FROM
WEATHER VANES
TO
SATELLITES

FROM WEATHER VANES TO SATELLITES

AN INTRODUCTION TO METEOROLOGY

HERBERT J. SPIEGEL
Miami-Dade Community College North

ARNOLD GRUBER
National Earth Satellite Service
National Oceanic and Atmospheric Administration

JOHN WILEY & SONS
New York   Chichester   Brisbane   Toronto   Singapore

Cover: Photography by Lowell Georgia/Photo-Researchers
Designed by Ann Marie Renzi

Copyright © 1983, by John Wiley & Sons, Inc.

All rights reserved. Published simultaneously in Canada.

Reproduction or translation of any part of
this work beyond that permitted by Sections
107 and 108 of the 1976 United States Copyright
Act without the permission of the copyright
owner is unlawful. Requests for permission
or further information should be addressed to
the Permissions Department, John Wiley & Sons.

Library of Congress Cataloging in Publication Data:

Spiegel, Herbert.
   From weather vanes to satellites.

   Includes index.
   1. Meteorology. I. Gruber, Arnold, 1940–
II. Title.

QC861.2.S64 1982   551.5   82-8349
ISBN 0-471-86401-3   AACR2

Printed in the United States of America

10 9 8 7 6 5 4 3 2 1

To our families
whose patience and understanding
made this book possible

PREFACE

Over the last few years, as the number of courses offered in meteorology has significantly increased, we have seen the need for a special type of meteorology textbook. In many two- and four-year colleges and universities, nonscience majors are enrolling in an introductory meteorology course in order to partially satisfy the science requirement for graduation. At the same time, these students realize that weather influences their daily lives and has a long range impact on their existence.

Ten to fifteen years ago most textbooks in meteorology made sense only to readers with strong mathematical backgrounds. However, recent books have leaned toward a nonmathematical approach to serve the needs of students in areas such as the arts, social sciences, and aviation. Although some of these publications have proven quite useful, we believe they have not really satisfied the needs of the general education oriented students. Specifically, these students were concerned with the applications of meteorology and its effects on their lives. This is a normal interest since most people are both familiar and concerned with their environment. It is this familiarity that encourages an understanding of meteorology from the applied viewpoint.

Our book presents the highly complicated science of meteorology in two distinct parts. The first part stresses basic concepts, which are the necessary ingredients in the learning of any science, while the second part contains Laboratory Exercises that integrate concepts with applied aspects of meteorology.

Chapter 1 is a survey of the uses of meteorology and a brief history of the science. Chapters 2 through 11 describe the elements that constitute the science of meteorology. These chapters stress a nonmathematical understanding of the physical concepts of the atmosphere. Where appropriate, we have included simple examples to help the reader grasp a concept easily. Chapter 12 integrates the previous chapters into a comprehensive view of weather observations, analysis, and forecasting, and reinforces the idea of a complex atmosphere with its many interacting forces. The last chapter on climate departs from the normal presentation of the subject. Most textbooks discuss in detail the climate classification schemes used by scientists. However, these schemes are static systems in that they are fixed and cannot adequately represent our changing climate even when they adopt a new scheme of classification. More important is an understanding of climate changes and how those changes can affect our lives. Especially important is the human influence on climatic changes, and the long range effect of global climate and its effect on planet earth in the future.

Following Chapter 12 are thirteen laboratory exercises that are keyed to appropriate text chapters. Each consists of exercises, experiments, and questions, and the entire section is perforated so that pages can be easily removed.

In keeping with the national trend toward the metric system, we have identified most physical units in the text using both metric and English systems. However, in a few selected instances, where common usage is predominant, only English units are presented.

It is hoped that as students read this book they will realize that weather influences all aspects of our civilization and provides the necessary information for our important and not-so-important everyday decisions.

Many people played an important role in the development of this book. First, we thank our families, whose constant encouragement made burning the midnight oil less of a chore than it might have been. We especially thank Phyllis Gruber, Toby and Karen Spiegel, who served as our grammarians, proofreaders, and typists; Don Ink, who prepared the chapter opening artwork with exacting care; Holly Rivera and Toni Ducassoux, who helped with typing the first draft and much of the correspondence. Their individual efforts sparked new life into our writing whenever it ebbed.

A special thanks to Joe Golden, Dick Jaffee, Mike Mogil, and Norman Mendelson for providing the photographs that in many cases appear here for the first time.

Herbert J. Spiegel
Arnold Gruber

CONTENTS

1.	**INTRODUCTION**	1
	Exposure to the Elements	2
	Special Adaptations of Weather Data	3
	Synoptic Meteorology	3
	Aviation Meteorology	4
	Agricultural Meteorology	4
	Weather Modifications	5
	Industrial Meteorology	6
	History of Meteorology	7
	Early Beginnings	7
	The Greek Culture	7
	The Age of Nonaccomplishment	8
	The Birth of a Science	9
	The Coming of Age	10
	History of the National Weather Service	13
2.	**THE ATMOSPHERE**	16
	Evolution of Our Atmosphere	17
	Composition of Our Present Atmosphere	17
	Vertical Structure of the Atmosphere	21
	Exploration of the Lower Atmosphere	23
	Exploration of the Upper Atmosphere	25
3.	**THE SUN-EARTH ENERGY SYSTEM**	26
	The Sun	27
	Prominent Solar Features	27
	Transfer of Heat Energy	28
	Heat Versus Temperature	29
	Electromagnetic Energy	30
	Solar Constant	30
	Earth Energy	32
	Energy Mechanisms for Understanding the Heat Balance	33
	The Heat Balance of Our Environment	33
4.	**TEMPERATURE OF THE AIR**	36
	Temperature Measuring Instruments	37
	Temperature Scales	40
	Temperature Conversions	40
	Temperature Variations	41
	Importance of Various Lapse Rates and Atmospheric Stability	46

5. WATER IN THE ATMOSPHERE — 49

- Changes of State — 50
- Moisture Parameters — 51
- Water Vapor Measuring Instruments — 53
- Evaporation, Condensation, and Precipitation Processes — 54
- Sources of Atmospheric Moisture — 56
- Hydrologic Cycle — 56

6. FOG, DEW, CLOUDS, AND PRECIPITATION — 57

- Water Droplet Size — 58
- Fog — 59
- Dew and Frost — 62
- Clouds — 62
- Precipitation Process — 67
- Precipitation Forms — 69
- Precipitation Measuring Devices — 70

7. ATMOSPHERIC PRESSURE — 73

- Pressure-Measuring Instruments — 74
- Pressure Units — 75
- Vertical Pressure Variations — 75
- Altimetry — 77
- Horizontal Pressure Variations — 77
- Pressure Terms — 77
- Semipermanent Pressure Patterns — 78

8. WINDS OF THE WORLD — 82

- The Three Laws of Motion — 83
- Primary Forces Affecting the Wind — 83
- Secondary Forces Affecting the Wind — 84
- The Balance of Wind Forces — 86
- Wind Systems of the World — 88
- The General Wind Pattern — 88
- Jet Stream Winds — 91
- Airflow Around Pressure Systems — 91
- Local Wind Systems — 92
- Wind Observations — 93
- Wind Chill Factor — 97

9. AIR MASSES OF THE WORLD — 98

- General Characteristics of an Air Mass — 99
- Source Regions — 99
- Air Mass Classification — 100
- Air Masses Affecting the United States — 101
- Modification of Air Masses — 103

10. WEATHER SYSTEMS — 106

- Frontal Systems — 107
- Weather Map Frontal Symbols — 112

CONTENTS xi

   Cyclones and Anticyclones   112
   Tropical Storms and Hurricanes   115

11. THUNDERSTORMS AND TORNADOES   **123**

   Thunderstorm Formation   124
   Lightning and Thunder   126
   Thunderstorm Weather   128
   Types of Thunderstorms   129
   Tornadoes   130

12. OBSERVATIONS, WEATHER MAPS, AND FORECASTING   **134**

   Observing the Weather   135
   Transmission and Collection of Weather Data   138
   Drawing Weather Maps   140
   Weather Forecasting   142
   The Future of Forecasting   143

13. CLIMATE   **145**

   Climate Controls   146
   Distribution of Climatic Elements   147
   Climate Variability and Change   150
   Climate's Effect on Society   153
   Man's Effect on Climate   155
   Global or Worldwide Effects   156

APPENDIX I   BLACK BODY RADIATION   159
APPENDIX II   GAS LAWS AND THE HYDROSTATIC EQUATION   161
APPENDIX III   DAILY WEATHER MAP DECODING SYMBOLS AND TABLES   163
APPENDIX IV   COMMON METRIC CONVERSIONS   169
GLOSSARY   171

LABORATORY EXERCISES   **177**

 1. MEASUREMENTS AND DIMENSION   177
 2. TEMPERATURE MEASUREMENTS   183
 3. THERMAL CONVECTION CELLS   187
 4. MOISTURE IN THE AIR   189
 5. ISOPLETHING   193
 6. OBSERVATION AND MAP PLOTTING   198
 7. SURFACE MAP ANALYSIS   205
 8. FRONTAL ANALYSIS AND FORECASTING   211
 9. UPPER ATMOSPHERE OBSERVATIONS   215
10. SUMMERTIME TEMPERATURE ANALYSIS   219
11. WINTERTIME TEMPERATURE ANALYSIS   223
12. ANNUAL TEMPERATURE RANGE ANALYSIS   227
13. STATISTICAL USE OF DATA   231

INDEX   235

FROM
WEATHER VANES
TO
SATELLITES

1 INTRODUCTION

Aristotle (384–322 B.C.), a Greek philosopher, wrote the first complete work on the subject of meteorology. His series of four books entitled *The Meteorologica* was a composite of meteorological data known at that time. It is from the title of his work that we derive the name *meteorology*.

LEARNING OBJECTIVES

After reading this chapter, you should be able to:

1. Describe the major advances in meteorology from its early beginnings to the present time.
2. List the various subbranches of meteorology and explain the importance of each to human life.
3. Trace the birth and growth of the National Weather Service of the United States.
4. Define *wind shear*, *hygrometer*, and *radiosonde*.
5. Explain how the word meteorology was adopted for a science dealing with weather changes in the atmosphere.
6. Describe the types of current weather satellites.
7. Identify when and with what developments meteorology became a true science.

The terms weather and climate have a universal meaning. All over the earth, people know they are completely dependent upon weather for their comfort and existence. There are no geographical boundaries to which our constantly moving atmosphere must adhere; thus, all countries of the world depend upon one another for information and cooperation in this most important scientific subject.

If one asks whether or not there exists some place on earth where a person could live and be safe from the fury of a natural disaster, the answer would have to be an emphatic no. In the case of atmospheric disasters, meteorologists can point to any one of dozens of hurricanes, floods, droughts, tornadoes, blizzards and thunderstorms that have raised havoc with the population centers of the world. However, weather and climate refer to more than just these natural disasters, for it is through their influences that people settle in a particular area, wear certain types of clothing, work in a specific occupation, participate in leisure time activities and, in particular, decide the type of food they eat. Strange as it may seem, weather also influences the prices people pay for goods and the outcome of an election.

EXPOSURE TO THE ELEMENTS

In recent years, the news media have reported on the vulnerability of people when exposed to the unleashed power of the atmosphere. The names of Anita (1977), Betsy (1965), Camille (1969) or Celia (1970) will be remembered by inhabitants who lived in the area in which these hurricanes caused death and destruction.

One of the most disasterous of these hurricanes, Camille, struck the Louisiana Gulf Coast with winds greater than 322 km/hr (200 mph) and tides exceeding 7.3 m (24 ft) in height. In her wake she left more than 300 people dead and millions of dollars worth of damage. However, Camille was a small tropical disturbance when compared to the cyclone that hit East Pakistan in November, 1970. Because of the population centers over which it moved, its 6.1 m (20 ft) storm surge resulted in over 300,000 deaths and a crop loss of 63 million dollars.

The word tornado strikes even more fear into the minds of those who hear it than the word hurricane. Carrying winds in excess of 483 km/hr (300 mph) in the most intense storm, this violent whirlwind is capable of picking up large objects within the center of its spiral and throwing them out great distances. Death and destruction can come swiftly from an atmosphere gone wild, such as in Birmingham, Alabama in April, 1977. Within 15 minutes, 22 people were killed, 130 injured, hundreds more left homeless, and millions of dollars of property destroyed. This example should vividly indicate what the atmopshere is capable of spawning given the right combination of conditions (Fig. 1-1).

Although the words hurricane or tornado carry the connotation of immediate disaster, other atmospheric uprisings, although more subtle, are just as dangerous. Short bursts of heavy rainfall might be advantageous for the farmer, but to people living in the floodplain of a confined river it could mean misfortune. In July, 1976, for example, the Big Thompson River in Colorado showed its strength. After a four-hour rainfall of more than 23 cm (9 in), a wall of water, fed by runoff from

Figure 1-1 The sparse remains after a tornado rolled through a section of Xenia, Ohio in April, 1974. (Courtesy National Severe Storms Laboratory, NOAA)

the surrounding mountains and imprisoned by vertical walls, roared down the canyon. In a few short minutes 135 visitors to this usually peaceful Colorado resort area were dead or missing. The road that runs beside the river was completely destroyed by what meteorologists call a *flash flood*. All rivers have the potential for flooding, and it is the sudden excessive rainfall or spring snowmelt that causes this disaster.

There are other natural disasters: drought conditions; wintertime snowfalls in excess of 1.2 m (4 ft); wind-biting, record-cold temperatures; oppressive, hot, humid summer weather; and rolling duststorms that remove fertile topsoil from agricultural land that create a set of hostile circumstances with which people must constantly cope (Fig. 1-2).

SPECIAL ADAPTATIONS OF WEATHER DATA

It is quite natural to think that meteorology involves just weather forecasting. Being constantly in contact with daily televised weather reports, hourly radio weather broadcasts, and a daily weather map in the local newspaper might give the impression that forecasting was the sum total of the science. Actually, forecasting is only one part of this very important science. There are at least 50 branches and subbranches of meteorology, most of which are useful to our daily well-being. It is relatively easy, while reading the following synopses, to think of a few applications in which people might use their knowledge of the atmosphere to benefit life on earth.

Synoptic Meteorology

Usually associated with weather forecasting, this branch of meteorology utilizes simultaneous weather observations to study the present conditions of our three-dimensional atmosphere. This is done by correlating surface and upper air observations into a set of weather maps. Utilizing basic physical concepts and extending the data into highly theoretical computations and analyses, the meteorologist can describe how the atmosphere will behave. Thus, the final output is a weather forecast for temperature, cloud cover, wind, pressure, and precipitation. Using this forecast, we can

Figure 1-2 Automobiles abandoned on the outskirts of Maryland after the great snowstorm of February, 1977. (Courtesy H. Michael Mogil)

make plans about what to wear, where to go, and how to get there.

Aviation Meteorology

In the United States alone, with 1600 daily commercial flights and hundreds of thousands of private pilots, it is necessary to have specialized weather reports geared strictly for aviation. A pilot not only wants to know the weather at departure and destination, but he or she must also be informed of inflight weather at flying altitude. Conditions such as icing, cloud cover, visibility, turbulence, wind shear, and thunderstorm activity are important to the safety of the aircraft. Commercial aircraft are not only concerned with safety, but also with the comfort of the paying passengers and the timeliness of arrival. If passengers could not fly in comfort and were subjected to long hours of rough flight, they might use other means of transportation. Therefore, forecasting areas of turbulence and the dissemination of actual turbulence reports assist the pilot in circumventing areas of particularly unstable conditions. Diverting aircraft because of poor weather conditions is quite expensive to the airlines and an inconvenience to the passengers. If an airplane has to be diverted from one city to another because of weather conditions, the airlines will usually find another mode of travel for the passengers or accommodate them for the necessary time period. If we multiply this situation by the thousands of passengers who might be diverted at any one time, we realize the high costs involved in such operations.

As a result of aviation's booming progress, meteorologists now specialize in specific forecasts and reports. Radar is constantly used to locate thunderstorms, such as that shown in Figure 1-3. Specialized instrumentation measure surface and upper atmosphere weather phenomena while satellites observe from above what we see from the surface. One recent advance in instrumentation is the *wind shear system*. Wind shear is a rapid change of wind direction or speed over a short distance. When it takes place, the lift of an aircraft usually decreases and on many occasions has resulted in aircraft accidents. By using observations from aircraft flight recorders, it is possible to determine that changing air temperature has much to do with wind shear. Therefore, temperature measuring devices are being installed in aircraft and around major airports to help identify these critical conditions. Thus, meteorologists participate in making flying a much safer and comfortable mode of transportation.

Agricultural Meteorology

At present, the United States is known as the "bread basket of the world." Raising more food than any other

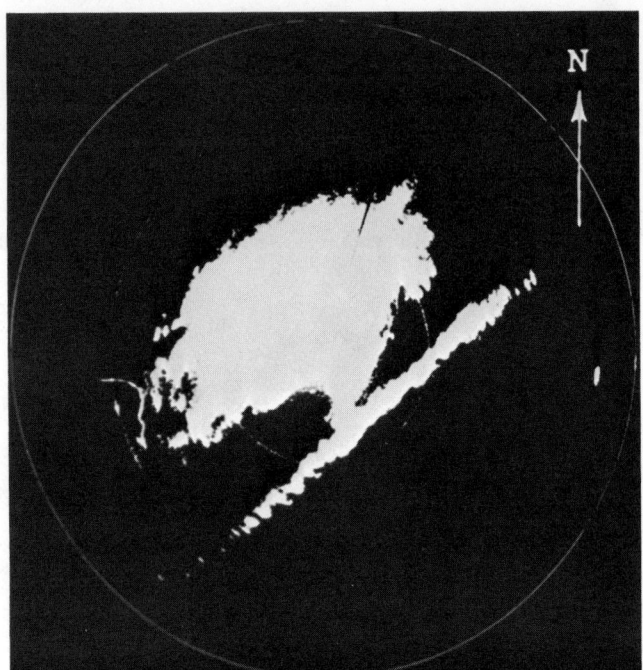

Figure 1-3 Thunderstorm squall line as seen on a weather radar scope. (Courtesy NOAA)

nation on earth is not just an accident. By applying scientific knowledge of soils, pesticides, fertilizers and weather, the American farmer has been able to continuously produce abundant crops resulting in the lowest food price of the entire world. After we compare the cost of food in places such as Japan, Sweden or England, we realize what American scientific technology has accomplished.

Because weather plays such an important part in our agricultural output, a branch of the National Weather Service of the United States is devoted exclusively to agricultural observations and forecasts. Since two of the most important agricultural weather elements are precipitation and solar radiation, observations are continuously made to determine available water and sunshine (Fig. 1-4). Tied closely to rainfall and temperature are other weather elements that have a direct effect on crop production. Frost conditions, wind flow, snow accumulation, and climatological data surrounding earliest planting and harvesting dates all have to be considered. The latter type of information is available from data archives at the National Weather Records Center in Ashville, North Carolina. If these data are used an earlier and more abundant yearly crop can result in greater profit for the farmer, while the costs for the consumer are held to a minimum. If we consider, for example, the higher price of watermelon at the beginning of the season and the lower price when it is in abundance, it is no wonder farmers use weather knowledge to be able to harvest as early as possible. Many other examples of

SPECIAL ADAPTATIONS OF WEATHER DATA 5

Figure 1-4 A solar-powered Remote Automatic Meteorological Observing Station (RAMOS) allows weather data such as solar radiation, temperature, wind, and precipitation to be measured in remote areas. This data is relayed via satellite to the National Weather Service and is used in part for agricultural forecasts. (Courtesy LeBarge, Inc., St. Louis, Missouri)

weather related agricultural benefits will be given in the text as we discuss each weather element separately.

Weather Modifications

Not only have meteorologists made great progress in understanding the atmosphere but, under certain conditions, they can also alter the environment. The branch of meteorology dealing with weather modification is just now coming into its prime. In the future it may be possible to order certain weather conditions just as we now order food from a restaurant.

At the present time, modest advances have been made in cloud seeding, hail suppression, fog dissipation, reduction of evaporation from a water surface, lightning suppression, and frost protection. Initial research is being conducted on hurricane seeding and tornado protection. Scientists have dreamed for many years of controlling the atmospheric environment: producing rain for a drought-ravaged world; suppressing rainfall from flood-threatened areas; steering hurricanes away from densely populated areas; preventing lightning-produced forest fires; and, in general, altering any life threatening or destructive situation (Fig. 1-5).

Figure 1-5 Typical multistroke lightning that causes hundreds of forest fires each year. (Courtesy United States Forest Service, Department of Agriculture)

Here is an abbreviated list of important weather modification projects.

1. **Reduction of Evaporation.** By the use of a harmless chemical such as acetyl alcohol, which is floated on the surface of a water body, evaporative loss can be reduced by as much as 30% under ideal conditions. This process can save much of our vital surface water supply. A second method is to construct shelterbelts, such as tree lines, around the water body. This will retard wind flow and reduce water loss due to evaporation.
2. **Increasing Precipitation.** Encouraged by the experiments of Drs. Vincent Schaeffer and Bernard Vonnegut, meteorologists have experimented with the cloud seeding process. At General Electric in 1946, these two scientists discovered that by introducing dry ice or silver iodide into supercooled clouds, they could produce large ice crystals. These large ice crystals would fall from the cloud, melt in the warm air below, and end up as raindrops hitting the ground. Although rainmaking does not have a major impact on available surface water, it can aid in local water problem areas. A second way of increasing precipitation is to add condensation nuclei to the air. Studies have shown that the addition of nuclei such as salt, dust, dirt, and smoke results in an increase of precipitation, especially over highly industrialized areas. Although this type of weather modification may be of value in the future, cloud seeding presently offers our best method of inducing precipitation.
3. **Hail Suppression.** Spurred by the fact that hail losses to agricultural crops amount to more than 600 million dollars yearly, the United States has embarked upon a hail suppression study. However, the Soviet Union has had the greatest success in this area. Experiments in the Caucausus Mountains have proved that hailstones can be reduced in size. Artillery shells filled with silver iodide have been fired into the supercooled liquid zone of a growing thunderstorm cloud. These crystals supply the seed for the formation of ice crystals and thus rob the available moisture from

the already formed water droplet. Thus, the water droplet cannot grow into a large hailstone. The seeding resulted in a large amount of rain or sleet, which was not devastating to the crops below. The success of this method has been excellent, and countries such as Kenya, Canada, and England have adopted similar systems (Fig. 1-6).

4. **Lightning Suppression.** Seeding of clouds with silver iodide has led the United States Forest Service to believe that lightning can be reduced. A specialized project called *Skyfire* was implemented to produce cloud to cloud lightning rather than cloud to ground lightning. However, a second method appears to be more promising. In this method, the cloud is seeded with aluminum-coated material. These bits of material prevent the buildup of maximum electrical potential by conducting electricity away from the potential center. This results in more frequent but generally weaker discharges.

5. **Frost Protection.** By the use of artificial devices to warm or stir the air, damage to agricultural crops can be held to a minimum. For many years smudge pots were used to reduce the amount of surface reradiation, but the dense smoke created air pollution problems. Current methods include smokeless orchard heaters to maintain temperatures above their critical level, and propeller-driven wind machines to stir up the air, thus reducing heat loss. In the cranberry bogs of New England, New Jersey, and Wisconsin, flooding of the low-lying areas is practiced. The flooding method is used because as water cools to its freezing point (0°C or 32°F), it releases stored energy into the air. With this release the rate of cooling of the air will slow down considerably. If the water has any appreciable depth, the slow release of this energy will result in the air remaining warmer longer.

There are many other areas of modification that meteorologists are researching. Among them is the slow release of water by controlling snowmelt; increasing snowfall rates over ski areas; and possibly melting the Greenland icecap. However, realizing that altering our atmosphere might produce an unwanted reaction that is not predictable, meteorologists are proceeding very cautiously. Understanding our atmosphere is the key to weather modification.

Industrial Meteorology

In the late 1940s, a few meteorologists established private weather firms to service special interest groups in industry and commerce. It was well known that weather forecasts provided by the United States Weather Bureau (now known as the National Weather Service) were not specialized enough for use by the private sector. At first, there was no ready-made market for this type of service, but it grew slowly and prospered into a very highly regarded subbranch of meteorology. Today some of the major areas served are land developments, agriculture, private aviation, maritime transport, petroleum companies, large retail chains, agricultural firms, local and state governments, and communications media.

The following examples are presented to show the information a special weather report provides beyond that supplied by the National Weather Service.

1. Since it is well known that weather affects the buying habits of the general public, commercial interests such as retail sales, sporting events, and manufacturing have a need for special weather services. In a recent study, it was determined that a person would walk only a block away from a rapid transit stop in inclement weather to purchase a needed commodity. In this case, it would be better to establish a bakery in a large urban city on a site within one block of a bus stop. In another instance a large retail chain hired a meteorology firm to produce a three-day forecast on a Friday so that they would know what specials to advertise in the Sunday newspaper. If rain was predicted for Monday, raincoats and umbrellas were featured in the Sunday advertisement. It would make more sense to advertise air conditioners when the temperature is in the 90s rather than in the 60s.

2. Local and state governments, as well as the private sector, need specialized information on snowfall and air pollution episodes. All major northern city or turnpike authorities have built into their budgets large amounts of funds for snow removal. Therefore, they must know when a storm will start, the location of the state in which it will first begin, how long it will last and the total snow depth. Using this information they can plan where to start their plowing and snow removal operations, and how many work crews to use (Fig. 1-7). All this preplanning adds up to many thousands of dollars in savings.

Just as important as snowfall forecasts is the need for a municipality to know when atmospheric conditions favor high air pollution problems. This knowledge may necessitate the closing of certain industrial complexes for short periods of

Figure 1-6 Hailstone damage to crops.

Figure 1-7 Snow removal is a high priority in most northern cities. Record snowfall in recent years has made the services of private meterologists quite important. (Courtesy NOAA)

time or, in extreme cases, the banning of all but emergency vehicles from the roads. Hospitals may be forewarned of the possible influx of patients who suffer from conditions that are aggravated by air pollution.

3. The petroleum industry has a specialized problem within its delivery system. Major gasoline producing companies must know where and when the weather will be conducive for extensive driving so that they may supply adequate amounts of fuel. Gasoline delivery is especially critical in the wintertime so as to avoid overstocking in areas where the roads are unusable. Heating oil also falls into this category. Because oil consumption is heavily dependent on outside air temperature, oil delivery firms must have a specialized forecast for their local area in order to ensure that an adequate quantity of heating fuel is available. This, of course, applies to the gas industry as well.

Other specialized information provided by the industrial meteorologist include forecasts for concrete pouring, marketing of perishables, the fishing industry, ski resorts, energy producing companies, insurance firms, and freight hauling.

These examples represent only a few of the many disciplines within the field of meteorology. Other applications exist within agriculture, forestry, health, and architecture. The field is continuing to grow into even more diversified areas, such as land, air and sea interaction, and the use of the satellite for global weather predictions. The future of meteorology lies within these newer fields with still other areas waiting to be discovered.

HISTORY OF METEOROLOGY

The study of any subject would not be complete without a brief history of how it began, what made it prosper, and what is in its future. The history of meteorology can be effectively divided into four parts:

1. Early beginnings
2. The age of nonaccomplishment
3. The birth of a science
4. The coming of age

Early Beginnings

It is quite evident from paintings and carvings on cave walls that early inhabitants were well aware of their environmental surroundings. They evidently understood that different types of weather affected their daily lives. Because they could not explain logically what was happening to their surroundings, they naturally turned to their gods with prayers for fair weather. To the ancients, storms must have indicated a punishment from an angry god who could control all physical events on the surface of the earth. All through our progress toward modern civilization, gods have played an important role in weather. *Boreas*, Greek god of the north wind; *Ra*, Egyptian god of the sun; and *Thor*, Norse god of thunder, are only a few of the many weather gods throughout history. Even today, many cultures still cling to the idea that praying to a weather god will change their environment (Fig. 1-8).

The Greek Culture

Greek scholars were among the earliest in Western civilization to make meteorological observations, and their keen interest in their environment led to the publishing of many theories on weather. Their love for abstract reasoning and theoretical deduction aided in remarkable advances in all the sciences. However, they were quite negligent in the experimental phase of the scientific process, relying only on observations within their natural surroundings. Therefore, many of their theories were later proven false.

Two prominent Greek philosophers during the Golden Age of Greece were *Aristotle* and *Hippocrates*. Aristotle, the author of *The Meteorologica*, the oldest known written work in meteorology, was able to compile everything known to that time on subjects such as rain, hail, lightning, thunder, snow, clouds, and wind. He included in his book a primary wind rose that indicated the prevailing direction of the wind (Fig. 1-9). Aristotle based his idea for the wind rose on both the seasons of the year and the daily path of the sun. His statements that lightning resulted from a dry wind ejected out of a cloud, and that thunder preceded lightning were two of many major errors in *The Meteorologica*. However, this four volume treatise stood as the undisputed authority on physics of the atmosphere for over 2000 years.

Figure 1-8 Both ancient and modern civilizations used prayers to placate weather gods. While the caveman prayed for good hunting weather, the American Indian performed a dance to bring on the rain.

Hippocrates (460–377 B.C.), contributed greatly to the knowledge of his environmental surroundings. Known as the "Father of Medicine," Hippocrates wrote a book on the medical well-being of man entitled *Airs, Waters, and Places*. This treatise, published around 400 B.C., explained how *climate* (long term weather) affected the health of people around the world. One of his outstanding contributions was his observation that large amounts of sunshine over a short period of time could lead to lesions on the skin. Many centuries later this statement was linked to skin cancer.

During the remainder of the Golden Age of Greece, only a few additional achievements in meteorology were made. Around 100 B.C. the Greeks, who were great sailors, invented the wind vane. This meteorological instrument was an important tool for the ever-increasing military and merchant fleets of this powerful nation. Accompanying the invention of the wind vane was the first attempt at a practical, general-type forecast and an explanation that the yearly flooding of the Nile River was due to excessive rainfall.

The Age of Nonaccomplishment

By 100 A.D. Greece was no longer a power, and most scientific investigation has ceased. The Romans had conquered the known world and, being a practically oriented people, did very little to further meteorology. Two notable exceptions were the weather predictions of Ptolemy (85–165 A.D.) based on astronomical data and the handbook for scientists entitled *Natural Questions*, written by Seneca (2–65 A.D.). In this treatise Seneca contradicted Aristotle's concept that wind was not a dry exhalation, but simply the movement of air.

Just as the Greek influence on the world came to an abrupt end, so did the Roman rule. Around 400 B.C. hordes of Germanic barbarians ravaged a disintegrating empire. Thus began the dark ages and a long period of scientific drought. During this time only a few technical advances were made, such as the plough and waterwheel. Around 1100 A.D. scientific theory once again had a slow, but steady emergence. But this period proved to be very difficult for scholars, since any findings contrary to the Church's teachings were consid-

ered heretical and punishable by death. The Church felt that nature was the work of God alone, and this attitude prevailed until the Scientific Revolution (1500–1900 A.D.).

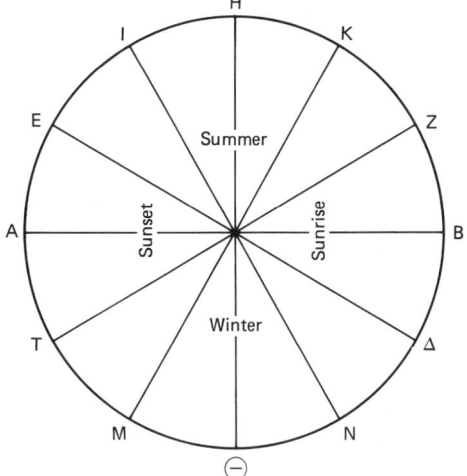

Figure 1-9 An early wind rose designed by Aristotle. It was based on astronomical reasoning rather than our more familiar directions.

The Birth of a Science

Until the seventeenth century most scientific findings were based strictly on sensory observations and influenced by Aristotle's treatise of the atmosphere. Up to this time, only two weather instruments, the rain gauge and wind vane, had been invented. Now we see the development, in rapid succession, of the hygrometer, thermometer, and barometer. With these three weather instruments meteorology advanced from a scientific subject to a true science. It has been stated that the invention of both the thermometer and barometer marked the beginning of meteorology as a science.

The first instrument developed was the *hygrometer*, used to measure the amount of water vapor in the air. Initially, the simple hygrometer consisted of a scale with wool and rocks in balance (Fig. 1-10). As the wool absorbed moisture or dried out, the scale would indicate the moisture content of the air.

Immediately following the hygrometer, Galileo Galilei (1564–1642) invented the first thermometer (Fig. 1-11). Called a *thermoscope*, the instrument was an air-filled glass cylinder inverted into a liquid surface. If the bulb were warmed before immersion into the liquid, air would be driven out of the glass tube. This cooling and contracting air caused the liquid within the res-

10 INTRODUCTION

Figure 1-10 The first hygrometer worked on the principle that an imbalance between two weights resulted from an absorption or evaporation of water.

ervoir to rise. By calculating the height of the liquid column, Galileo could measure relative degrees of hotness or coolness.

The barometer was developed, purely by chance, soon after the thermometer was invented. In 1644 Evangelista Torricelli (1608–1647), a noted physicist and mathematician, invented this pressure measuring device. Torricelli was greatly interested in the ability of suction pumps to lift water to a level no higher than 34 feet above the free water level. It had previously been assumed that after rising higher than 34 feet a column of water would break due to its own weight. However, Torricelli theorized that the weight of the air pressing down on the open water surface pushed the column of water upward. To prove his theory, Torricelli constructed a primitive barometer fashioned after an earlier experiment of Gasparo Berti (Fig. 1-12). It consisted of a one meter glass tube closed at one end and filled with mercury. With this instrument Torricelli was able to show that air indeed had weight and that it exerted a pressure that changed with temperature.

In the eighteenth and nineteenth centuries these three weather instruments were improved upon. H. B. deSaussure increased the hygrometer's accuracy by using human hair to measure the atmosphere's moisture (Fig. 1-13). Thermometers were also improved greatly. They now had a scale of numbers that used the fixed freezing and boiling points of water as a standard. The thermometer's actuating element went from water to alcohol to mercury, and for the first time could be standardized to a known scale. The barometer also became much improved. A portable pressure measuring instrument called an *aneroid* was developed. It used no mercury and was much less fragile.

Other notable developments in the eighteenth and nineteenth centuries were the theories to explain evaporation and condensation of water, the identification of the four major gases of the atmosphere, the discovery of atmospheric electricity by Benjamin Franklin, and the classification of cloud types by Luke Howard. Observational networks were set up to acquire surface weather data, kite flying stations provided upper air observations, and the telegraph provided a rapid means of weather data dissemination. With the invention of the telegraph a tremendous leap forward took place in *synoptic meteorology*, thus allowing meteorologists to rapidly analyze current weather data. At last, the necessary ingredients were present for a mushrooming scientific revolution.

The Coming of Age

Undoubtedly, the major advances in both theoretical and applied meteorology have been and are still being made during the twentieth century. In the early 1900s Norwegian meteorologists led by Jacob Bjerknes devel-

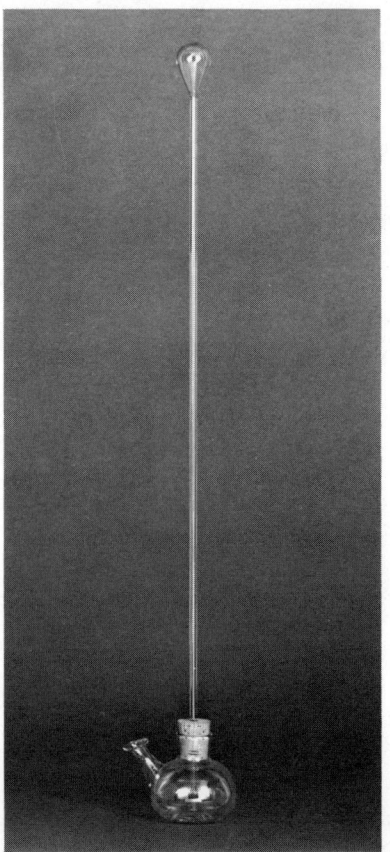

Figure 1-11 The Galileo thermoscope. (Smithsonian Institution Photo No. 61900)

HISTORY OF METEOROLOGY 11

the radiosonde, the use of weather radar, the observation of the world by weather satellites, and the development of the high speed computer.

Radiosonde. The radiosonde is essentially a radio instrumented package carried aloft by a lighter-than-air balloon (Fig. 2-6). This instrument is capable of either changing the frequency of a radio signal or transmitting a specialized code, which is then reinterpreted at the receiving station as weather observations. With this invention meteorologists had at their disposal the capability of picturing the atmosphere in three dimensions, a necessity if modern forecasting techniques were to be employed. Additional information on the radiosonde is given in Chapter 2.

Weather Radar. Although radar was not originally developed for use in weather, its availability allowed meteorologists to detect and track severe storms. With today's sophisticated equipment weather can be seen 250 miles in all directions and the intensity can be transformed into different colors. As in Figure 1-14, radar can track hurricanes throughout their life cycle. With this capability, meteorologists can predict areas of most probable landfall and issue advance warnings. As a re-

Figure 1-12 The famous Gasparo Berti experiment that eventually led to the proof of the existence of a vacuum and the invention of the mercurial barometer by Torricelli (Smithsonian Institution Photo No. 60824)

oped the now famous *cyclone* and *polar front* theories. They described conditions in which two huge air masses with different temperature and moisture characteristics collided. The boundary zone of this collision was called a front. They also went a step further and theorized how this front could develop into a wave cyclone around a low pressure area. These two theories are still used today in weather forecasting.

With the invention of the airplane, the need for meteorological data became a necessity. With this need came the onset of explosive discoveries in weather observation and forecasting. Modern technological discoveries were used to improve our capabilities to foretell the weather. Equations describing the behavior of the atmosphere were developed and incorporated within the forecasting scheme.

Out of all the outstanding achievements in the twentieth century, four are extraordinary: the invention of

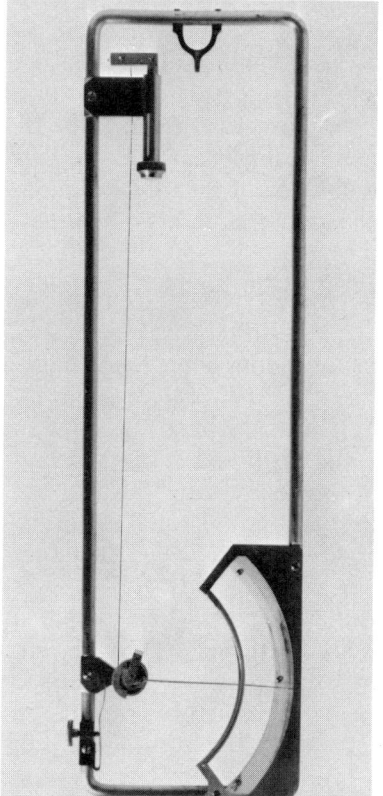

Figure 1-13 H. B. deSaussure's original hair hygrometer. (Smithsonian Institution Photo No. 61901-B)

12 INTRODUCTION

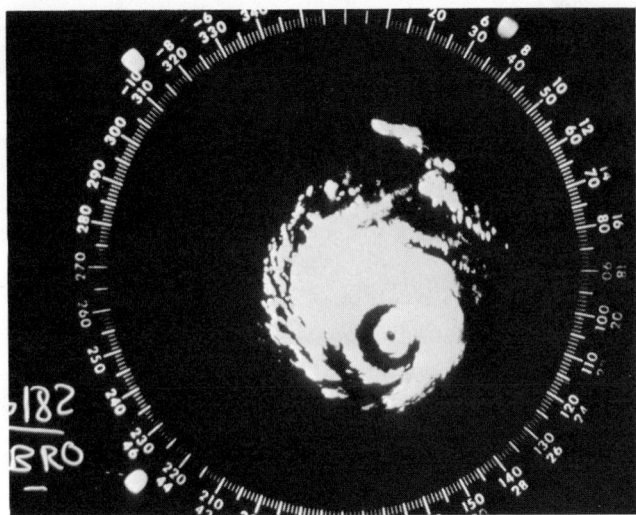

Figure 1-14 Hurricane as seen on a radar scope. (Courtesy NOAA)

sult, many lives can be saved and millions of dollars in damages prevented.

High Speed Computers. In the 1950s an electronic revolution began and is still going on today. Simple computers developed during that time have given way to high speed computers and microprocessors. With this progress came the meteorologist's ability to solve the physical equations governing the behavior of the atmosphere. Utilizing raw observational data, weather maps are printed out by machine (Figure 1-15) and weather forecasts made by numerical prediction. By using the computer, scientists finally have a tool that can handle the vast amounts of raw data that are constantly funneling in from worldwide observational networks.

Weather Satellites. On April 1, 1960, the National Aeronautics and Space Administration (NASA), successfully launched the first of a series of operational meteorological satellites. Named *TIROS* (Television Infra-Red Observation Satellite), these satellites were used to provide pictures of both the earth and its cloud structures. With these pictures meteorologists now had the ability to observe, for the first time, large scale cloud patterns, particularly in areas not covered by other types of observations.

Tiros was a relatively simple satellite when compared to our present day launchings. It had 9000 solar cells embedded in its exterior, which provided the power to operate its onboard systems. The pictures were taken principally by a television camera and in the 79 days of successful operation, it sent back over 20,000 photographs. In rapid succession, a series of nine additional Tiros satellites were launched, each with a much better camera storage and transmittal system.

Following the Tiros series, a second generation of nine operational satellites was launched. Called *ESSA* (Environmental Survey Satellite), this series provided global coverage with 156 pictures taken every 24 hours. The ESSA satellites first began in 1966 and the last of the series was deactivated on November 29, 1973. It provided a low cost method of obtaining much needed weather data.

Other operational satellites were still to come. In 1966 the first of the GOES (Geostationary Operational Environmental Satellites) were launched. These satellites had an orbital speed equal to the earth's rotation and, therefore, once put into orbit above the equator they stay positioned at a fixed location. Improvements continued and with the SMS/GOES series the ability of meteorologists to collect weather and other scientific data from widely dispersed areas became a reality. These orbiting platforms can collect data from up to 10,000 observation sites every 6 hours and transmit this data to weather stations around the world. Satellites now operating are capable of sending back photographs such as those in Figures 1-16 and 1-17.

Named after the Latin word for cloud, the *Nimbus* satellite was launched by NASA. It was designed to be a research and development spacecraft to test improved sensors. The Nimbus series has proved its worth with many of our current satellites. It used equipment origi-

Figure 1-15 A computer-drawn map of Relative Humidities at approximately 1.5 km (5000 ft) from the earth's surface. (Courtesy NOAA)

Figure 1-16 A global view of clouds depicting weather systems taken by the ATS 3 satellite. Note the hurricane with its distinctive eye and spiral bands in the Atlantic Ocean. (Courtesy NOAA)

nally tested on spacecraft similar to the one pictured in Figure 1-18.

HISTORY OF THE NATIONAL WEATHER SERVICE

Prior to the establishment of the first network of weather reporting stations in the United States, many problems had to be overcome. Instruments had to be developed that were accurate enough to measure instantaneous environmental conditions, then a means of collecting and transmitting these simultaneous observations had to be established, and government officials in Washington had to be made aware of the importance of such a service.

Instruments such as the barometer and thermometer had been developed earlier in the seventeenth century. These technological developments, coupled with the nineteenth century invention of a moisture measuring device called the sling pschrometer, and the ability to take upper atmosphere observations using kites and balloons, gave scientists the base from which a major network of observing stations could be established. However, it was not until the telegraph was invented in 1837 that scientists had the necessary tool to collect these observations from distant locations. Once collected, the data could be used to draw weather maps and issue storm warnings to the general public. The final step came on February 2, 1870 when Congressman Halbert Paine of Milwaukee, Wisconsin introduced a resolution in Congress to require the Secretary of War to set up a system for taking meteorological observations at all military stations. Thus, under the United States Signal Service, a division of meteorology commonly called the Weather Bureau was born.

Immediately, the Signal Service began to take and transmit observations from 24 military bases around the United States, and in November, 1870, it issued its first weather warning for the Great Lakes area.

During the ensuing 21 years the Weather Service

14 INTRODUCTION

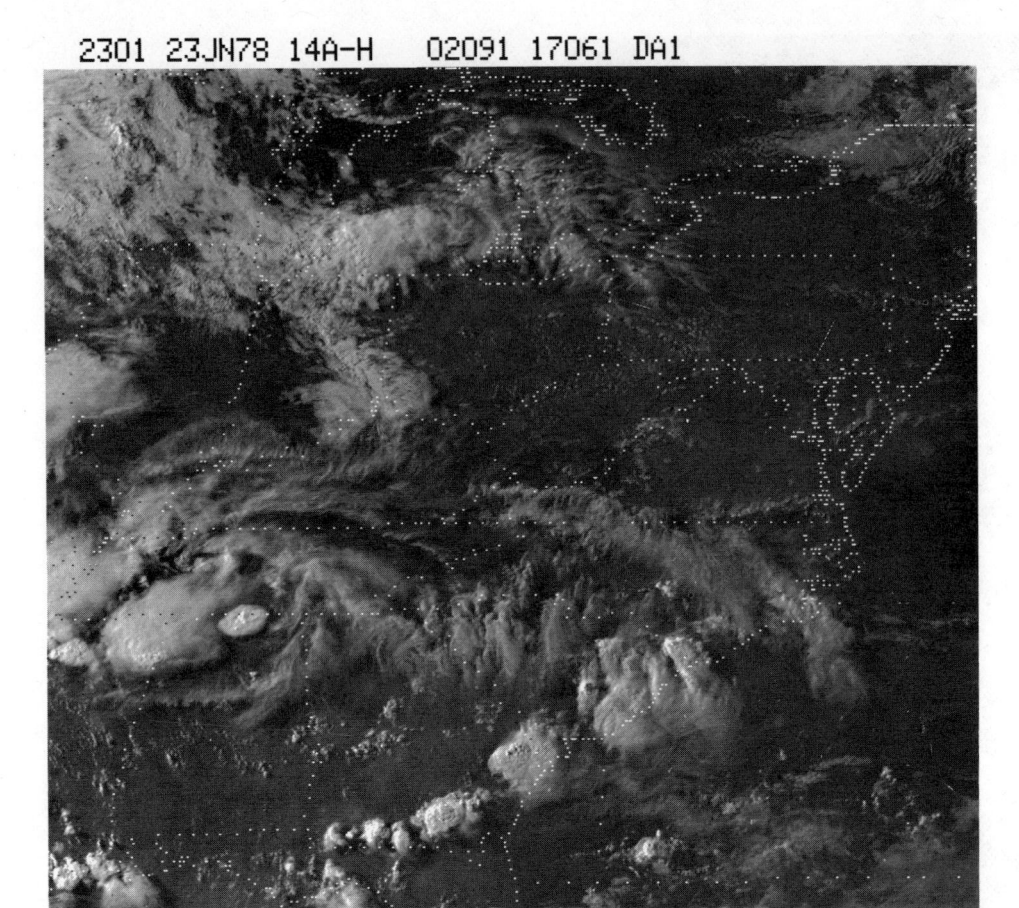

Figure 1-17 High resolution visible imagery from satellites can monitor small (mesoscale) weather patterns. Cloud tops can be easily seen as convective thunderstorm activity increases along the coastline of South Carolina, Georgia, and on into northern Florida. (Courtesy NOAA)

prospered and grew. Unfortunately, the Army Signal Service had more important missions to accomplish for the War Department, which led to the transfer of this service to the Department of Agriculture, where it was renamed the United States Weather Bureau. By this time daily forecasts and flood warnings were being issued; a network of 284 observing stations had been established; observations were being taken on naval and merchant vessels; and specialized forecasts for agriculture were being distributed. These services were minor, however, compared to those that followed the invention of the wireless telegraph and the airplane.

Rapid Growth of the Weather Bureau

Guglielmo Marconi's development of the wireless telegraph in 1895 greatly enhanced the ability of the Weather Bureau to expand its observing network. Not only could additional weather stations be set up in remote areas, but instantaneous observations from foreign countries and vessels at sea were now realities. Forecasts could be transmitted to all parts of the world, and international cooperation flourished.

It appears that no other event in history was more important to the growth of the Weather Bureau than the invention of the airplane. From its inception at Kitty Hawk in 1903 to its phenomenal growth through two World Wars and into the jet age, the airplane and meteorology have progressed together. As early as 1926 Congress realized how important meteorology was to aviation, and, passed the Air Commerce Act, which made the Weather Bureau the agency responsible for all weather services of civilian aviation. With this responsibility came additional funds and manpower that enabled the Weather Bureau to branch out into research, improvement in forecasting techniques, and the use of aircraft for upper air observations. Additional meteorological importance was reflected in the need for data and forecasts during World War II. This was especially true for the Normandy Invasion of France. It was the forecasting of the right weather conditions that allowed the Allied High Command to choose the best

HISTORY OF THE NATIONAL WEATHER SERVICE 15

Figure 1-18 The Nimbus spacecraft receiving final adjustments before launch. (Courtesy NASA)

time period for the invasion that caught the Germans by surprise and ended in a victory for the Allies.

In 1940 the Weather Bureau was transferred to the Department of Commerce. By this time all airplane upper air observation had been replaced by radiosonde stations. For the first time meteorologists had an inexpensive way of gathering upper air data, which led to a better understanding of the atmosphere and the development of new forecasting techniques.

In rapid succession both 5 day and 30 day outlooks were issued; the National Weather Records Center was established at Asheville, North Carolina; and a numerical weather prediction unit was established at Suitland, Maryland. But this postwar progress was nothing compared to what was about to take place.

In 1965 the Weather Bureau was again transferred, this time to the *Environmental Science Service Administration*. ESSA, as it was called, merged with the Coast and Geodetic Survey and a few other smaller agencies. However, three other components were created: ESSA Research Laboratory, Environmental Data Service, and the National Environmental Satellite Center.

In 1970 another reorganization took place and a new agency, NOAA (National Oceanic and Atmospheric Administration) was created within the Department of Commerce. This merged the disciplines of meteorology, oceanography, and earth science and enabled these agencies to use common facilities to study our environmental processes. In addition to being reorganized the United States Weather Bureau, which was only one segment of the NOAA family, was renamed the *National Weather Service*, and its main tasks were to:

1. Supply all the United States meteorological data using observing and reporting sites that have been established both on land and sea.
2. Provide for the safety of all concerned by issuing weather forecasts, hazardous weather warnings, and flood warnings to the general public.
3. Provide specialized forecasts to aviation, agriculture, and air pollution facilities.
4. Establish an automated data collection center, called the *National Meteorological Center* (NMC), that would collect information via teletype, store and process by computer; analyze and forecast both manually and by computer; and distribute the finished product by teletype and *facsimile*.

A second component of NOAA is the *National Environmental Satellite Service* (NESS). It was formed to operate the environmental satellite system of the United States, and to prepare satellite cloud pictures, temperature soundings, wind estimates, and other quantitative data for use by the National Weather Service.

A third branch of NOAA is the EDIS (*Environmental Data Information Service*). One of its components, the National Climatic Center, located in Asheville, North Carolina, is the storage site for meteorological data and, therefore, is responsible for the publication of local, national and worldwide climatological reports and summaries. In addition the EDIS operates the National Oceanographic Data Center and provides data support for other data gathering agencies.

The last segment of NOAA's meteorological component is ERL (*Environmental Research Laboratory*) headquartered in Boulder, Colorado. The laboratories are spread throughout the United States with the five most important ones being the following:

1. National Severe Storms Laboratory, Norman, Oklahoma.
2. Atlantic Oceanographic and Meteorology Laboratory, Key Biscayne, Florida.
3. Pacific Oceanographic and Meteorology Laboratory, Seattle, Washington.
4. Wave Propagation Laboratory, Boulder, Colorado.
5. Geophysical Fluid Dynamic Laboratory, Princeton, N. J.

It is obvious that meteorology has made great progress since the time of the Greeks and Romans. The National Weather Service, with its related organizations, have formed a complicated worldwide network capable of providing a most necessary service to all humanity.

2 THE ATMOSPHERE

L. P. Teisserenc DeBort (1855–1913), a French meteorologist, was the first person to develop the *meteorograph*, which was sent aloft by an unmanned balloon. After the balloon burst the meteorograph descended by parachute. Upon retrieval the upper air data recorded on the instrument could be analyzed. It was through such observations that DeBort first discovered that temperatures in the atmosphere did not continuously decrease.

LEARNING OBJECTIVES

After reading this chapter, you should be able to:

1. Describe the composition of the atmosphere and list the four major gases.
2. Describe the four steps in the creation of our present atmosphere and pinpoint the time when oxygen first appeared.
3. Explain the importance of ozone, water vapor, and the four major gases of the atmosphere.
4. List the major layers of the atmosphere and describe their important features.
5. Define *water vapor, aerosols, troposphere, stratosphere, mesosphere, thermosphere, ionosphere,* and *ozonosphere*.
6. Describe how a radiosonde and rocket are used for upper air observations.

The atmosphere is a mixture of invisible gases extending from the earth's surface to an indefinite height. Although it is difficult to determine the upper limit of the atmosphere with any precision, 500 km (300 miles) is sometimes taken as its average height. At this point the earth's gravitational attraction cannot keep a gas molecule in orbit. Thus, gas molecules above this level can reach their escape velocities and be lost to outer space.

THE EVOLUTION OF OUR ATMOSPHERE

The development of our atmosphere is closely linked to the formation of the earth itself. Theories of earth formation are numerous, and change as geologic evidence accumulates. Current theory indicates that the initial formation of the earth occurred about 4.8 billion years ago when relatively small cold objects that had little or no gaseous envelope associated with them were attracted to one another. Any gaseous component that may have been linked with this formative stage was probably lost to outer space since its mass was too small for gravity to contain it within the region that was to form the earth.

As the earth's interior heated up due to compression, radiative heating, and perhaps other chemical reactions, expulsion of volatile components took place to form our oceans and atmospheres. Geological evidence suggest that this process took place about 3.6 billion years ago, about 1 billion years after the initial formation of the earth. Furthermore, it is theorized that the principle gaseous components at that time were water vapor, (H_2O), carbon dioxide (CO_2), carbon monoxide (CO), nitrogen (N_2), and substantial quantities of methane (CH_4), and ammonia (NH_3). It should be noted that there was no free oxygen (O_2) in the earth's primitive atmosphere.

Free oxygen appeared as an atmospheric component roughly 2 billion years ago. There are two principle mechanisms for the production of O_2. One is liberation of oxygen through plant *photosynthesis*. Photosynthesis is a plant process that combines carbon dioxide, water, and sunlight to produce sugar and oxygen. The second method is through *photolytic dissociation*. This is a process of splitting molecules through the absorption of light. When water vapor at high altitudes absorbs ultraviolet light, the water molecules splits into hydrogen atoms (H), which are light enough to escape to outer space, and the O_2 molecule, which remains in the earth's atmosphere as free oxygen because it is heavier.

COMPOSITION OF OUR PRESENT ATMOSPHERE

The atmosphere basically consists of three components: dry gases, water vapor, and solid particles.

Because of diffusion and mixing by wind, the mixture of dry gases is relatively constant up to about 80 km (50 miles). The four most abundant gases by volume are nitrogen, oxygen, argon, and carbon dioxide. Together they account for 99.9% (by volume) of the dry air. Nitrogen and oxygen alone account for 99.03% of the volume, while argon and carbon dioxide consist of under 1% of the mixture.

Above 80 km (50 miles), where large scale vertical mixing does not readily occur, the gases are distributed according to their densities, the heavier gases being on the bottom. At these levels and higher the process of ionization takes place. This occurs when ultraviolet radiation from the sun produces charged particles called ions by stripping electrons from oxygen and nitrogen molecules. The maximum concentration of these charged particles is at about 400 km (250 miles) where they form part of the *ionosphere*.

18 THE ATMOSPHERE

Meteorologists are primarily concerned with the lowest 50 km (about 30 miles) of the atmosphere. This region contains the layers known as the *troposphere* and *stratosphere* and contains 99% of the entire mass of the atmosphere.

Nitrogen

Nitrogen, the most abundant gas in the atmosphere, was first identified in 1722 by a Scottish botanist Daniel Rutherford. It is a tasteless, odorless, and colorless gas that is for all practical purposes chemically inactive. Neither people nor other animals can use atmospheric nitrogen directly even though it produces protein, which is vital to both growth and life. Since we cannot manufacture protein from nitrogen, we must depend on plant life to produce usable nitrogen through the *nitrogen cycle* (Fig. 2-1).

The cycle begins with an unlimited supply of atmospheric nitrogen. This gas is transformed into simple nitrite and nitrate compounds by bacteria that live among the roots of certain cover crops such as clover, soybeans, and alfalfa (Fig. 2-2). A small amount of nitrogen is also transformed into these simple nitrogen compounds during thunderstorms, when lightning produces the energy needed to initiate the process. Such nitrogen compounds are carried down to the earth by precipitation, and they replenish the earth's nitrogen supply. The cycle continues when plants convert the nitrogen to protein, which is then ingested by animals and man. Eventually the nitrogen is returned to the earth either through the elimination of waste products or by decomposition after death. Finally, the earth gives up nitrogen to the atmosphere through the

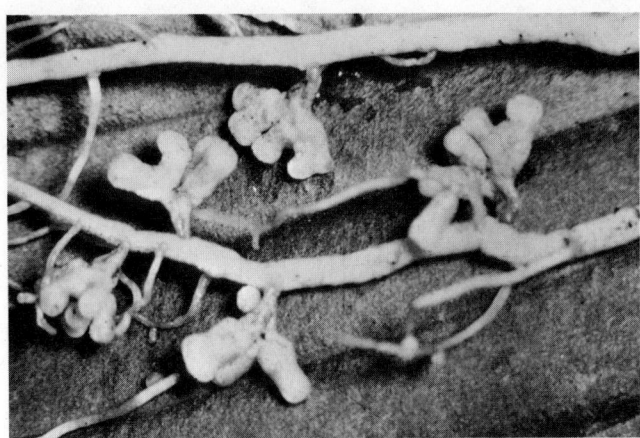

Figure 2-2 Nitrogen-fixing bacteria grow in the nodules of the clover roots. (Courtesy Carolina Biological Supply Co., Burlington, N.C.)

work of the ever-present bacteria, which completes the cycle.

The rate at which the atmosphere loses nitrogen to the soil is essentially balanced by the rate at which it is returned to the atmosphere by *denitrification*. The widespread use of fertilizers however, may be upsetting this balance.

Oxygen

The second most abundant gas in the atmosphere (21% by volume) is oxygen, which was discovered by Joseph Priestly, an English clergyman. Oxygen is a gas that is essential to all air-breathing animals and accounts for 89% by weight of all the water found on the earth or in the atmosphere.

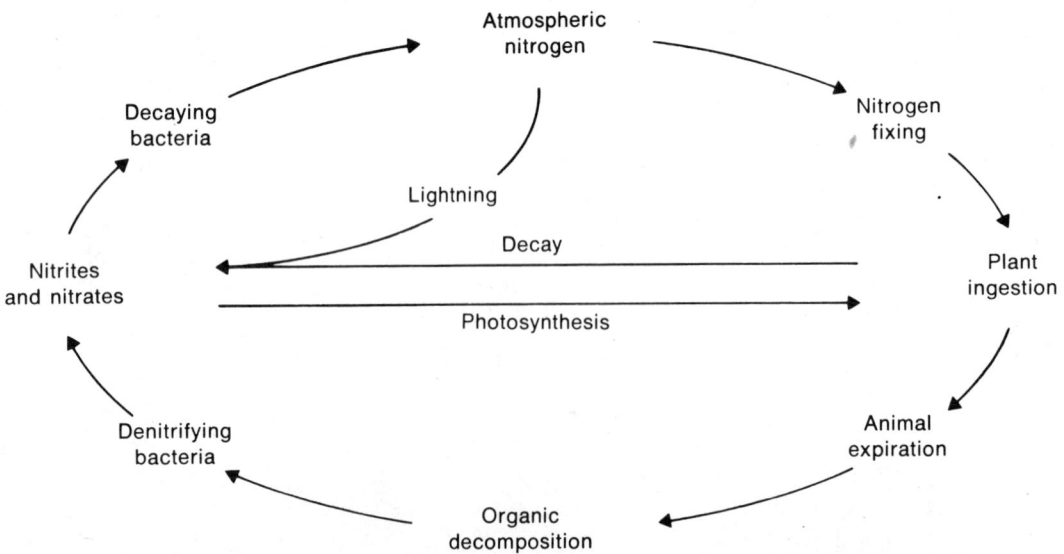

Figure 2-1 The "nitrogen-cycle" process consists of converting nitrogen into nitrogen compounds and then into nitrogen again. (From Thomas L. Burrus and Herbert J. Spiegel, *Earth in Crisis*, 2nd Ed., St. Louis, 1981, The C. V. Mosby Co.)

This colorless, odorless, tasteless gas is highly reactive and is capable of combining with almost all the other elements. This ability to combine easily results in the formation of many oxides. In addition to the previously mentioned *nitrogen oxides*, oxides of sulfur and carbon can also produce major pollution problems for urban areas.

Sulfur is an abundant by-product of the incomplete combustion of fossil fuels; it is released into the atmosphere during combustion. There it combines with free oxygen (O_2) to form *sulfur dioxide* (SO_2). This colorless, acrid-smelling gas can in turn be easily transformed into *sulfuric acid* (H_2SO_4) by combining with the water droplets in the air. The end result is usually the yellowing of vegetation, corrosion of iron or steel and upper respiratory diseases in humans. Breathing a mist of this highly toxic substance can severely injure lung tissues.

Argon

Argon is one of the six inert gases found in the atmosphere. It is relatively abundant compared to the remaining gases in the air. Since argon does not combine with other elements, its usefulness is quite limited. The gas is mainly used in prolonging the life of tungsten filaments in light bulbs and in welding.

Carbon Dioxide

The last of the four most abundant constituents of the atmosphere is carbon dioxide. This is a colorless, odorless, gas that has a sour taste when mixed in sufficient quantities with water. The most important function of CO_2 is its role in photosynthesis. In this process it is absorbed by plants that combine it with water in the presence of sunlight to produce sugar and free oxygen. The sugar is used by the plant producer, while the oxygen is released into the atmosphere (Fig. 2-3).

One of the most familiar properties of carbon dioxide is its ability to change directly into a solid without passing through the liquid stage. This process, called *sublimation*, is reversible, and this makes solid CO_2 a valuable refrigerant, since there is no messy liquid left as an end product. DRY ICE, the trade name of solid CO_2 is an important tool in scientific research, where very low temperatures are often needed.

Carbon dioxide is also important as a major pollutant of our atmosphere. This is because CO_2 is released as a result of the incomplete combustion of fossil fuels. Once again automobiles are among the main contributors to the formation of this gas. Since its relative abundance is dependent on the extent of urbanization in any given area, extremely high concentrations of carbon dioxide are found in metropolitan areas.

Carbon dioxide also prevents radiation emitted by the earth from escaping into space. Thus, it is believed that increases in the CO_2 content can lead to worldwide temperature increases and an alteration of world climate. Fortunately, some of the increased CO_2 produced by the combustion of fossil fuels is dissolved in the oceans and some is probably used in plant growth. Thus, the *biosphere* and *hydrosphere* accommodate much of the increased amount of CO_2 resulting from combustion, and only about 40 to 50%* of the amount released by burning fossil fuels enters the atmosphere directly.

* J. Murray Mitchell, Jr., "Carbon Dioxide and Future Climate," *EDIS*, Reprint from Environmental Data Services, NOAA, March, 1977.

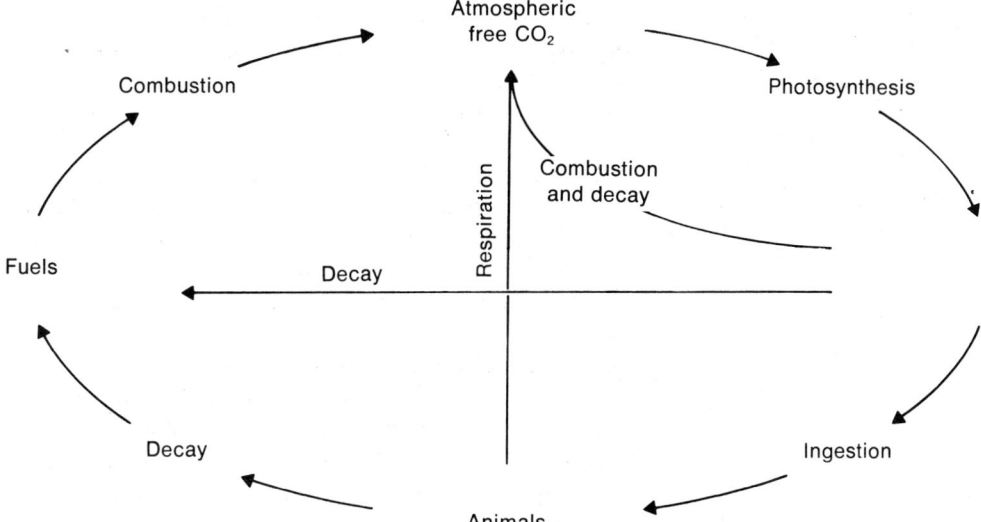

Figure 2-3 Path of carbon dioxide from atmosphere to plant and animals that use it in their life cycles to fossil fuels and back to the atmosphere. (From Thomas L. Burrus, and Herbert J. Spiegel, *Earth in Crisis*, 2nd Ed. St. Louis, 1980, The C. V. Mosby Co.)

Ozone

Ozone is found in the atmosphere in varying but very minute quantities. Although this gas is relatively rare, scientists consider ozone critical to the existence of life on this planet. If this bluish-tinged gas did not exist at heights of 20 to 30 km (12 to 18 miles) in the atmosphere, highly concentrated amounts of ultraviolet radiation would strike the surface of the earth, causing severe damage to plant and animal life. Fortunately, the ozone layer absorbs most of this potentially lethal radiation. The small amount of ultraviolet radiation that does eventually penetrate the earth's surface is still enough to produce a painful sunburn, and large doses of ultraviolet light can result in skin cancer.

However, human activities threaten to reduce the ozone layer. The widespread use of *chlorofluoremethanes* (CFM) commonly used in spray cans as propellants and as working fluids in refrigeration systems has resulted in a high concentration of CFM in the stratosphere. Through photochemical reactions these high concentrations react to break down the ozone. It is for this reason that chlorofluoremethanes have been prohibited in the many spray can products that we use every day.

The widespread use of chemical fertilizers, while a boon to agriculture production, has also increased the amount of nitrogen oxides (NO_x) in the stratosphere which, in turn, affects the production and removal of ozone. The above activities of society and their possible effect on the stratosphere are fairly recent discoveries but, because they can have such a serious impact on our well-being, are receiving considerable and well-deserved attention.

At the earth's surface ozone presents a definite threat to human health. This pungent, garlic-smelling gas is one of the most powerful air pollution irritants known. At low levels it can be created by lightning or by the reaction of elements in automobile exhaust. The production of nitrogen dioxide (NO_2) and its subsequent reaction with sunlight yields ozone as an end product. This photochemical reaction sometimes leads to smog problems in the greater Los Angeles, California area. The possible effects of ozone on humans include eye irritations, headaches, coughing, and severe lung diseases. The effects on agriculture include reduction of crop yields and damage to forests while ozone affects industry mainly through rubber deterioration.

Water

Water vapor, just as the name implies, is water in gaseous form. It enters the atmosphere through the processes of *evaporation* from water surfaces and *transpiration* by plant life. Since the water available for conversion into water vapor is found on the earth's surface, it is quite understandable that 90% of all water vapor in the atmosphere is found in the lower several kilometers (2 to 3 miles).

Water vapor is quite variable, ranging from almost zero to about 4% by volume. Part of the importance of water vapor is that over the range of temperatures found on earth and in the atmosphere, water can exist in a liquid and solid state as well as gas. Thus it is one of the most important variables in the atmosphere. More about the properties of atmospheric water in all its phases will be discussed in Chapter 5.

Solid Particles

Spread throughout the lower troposphere are solid particles called *aerosols*. The major sources of aerosols are sea salt from evaporated ocean spray, wind-blown dust, and debris from forest fires and volcanoes. Human activity also contributes to the particle load of the atmosphere through the burning of fossil fuels, agricultural operations, and manmade forest fires. Nevertheless, it has been estimated that worldwide human activity accounts for only about 10% of the aerosol burden of the atmosphere. Because manmade sources are concentrated in relatively small areas, on a local level human activities can account for nearly all of the aerosols present in the atmosphere.

The residence time of aerosols in the atmosphere is on the order of days to weeks. Most aerosols are washed out in precipitation and some of the larger particles tend to settle gravitationally.

Aerosols can have an important impact on the surface temperature; however, scientists are uncertain whether the overall effect should be one of warming or cooling. The dilemma comes about because aerosols, like CO_2, can trap energy emitted from the earth's surface, which results in a warming effect. However, the solid particles can also reflect more of the incoming solar energy back to space yielding a cooling effect. Research shows that warming or cooling depends on a variety of factors, such as the physical properties of aerosols, and altitude at which they reside. The violent volcanic eruption of Krakatoa (near Java) in 1883, for example, produced volcanic ash that remained suspended for nearly 2 years before settling back to earth. With this quantity of volcanic dust and ash in the air over such a long time period, and a decrease of solar energy reaching the surface, it is quite obvious that temperatures around the world could have been affected. In fact, the year following the eruption of Krakatoa was known as "the year without a summer." Temperatures fell drastically, rivers froze over in the late spring, crops were ruined, and the economies of many nations suffered. Although the following winter was extremely bitter, normal temperatures finally returned two years later.

Cloud Condensation Nuclei

Many of the aerosols act as centers on which condensation of water vapor takes place. Those aerosols are also called cloud condensation nuclei. These nuclei are extremely important in the formation of clouds and subsequent rainfall. For example, during METROMEX,* which is a research project designed to investigate city induced weather effects at St. Louis, Missouri, it was observed that a significant increase in summer rainfall and a large increase in thunderstorm and hailstorm activities occurred just east of the city. Much of the increased intensity and amount of precipitation is related to the increase of aerosol and condensation nuclei present over and downwind of the highly industrialized city. The observed increase in summer precipitation (10-30%) has had both positive and negative impacts. For example, agricultural yields in the area have increased while annual crop losses due to hail have been three times greater than elsewhere in areas further east of St. Louis since 1950.

VERTICAL STRUCTURE OF THE ATMOSPHERE

Although the atmospheric gases are well mixed to heights of 80 km (50 miles), the physical characteristics of the atmosphere in the vertical direction are quite distinct resulting in a well defined layered structure. The various layers or shells are most commonly identified by their characteristic temperature structure, shown in Figure 2-4.

Troposphere

The layer closest to the earth's surface is known as the troposphere (from the Greek *tropo*, meaning to turn). It is characterized, on the average, by a uniform decrease of temperature with height, and has an average thickness of about 11 km (7 miles). It reaches its greatest height of 18 km (11 miles) at the equator and its shallowest point of 8 km (5 miles) at the poles. These of course represent conditions obtained by averaging many years of observations. An individual observation of the vertical structure of the troposphere may show large departures from the average. The troposphere contains about 75% of the total mass of the atmosphere and due to its closeness to the earth's surface, contains essentially all the water vapor and particulate matter. It also is the layer where most of the heat exchange between the earth and atmosphere takes place. An important aspect of this layer is the over-turning of air that occurs here. This is the mechanism which distributes heat, moisture and solid particles through the troposphere. Obviously, this layer is of primary concern to meteorologists since weather, as we know it, and the processes that affect the weather, take place in this lowest layer.

Stratosphere

The second layer in the atmosphere is called the stratosphere (from the Greek *strato* meaning horizontally layered). It extends on the average from 11 to 45 km (7 to 28 miles) above the earth's surface and, as can be seen in Figure 2-4, its temperature is nearly constant in the lower portion of the layer and increases with height elsewhere. As a consequence of this temperature distribution vertical motions and overturning are greatly reduced. In fact, the stratosphere acts as a gigantic lid on the overturning of the troposphere. This results in the following characteristics.

1. There are only minute amounts of water vapor in the stratosphere, put there by the giant thunderstorm clouds that occasionally penetrate into the lower stratosphere.
2. Dust and other solid particles are also rarely found there. However, powerful volcanic eruptions such as Krakatoa are capable of injecting solid particles into the stratosphere and because of the lack of overturning, they tend to remain there a long time.
3. There are a few clouds and no precipitation in the stratosphere. However, a certain type of cloud called *nacreous* or *mother of pearl* cloud, which is named for its irridescent appearance, is found there. The composition of nacreous clouds is unknown, and it is rarely observed.
4. Most of the ozone found in the atmosphere is located in the stratosphere, with a maximum concentration centered at approximately 25 km (15 miles). It is in this layer that most of the ultraviolet energy from the sun is absorbed. This absorption of energy plays an important role in determining the temperature structure of the stratosphere.

The stratosphere contains about 24% of the mass of the atmosphere. Thus the troposphere and stratosphere combined contain 99% of the total mass of the atmosphere. This layer is also of importance to meteorologists particularly with the advent of supersonic aircraft that regularly fly in the stratosphere, and require meteorological support such as wind and temperature forecasts.

Mesosphere

The layer located above the stratosphere is called the mesosphere (from the Greek *meso* meaning middle). It lies 45 to 92 km (27 to 55 miles) above the earth. The mesosphere is the third of the four major layers of the atmosphere and is characterized by sharply declining temperatures that eventually reach −93°C (−137°F) at the top.

---

* For more information see *Metromex Update*, Bulletin of the American Meteorological Society, Vol. 57, No. 3 (1976), pp. 304-308.

22 THE ATMOSPHERE

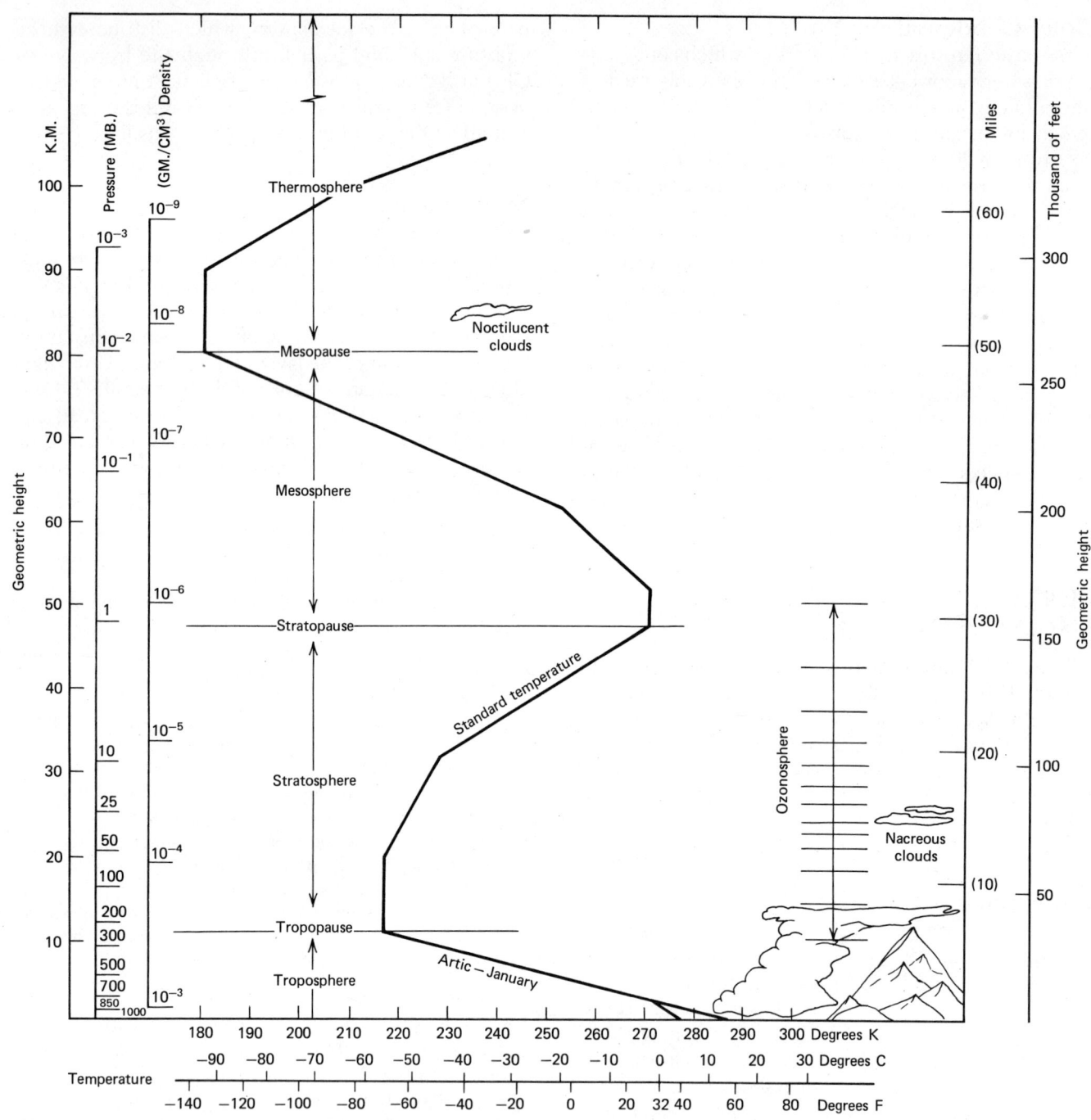

Figure 2-4 The average vertical temperature structure (standard atmosphere) and its relationship to the various atmospheric layers. (Adapted from U. S. Navy Weather Research Facility, Monterey, California)

Thermosphere

The last layer of the atmosphere, which begins at 92 km (55 miles) above the earth, is called the thermosphere (from the Greek *Therme*, meaning heat). The thermosphere is characterized by high temperatures due to its extremely thin air. Temperatures in the thermosphere are estimated to be as high as 1370°C (2500°F). However, this is a measure of molecular activity and because of the very low density at this altitude would not at all be like temperature near the surface.

Two rare and peculiar phenomena take place in the thermosphere. The first, the *Aurora Borealis* and *Aurora Australis* (the northern and southern lights) are glowing brilliant yellow, green, and red colors seen in the skies of the polar regions at night. These auroral displays result from showers of charged particles being thrown out from the sun into the upper atmosphere during solar magnetic storms (Fig. 3-2).

The second rarity is that of cloud formations called *noctilucent clouds* that are thought to be composed of cosmic dust particles. These clouds can be seen moving swiftly across the sky on long clear summer nights. Twilight is usually the best time to observe them.

Zones of Transition

Separating the layers of the atmosphere are zones of transition. These are regions no more than 1 mile thick that are called *pauses*. Thus the zone marking the transition from troposphere to stratosphere is called the *tropopause*. Other transition areas are the *stratopause* and *mesopause*.

Ionosphere and Ozonosphere

Superimposed on the atmosphere are two more important shells or layers. The highest of the two is the ionosphere, which is located above the thermosphere. Here unfiltered x-ray and ultraviolet energy from the sun ionize the atoms of the gases located there. These charged atoms are called *ions*, hence the name "ionosphere." These ions interfere with shortwave radio transmission and repeatedly reflect these short waves and broadcast them back to earth. As a result it is possible to transmit and receive voice communications over great distances. Ham radio operators express this phenomenon as the radio wave skip. This skip is more pronounced at night than at any other time (Fig. 2-5).

The second important layer is the ozonosphere. This layer extends from the bottom of the stratosphere to about 50 km (30 miles) just above the stratosphere. All the ozone of the upper atmosphere is found within the ozonosphere. It should be noted that if all the ozone within the ozonosphere were compressed and brought down to sea level, a layer only a fraction of an inch thick would result. This may seem to be only a trivial quantity, but, as previously mentioned, without this all-important gas excessive amounts of ultraviolet energy would penetrate to the earth's surface, which could prove fatal to plant and animal life on earth.

EXPLORATION OF THE LOWER ATMOSPHERE

At about the beginning of the twentieth century the most commonly used methods for making measurements in the lower atmosphere were kites and manned balloons. Manned balloons were dangerous and limited in the heights they could reach, rarely exceeding 5 to 8 km (3 to 5 miles). Kites, while being able to achieve somewhat higher altitudes, were dependent on sufficient ground wind for flying.

With the perfection of radio communication the unmanned balloon became the main method of measuring the lower atmosphere. The balloon system, known as a radiosonde, is shown in Figure 2-6. The balloon is generally filled with either hydrogen or helium and can reach altitudes of about 30 km (18 miles) before bursting. The instrument package measures temperature, pressure, and humidity as a function of height. The measurements are transmitted via radio signals to a ground station, which also tracks the balloon using a radio theodilite to obtain wind data. This system is used by weather services throughout the world to provide routine (usually twice daily) observations that are utilized in regional and global weather forecasts. A major deficiency of this system is that it is primarily limited to

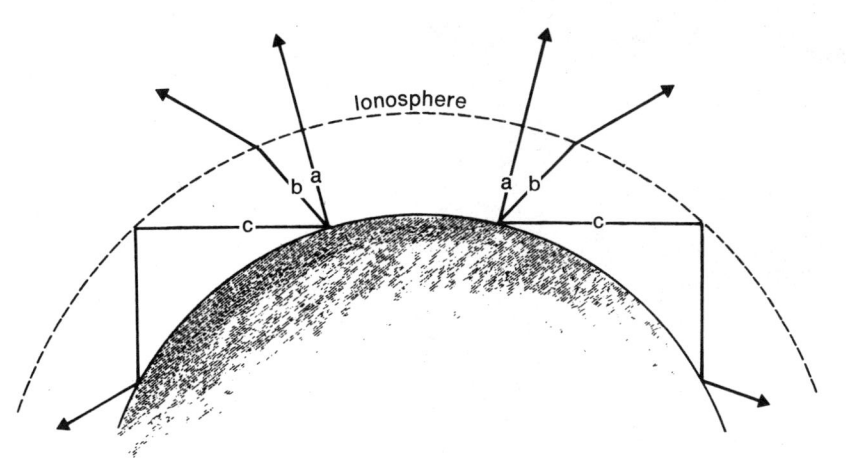

Figure 2-5 Only radio waves of certain angle will be reflected by the ionosphere. Vertical or near vertical rays (*a* and *b*) pass through unaltered or slightly changed in direction. Waves that hit the ionosphere at critical angles (*c*) are reflected. (From Thomas L. Burrus and Herbert J. Spiegel: *Earth in Crisis*, 2nd Ed., St. Louis, 1980, The C. V. Mosby Co.)

24 THE ATMOSPHERE

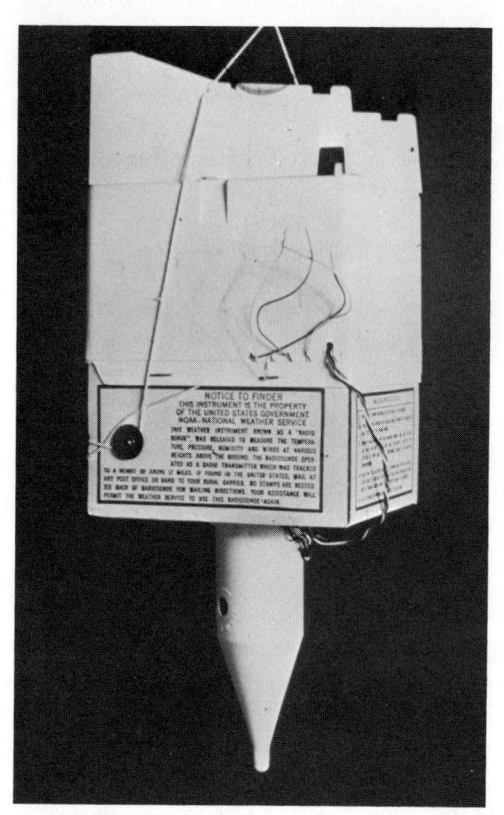

Figure 2-6 A radiosonde consists of an instrument package suspended from a balloon that is usually filled with helium to achieve buoyancy. (*left*) Instrument (*right*) entire system. (Courtesy NOAA)

Figure 2-7 TIROS N spacecraft used to obtain global meteorological data. There will be two such satellites orbiting the earth at the same time at altitudes of 840 km (520 miles) and 870 km (540 miles). (Courtesy National Environmental Satellite Service, NOAA)

observations over land. Consequently, the oceanic areas are poorly measured, thus leaving large areas unobserved.

This problem of large unobserved areas, has been reduced to a large extent with the advent of meteorological satellites (Fig. 2-7). These satellites orbit the earth at altitudes of 900 km (540 miles) or higher and provide data on the vertical temperature and pressure of the atmosphere by the use of remote sensing techniques. These techniques are called remote sensing since they do not physically measure the temperature and pressure at a specific height, but receive radiation data from the atmosphere. These data are transmitted to ground stations and then analyzed to produce a temperature and pressure sounding. The satellites observe the entire globe and, therefore, can provide the important information over the oceanic areas where there are no radiosondes. Satellites are also capable of providing wind data, by tracking the motion of clouds over time intervals of 30 minutes to a few hours. However, they provide the information only at places and heights where clouds exist.

Other techniques are also employed to measure the lower atmosphere. One procedure is to drop an instrument package on a parachute from an aircraft. The aircraft in this instance is the receiving station for the radio signals. This procedure is used most frequently in special investigations such as aircraft surveillance of hurricanes and storms over nearby coastal area.

Constant level balloons are sometimes employed in investigations of the atmosphere. These balloons, as their name indicates, are designed to remain at a constant pressure level in the atmosphere. They are tracked for periods of months and are used mostly for investigations of wind-flow characteristics at a given altitude.

EXPLORATION OF THE UPPER ATMOSPHERE

As in the lower atmosphere, the upper atmosphere is explored by radiosondes and satellites.

Because of the balloon bursting height, radiosondes are limited to altitudes of about 30 km (18 miles) or less. Thus they are capable of providing information only in the lower portion of the upper atmosphere. This deficiency is overcome by meteorological rockets. These rockets are launched from a few locations around the

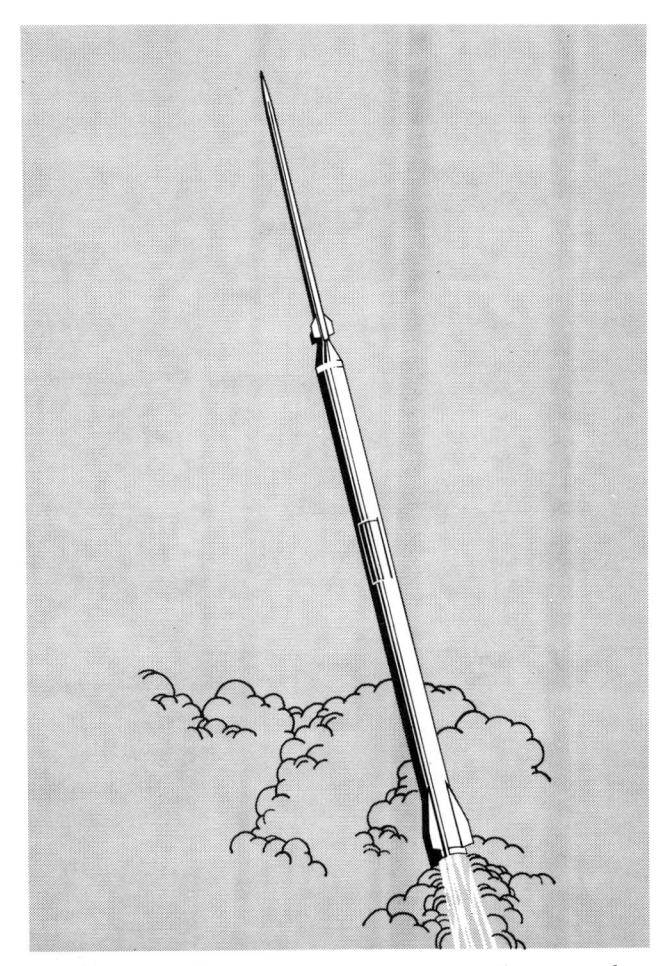

Figure 2-8 Rockets are used to investigate the atmosphere between 20 km (12 miles) and about 70 km (42 miles).

world and reach altitudes of about 65 to 70 km (39 to 42 miles). The rocket (Fig. 2-8) carries a package that is ejected at a predesignated height, and descends on a parachute much like a dropsonde. Temperature and pressure observations are radioed to a ground station where radios or radars are also employed to track the instrument to obtain wind information. The rockets provide useful information to about 20 km (12 miles), thus providing about 10 km (6 miles) overlap with radiosonde data.

The combination of radiosondes, satellites, and rockets are providing meteorologists with a good picture of the horizontal and vertical structure of the upper atmosphere.

3
THE SUN-EARTH ENERGY SYSTEM

C. G. Abbot (1872–1919) an American astrophysicist who, in the early 1900s, as the Secretary of the Smithsonian Institute, pioneered the measurement of solar radiation. Abbot devised a method of measuring solar radiation and, with the aid of other scientists, calculated an acceptable value for the *solar constant*.

LEARNING OBJECTIVES

After reading this chapter, you should be able to:

1. Describe the prominent features of the sun.
2. Compare the three major processes for transferring energy.
3. List the major wavelength ranges of energy in the electromagnetic spectrum and describe their effects on people.
4. Given the incoming and outgoing energy, compute an energy balance of the earth-atmosphere system.
5. Compare earth energy to solar energy.
6. Describe the greenhouse effect.
7. Define *albedo, net radiation, black body, reflection, absorption, scattering, refraction,* and *specific heat*.
8. Explain why the sky is blue.

Nearly all energy necessary for people and the atmosphere comes ultimately from the one star in our solar system, the sun. This energy source fuels our atmospheric circulation, drives the earth's weather systems and, most importantly, is responsible for the *evaporation-precipitation cycle*. This cycle not only has meteorological significance but also an economical, agricultural, and social impact. However, this solar energy is not always used directly even though one thinks immediately of a surface being heated as a result of exposure to visible sunlight. There are several different processes and energy transformations that occur before the energy can be used to drive the winds or cause weather changes. This chapter describes some of the more important processes in order that you may fully understand one of the major ingredients that make up the weather.

THE SUN

Approximately 1.5×10^8 km (93 million miles) away from earth is a huge rotating sphere of hot gases. Called through the ages by many names such as *Ra* and *Apollo*, it has been both feared and adored. The earth, like the other eight planets, orbits the sun in an elliptical path being closest to it about January 3 and furthest away six months later around July 3.

Why is it the sun, and not any of the other celestial bodies of our solar system, that is the source of our energy?

The sun is our solar system's most massive body, composed of mostly helium and hydrogen gases. In the *core* or center of the sun a natural nuclear reaction takes place each second in which huge quantities of hydrogen gas are transformed into helium gas. This reaction also converts some hydrogen mass into energy, resulting in extremely high interior temperatures. The energy is transferred to the photosphere where it is radiated away.

The energy we receive from the sun exits from the *photosphere*, a shell of gas, 500 km (300 miles) thick. At its outer surface the photosphere extends into a second region known as the *chromosphere*.

The region above the chromosphere is called the *corona*, which is a layer of variable thickness. Figure 3-1 illustrates the various layers and some of the features discussed below.

PROMINENT SOLAR FEATURES

Although there are many solar features, two of them are of particular interest to meteorologists. The best known are the dark features called sunspots, which are observed on the photosphere (Fig. 3-1). They are dark because they are slightly cooler than the surrounding gases, and represent disturbances that have penetrated the photosphere from below. Sunspots have been observed for hundreds of years, and scientists have established a definite cycle in the number of sunspots. This cycle, known as the "*sunspot cycle*" has on the average an 11-year period between maxima. Many attempts have been made to relate sunspot activity to weather changes, and while a considerable amount of statistical evidence has been presented suggesting such relationships, no clear-cut physical mechanisms have yet been established, and scientists are still debating the issue. Nevertheless, there are changes in thermosphere temperatures related to sunspot activity.

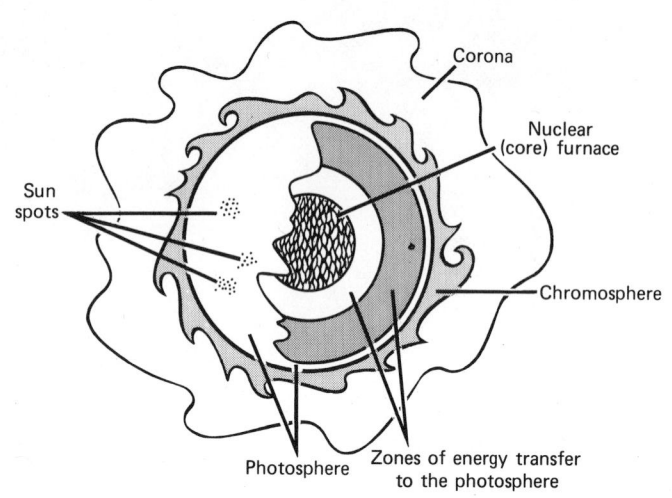

Figure 3-1 Cross section of the sun showing its various layers and features.

The second feature is that of *solar flares*. These are short-lived, very intense bursts of energy that occur in the chromosphere. Large amounts of x-ray, radio, and ultraviolet energy are emitted during solar flares and this, in turn, can affect radio transmissions on earth. During these eruptions material particles such as electrons and protons are emitted, which in turn produce the bright *aurora* of the upper atmosphere. These are called the *aurora borealis* or *northern lights*, and *aurora australis* or *southern lights*. Figure 3-2 is a photograph of the Aurora Borealis as seen from the University of Alaska during the International Geophysical Year, 1955.

TRANSFER OF HEAT ENERGY

To gain a better understanding of how radiation received from the sun affects our earth-atmosphere system, it is necessary to examine how heat is transferred. There are three principal ways of transferring heat energy: *conduction*, *convection*, and *radiation*.

Conduction is a process of heat transmission accomplished by the transfer of energy during the collision of rapidly moving molecules within the substance itself. A common example is the heating of a metal rod at one end (Fig. 3-3). Solids such as the rod are good conductors and after a short time the far end of the rod gets hot. Conduction is used when an electric stove is turned on with a pot of water on it. The coils heat up, heating the pot, which then heats the water, all due to this same process. Another example is the use of a soldering iron. The melting of the solder is caused by the transfer of heat by conduction to both the solder and the material being soldered.

Since gases and liquids are both poor conductors and the conduction in air is so slow, this process is only important in heat flow from the earth's surface to the air immediately above it.

Convection is the transfer of heat by the actual movement of the heated substance. (In meteorology convection usually is reserved for vertical movement.) This process occurs in substances with molecules that are free to move such as a liquid or gas. An example of heat transfer by convection is the rising of less dense warm air and the sinking of denser cold air (Fig. 3-4). This of course is the principle used to heat homes. Convection is one of the most important mechanisms for

Figure 3-2 Multiple-rayed arcs of the Aurora Borealis. (Courtesy V. P. Hessler, Geophysical Institute, University of Alaska)

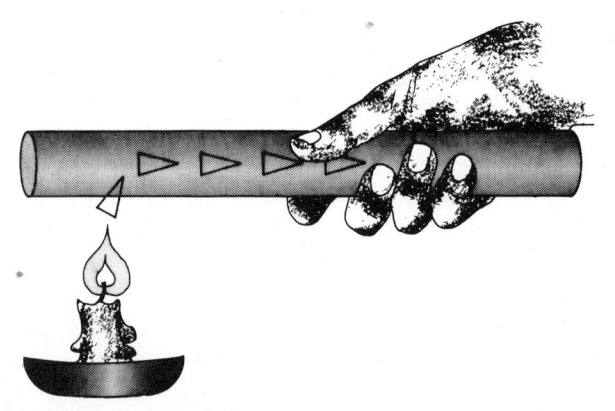

Figure 3-3 The collision of molecules results in the direct transfer of energy and is called conduction. (From Thomas L. Burrus and Herbert J. Spiegel, *Earth in Crisis*, 2nd Ed., St. Louis, 1980, The C. V. Mosby Co.)

Figure 3-4 Air rises up a chimney because it is heated from below, becoming less dense and, therefore, more buoyant than the surrounding air. (From Thomas L. Burrus and Herbert J. Spiegel, *Earth in Crisis*, 2nd Ed., St. Louis, 1980, The C. V. Mosby Co.)

heating the atmosphere. It should be noted that liquids and gases can also move horizontally and when heat is transferred in this way the process is called *advection*.

Radiation is the process of heat transfer by wave motions. Radiation is uniquely different from conduction and convection because it requires no molecular motion or intervening medium. Therefore, energy can be transferred through a vacuum such as outer space. This is a very important form of heat transfer since it is how the sun's energy reaches the earth. Heating by radiation only takes place when the radiation is absorbed by some medium, such as air, water, or earth. All substances radiate energy, the amount and wavelength of radiation emitted being determined by their temperature. (Wavelength is the distance between successive peaks of the wave motion, as shown in Fig. 3-5) When a substance absorbs energy, it heats up and immediately starts to emit energy. Therefore, the total amount of heating depends upon the amount of energy absorbed versus the amount emitted. Most radiation is not visible to us, but the heat we feel when we stand next to a very hot object is mainly a result of radiation. Infrared film is a good example of how use is made of the radiative properties of substances. It acts in much the same way as ordinary film except that it responds to wavelengths that are outside the range of wavelengths that the

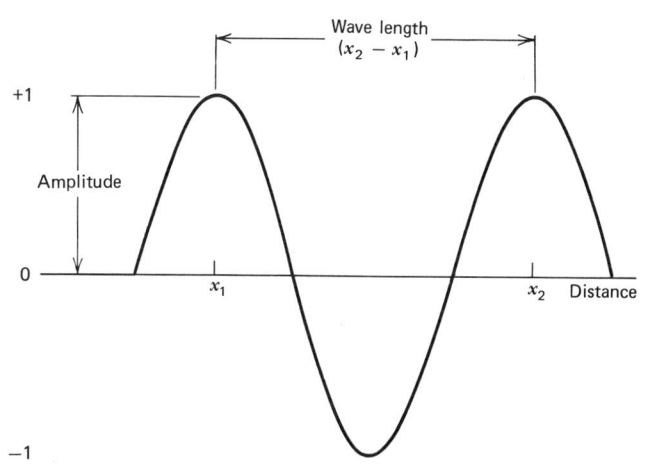

Figure 3-5 Wavelength is the distance $(x_2 - x_1)$ measured from crest to crest while frequency is the measurement of how many peaks pass a point in a set time (e.g., 1 second). Period is the time it takes 1 wavelength to pass a given point. Amplitude of the wave (A) is $\frac{1}{2}$ the distance between trough and crest.

human eye can see. The science of remote sensing, that is, measuring and recording information from a distance, depends on instruments that are sensitive to various wavelengths of radiation.

For many years it has been known that *clear air turbulence* (irregular and random motions in cloud-free air) has been the cause of numerous aircraft accidents. Recently, scientists have discovered that by using an *infrared radiometer* they could remotely measure the temperature fluctuations associated with this phenomena. Therefore, they could detect the most probable areas of occurrence and provide an early warning for aircraft. A more detailed discussion on the various wavelengths of radiation is presented later in this chapter.

HEAT VERSUS TEMPERATURE

Heat and temperature are closely related quantities that are frequently confused. Heat is best thought of as the amount of energy absorbed by a substance. For example, water boils because it has absorbed a certain amount of heat. Temperature, on the other hand, represents a measure of the motion of the molecules of the substance being heated. Different substances require different amounts of heat to achieve the same molecular motion or what we call the same temperature. Conversely, different substances having the same amount of heat applied to them will experience different temperature changes.

A measure of the amount of energy required to raise the temperature of one gram of a substance is given by a quantity called the *specific heat*. In gases, specific heats are generally measured at constant pressure or constant volume. Water has a specific heat about four times larger than air at constant pressure. Thus, it requires four times as much heat (energy) to raise the temperature of water 1 degree than it does to raise the air temperature 1 degree. The specific heat of soil is also quite low, about $\frac{1}{2}$ to $\frac{1}{3}$ that of water. This is why land heats up faster than water with the same amount of energy. When we consider that about 75% of the earth's surface is covered by water, the different specific heats have far-reaching consequences in meteorology, particularly in helping shape the surface temperature distribution.

ELECTROMAGNETIC ENERGY

Energy from the sun is known as *electromagnetic energy*. This *solar radiation*, as it is sometimes called, travels at the speed of light 3×10^8 meters/sec (186,000 miles/sec) and takes 8 minutes to reach us. Scientists have been able to identify radiant energy of different wavelengths, each having its own unique characteristics. This collection of wavelengths is known as the electromagnetic spectrum and ranges from the very short gamma rays to the very long radio waves. Figure 3-6 shows the different regions in the electromagnetic spectrum.

Because of the high temperature of the sun's surface, almost 99% of the solar energy reaching the top of our atmosphere arrives in the form of short wavelengths of energy. In fact, approximately 40% of the energy lies within an area sensitive to the human eye. This is called the *visible range* and lies between 0.4 and 0.7 micrometers. A *micrometer* (μm) or micron, as it is sometimes called, is a measurement of length and is equal to 10^{-6} meters, which is a very small unit. Other units commonly used to measure wavelength are the Angstrom (10^{-10}m), millimeters (10^{-3}m), and kilometer (10^3m).

The visible spectrum consists of wavelengths that can be discriminated by the human eye as different colors. These range from violet at the short end to the much longer red wavelengths. Very little of this visible energy is absorbed by the earth's atmosphere so that while a major portion of the total incoming energy falls in this range, the atmosphere is not directly heated by it.

Other wavelengths of energy shorter than the visible consists of *ultraviolet*, *x-rays*, and *gamma radiation*. These three account for 9% of the energy reaching the earth's atmosphere but very little ever reaches the earth's surface. Gamma and x-rays are mostly absorbed by the high altitude's oxygen and nitrogen molecules that become ionized upon absorbing these rays and form the ionosphere. Ultraviolet (U-V) energy is mostly absorbed by the ozone in the stratosphere. That small amount of U-V radiation that does get through is intense enough to cause the radiation burns commonly called sunburn. Prolonged exposure to this energy is the leading cause of skin cancer. Environmentalists are concerned that pollutants may decrease the amount of ozone in the atmosphere and allow more ultraviolet radiation to reach the ground.

On the other side of the visible spectrum is the near infrared. A small part of this energy is strongly absorbed by the carbon dioxide and water vapor in the atmosphere. This absorption, which represents a small fraction of the sun's total energy, heats the atmosphere very little. It is the energy absorbed at the earth's surface and then retransmitted that is responsible for the air's temperature.

The distribution of the incoming solar energy will be considered later when the heat balance of the earth—atmosphere system is discussed.

SOLAR CONSTANT

If we were to measure the energy received at the top of the atmosphere on an area perpendicular to the sun's rays (Fig. 3-7), we would find that the amount of energy remained constant except for variations in the earth–sun distance. This energy, known as the *solar constant*,

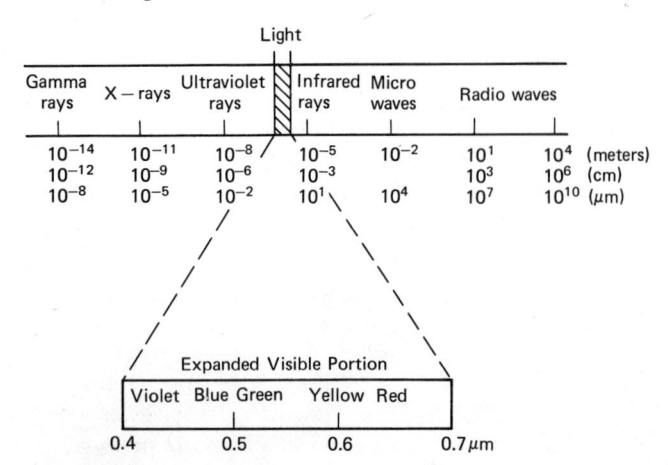

Figure 3-6 The electromagnetic spectrum.

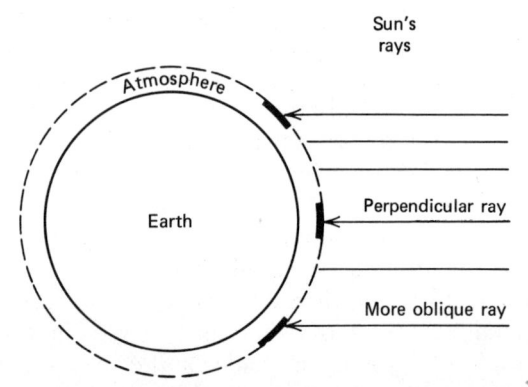

Figure 3-7 There is only one ray that is perpendicular to the top of the atmosphere at any one time. Notice the larger angles of incidence toward the poles.

has been measured from rockets, balloons, and most recently by satellites. Since the solar constant represents the amount of energy per unit time, per unit area at the top of the atmosphere, perpendicular to the sun's rays at the average earth-sun distance, its dimensions can be written as (cal/cm²/min). This label is given the additional name of the *langley*/min in honor of Samuel P. Langley, an American radiation expert. Latest satellite measurements indicate that this constant value is 1.97* langleys/min. It is interesting to note that the average earth-sun distance was chosen for the definition of the solar constant. This was needed simply because the earth's path around the sun is elliptical and the amount of energy received varies by about ±3.3% at its closest and farthest points from the sun.

To gain some idea of how much energy this represents, it has been estimated that the amount of energy falling on a football field in 12 hours is enough to run the street lights of New York City for about 2 days. The daily amount of solar radiation actually incident at any given location at the top of the atmosphere depends on several factors in addition to the varying earth-sun distance. These principal factors are the angle of incidence of solar radiation and the duration of daylight.

Angle of Incidence

The angle of incidence is a very important control on how much energy is received at the top of the atmosphere. A ray of energy perpendicular to a surface (Fig. 3-8) describes a circular area, while a ray at an oblique angle (>90°) produces an ellipse. If both rays represented the same amount of energy, the total energy per unit area would be greater for the circle because it represents a smaller area. In other words, the energy at an oblique angle is spread over a bigger area. This is exactly the situation encountered when we consider the energy streaming toward the earth. Because the earth's surface is curved the solar beam is more oblique at the polar regions than at the equator. This results in a smaller amount of energy per unit area at the poles.

Duration of Daylight

The duration of daylight is closely related to the season of the year, which in turn is related to the position of the sun relative to the earth. During the northern hemisphere summer the North Pole is tilted toward the sun, resulting in a longer period of daylight and more perpendicular rays. During the winter the North Pole points away from the sun, resulting in shorter days and more oblique rays. Interestingly, during the northern

---
* R. C. Willson, S. Gulkis, M. Janssen, H. S. Hudson, and G. A. Chapman, Observations of Solar Irradiance Variability, *Science*, Vol. 211 (1981), 700–702.

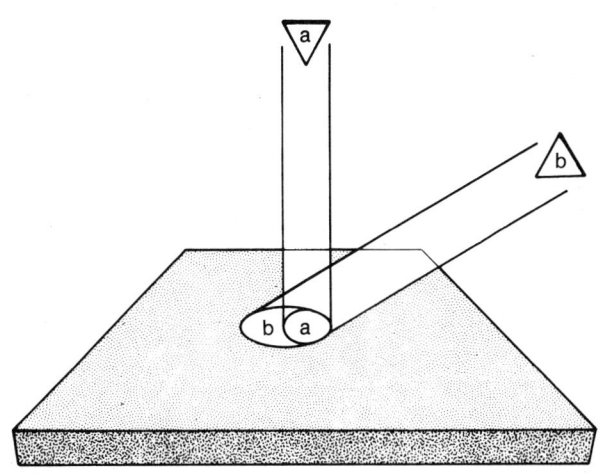

Figure 3-8 A sketch of angle of incidence and the area it describes. Ray *a* circumscribes a circle while ray *b*, which has a large angle of incidence, describes an ellipse. You can perform a simple experiment showing how the area changes with incidence angle by using a flashlight and a flat surface. (From Thomas L. Burrus and Herbert J. Spiegel, *Earth in Crisis*, 2nd Ed., St. Louis, 1980, The C. V. Mosby Co.)

hemisphere summer the earth is farthest from the sun. The longer period of daylight and the more perpendicular rays overcome the effect of distance.

The *daily* amounts of incoming solar energy at the top of the atmosphere have been calculated and are shown in Figure 3-9. The largest values, in excess of

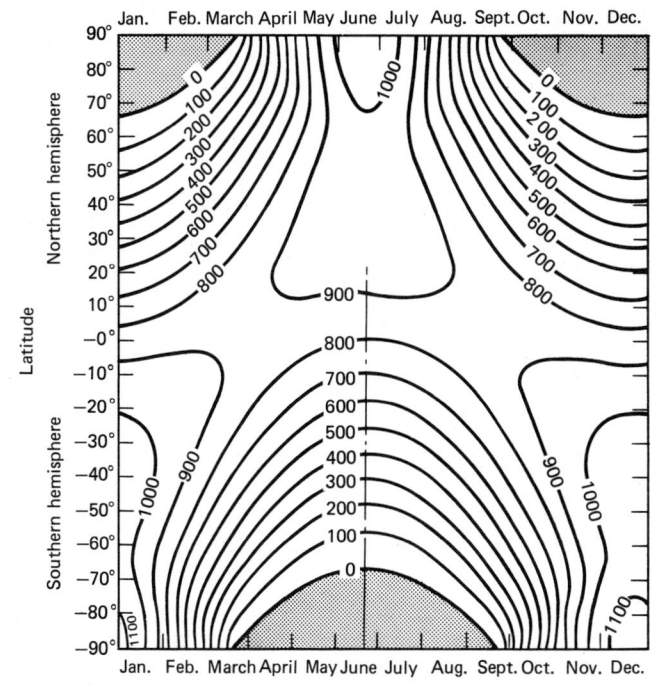

Figure 3-9 Total daily solar radiation on a horizontal surface in langleys. The shaded areas represent no incoming solar energy and indicates total darkness. (Adapted from Smithsonian Meterological Tables)

1100 langleys per day, are found at the poles during the summer because the sun never sets. Of course, during the winter the sun never rises and the polar regions receive no solar energy (the shaded areas).

The amount of energy that reaches the surface of the earth is vastly different from the amount incident at the top of the atmosphere. The four most important processes that reduce the amount of solar energy arriving at the earth's surface are *reflection*, *absorption*, *scattering*, and *refraction* (Figures 3-10a to 3-10d).

Reflection, as shown in Figure 3-10a, is the process that turns back a part of the energy striking a surface. It is the principle upon which a mirror operates, turning back essentially all the light incident upon it. Clouds in the atmosphere reflect up to 80% of the incident solar energy. At the earth's surface snow and light-colored sand are good reflectors.

In absorption (Fig. 3-10b) incident energy is retained by a substance. Bodies that absorb this energy heat up, leading to an energy transformation. Such a transformation from radiant energy to heat is accompanied by a temperature change. In the atmosphere x-rays and gamma rays are absorbed in the ionosphere while ozone, found in the stratosphere, absorbs nearly all the ultraviolet radiation. Other gases and particulates in the troposphere absorb small amounts of the remaining energy.

In scattering, energy undergoes a change in direction. As shown in Figure 3-10c scattering is distinct from reflection, because energy disperses in all directions. Since the energy can travel in all directions, part of it may be lost as it returns to outer space, while some of it may hit the earth's surface but only after a change in the direction of path. Air molecules preferentially scatter the blue wavelengths of solar energy. Consequently, blue light is scattered downward and makes the sky appears blue. Scattering is also the reason the sun takes on a reddish color at sunrise and sunset. Since the sun's path through the atmosphere is much longer at these times than at any other daylight time, more blue light is scattered out of the beam, leaving more red light.

Refraction (Fig. 3-10d) results in a change of energy's direction. When light travels between two mediums with different densities, such as water and air, the beam bends (refracts) without losing or gaining any energy. Even though there is no energy loss, refraction processes are important in producing certain phenomena such as rainbows, which are a result of both refraction and reflection by water drops in the atmosphere.

EARTH ENERGY

Just like the sun, the earth emits radiation although it is concentrated in different wavelengths. The wavelength of emitted radiation can be described based upon the properties of a *black body*.

A black body is an ideal radiating body. It emits radiation in a way that depends only on its temperature. Although we don't expect to encounter such a body in nature, the earth-atmosphere system often behaves as if it were a black body. From previous research it is known that the lower the temperature, the longer the wavelength of maximum radiation.

The sun radiates at a temperature of about 6000°C, and it emits a large amount of energy in the short wavelength visible range. On the other hand, the combined earth-atmosphere system has a fairly low radiation temperature (about $-18°C$), and the wavelength of maximum emission is at longer wavelengths, called the infrared part of the spectrum (Fig. 3-6). Thus, the radiation received from the sun is concentrated around short wavelengths, while the radiation emitted by the earth and atmosphere is in the longer wavelengths. Because of this clear distinction in wavelengths between the incoming and outgoing radiation, the incoming energy is commonly identified as shortwave radiation while energy emitted by the earth-atmosphere is known as longwave radiation. For additional information on black body radiation turn to Appendix I.

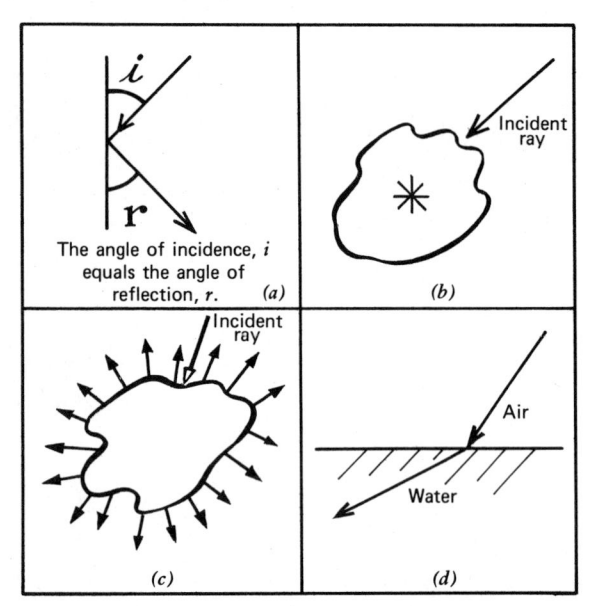

Figure 3-10 (a) Reflection—Angle of incidence equals angle of reflection. (b) Absorption—Objects that absorb energy must also reradiate it away or else they would heat up indefinitely. (c) Scattering—Without scattering the sky away from the sun would appear black similar to outer space and the planets would have no atmosphere. (d) Refraction—Mirages that result from terrestrial refractions and objects seen in water are not exactly in the location where they seem to be.

ENERGY MECHANISMS FOR UNDERSTANDING THE HEAT BALANCE

The heat balance of our environment is dependent on an equal amount of absorbed and emitted radiation over a very long time period. This balance must exist between incoming solar and outgoing terrestial radiation. Therefore, all energy sources and sinks must be accounted for if we are to understand what drives the wind or causes global temperature and weather distribution.

Greenhouse Effect

Earlier in the chapter it was stated that most of the visible short wave energy was transmitted freely to the surface of the earth. However, the atmosphere is not transparent to longwave radiation emitted from the earth's surface. Water vapor, carbon dioxide, and to a small extent oxygen absorb long wave radiation and reemit some of it back to the earth's surface. Thus, we have a recycling of terrestial radiation called *counter radiation*. This mechanism is one of the important ways the atmosphere's temperatures remain within a livable range. In effect, the atmosphere traps some of the emitted long wave radiation. This explains why it is warmer on a cloudy night and colder under clear skies since clouds trap and return more energy to the surface. It is also one of the reasons why it is warmer in the cities (more CO_2) than in the country. The glass in a greenhouse acts in a similar way, that is, it absorbs and reemits radiation downward, thus trapping the energy and tending to help keep temperatures in a greenhouse warmer than they would be normally. Thus the name greenhouse effect is applied to the atmospheric process that tends to trap longwave radiation.*

Latent Heating

Water can exist in three phases: liquid, gaseous and solid. In order for transformation from one phase to another to occur, energy is either given off or absorbed. For example, when a liquid is transformed to a gas, (a process known as *evaporation*), energy is required and this leads to cooling. A common example that many of us experience is the evaporation of perspiration from our bodies, which tends to cool our skin and, incidently, is one of the main mechanisms of cooling our bodies. Conversely, when water vapor *condenses* to a liquid, energy is given off and warming can occur. This warming is caused by the release of latent heat. For a fixed quantity of water (liquid or vapor) the amount of energy given off in condensing is exactly the same amount of energy required to evaporate the water. Thus, the evaporation process cools while the condensation process warms. This process is extremely important in maintaining the heat balance of the earth and atmosphere separately. The melting and freezing of water also involves latent heat, with the same energy being absorbed during melting as is released during freezing.

Albedo

The albedo is the ratio of the amount of radiation reflected by a body to the amount it received, and usually is expressed as a percent. For the atmosphere, clouds make the largest contribution to the albedo while on the earth's surface sand, snow, and ice covered regions have the highest albedo. The albedo of the earth–atmosphere, recently determined by satellite measurements, is about 31%. This means that 31% of the solar energy incident at the top of the atmosphere is sent back to space leaving 69% to be absorbed.

THE HEAT BALANCE OF OUR ENVIRONMENT

We are now in a position to examine and understand some of the details of the heat balance. As indicated, when we considered the earth-atmosphere system, the heat balance is between the absorbed solar radiation and the earth's emitted radiation. Over a time period of about a year the amount of energy absorbed by the earth-atmosphere over the entire globe is equal to the amount emitted by the system. Consequently there is no net gain or loss of energy and no increase or decrease of temperature. This is not true if we examine the atmosphere and surface separately, or even if we examine the combined system at different latitude belts.

In order to obtain a better insight into the energy balance, we shall discuss the earth's surface and its atmosphere separately. If we were to start off with 100 units of solar radiation hitting the top of the atmosphere, what would be left when it finally reached the surface? Utilizing a combination of calculations and measurements made by satellites, meteorologists have been able to draw a heat balance diagram, such as Figure 3-11, which shows the distribution of 100 units of incoming energy. The left side of the diagram indicates what is *reflected*, *scattered*, and *absorbed* as the shortwave energy makes its way to the earth's surface. The amount reflected and scattered, which is called the *albedo*, totals

---

* A greenhouse actually keeps the air warmer in two ways: (1) by preventing heat transport by convection and (2) by trapping radiation. The first is believed to be much more important than the second. Nevertheless, the process of trapping radiation in the atmosphere is still termed the *greenhouse effect*.

34 THE SUN-EARTH ENERGY SYSTEM

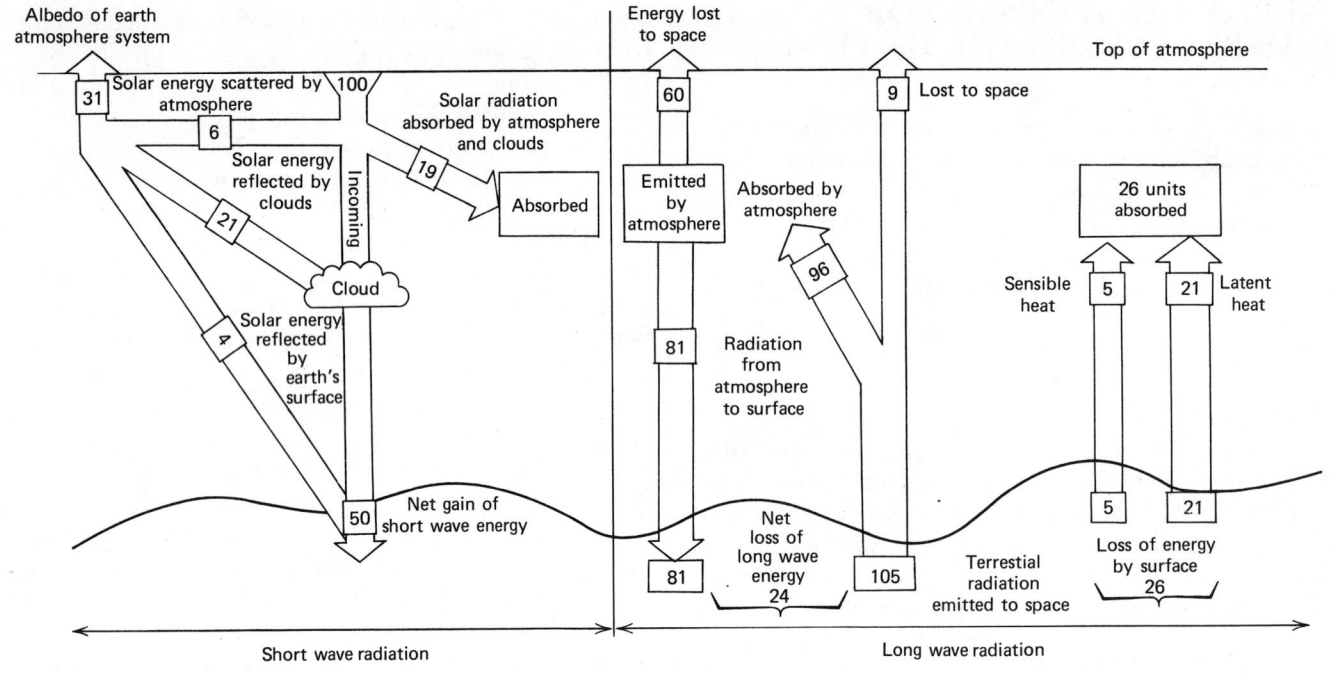

Figure 3-11 Heat balance diagram. Notice that the earth atmosphere and outer space are all in balance.

31%.* Of this amount, 21% is reflected from clouds, 6% is scattered by the atmosphere back to outer space while the remaining 4% is reflected by the ice, snow, water, and other types of material on the ground. Absorption, which also takes place within the atmosphere and clouds total 19%. After you add up the three depleting processes, 50% of the energy remains, which is the energy absorbed by the ground.

Now that all the shortwave energy has been accounted for, the longwave radiation must be summarized. This is done immediately to the right of the shortwave balance of Figure 3-11. The earth's surface emits 105 units of longwave radiation that it had previously received from the sun as shortwave energy and the atmosphere as longwave radiation. Because of the *greenhouse effect* most of this radiation, (96) is absorbed by the atmosphere while 9 units are transmitted directly to space. The atmosphere gives off 141 units previously absorbed from incoming solar radiation and longwave ground reradiation. Of this amount 60 units are lost to space while 81 are absorbed by earth and called *counter radiation*.

Separating outer space from the earth-atmosphere system, we see that the top of the atmosphere has 100 units incoming and a balance of 100 units outgoing. The atmosphere has a total net loss of 26 units of energy (115 absorbed, 141 units emitted), but the earth's surface makes up for the deficit by a 26 unit surplus (131 absorbed, 105 units emitted). If the atmosphere and earth are considered separately, the earth should heat up while the air would continue to get colder. However, temperatures observed over long periods of time indicate that this is not so. Therefore, energy must be transferred from the earth's surface to the atmosphere. How is this done?

It is accomplished by latent heat transfer, which accounts for 21 units lost by the earth's surface by evaporation. This results in a subsequent gain by the atmosphere. A second process known as *sensible heat transfer* also cools the earth's surface and heats the atmosphere. This is accomplished by convection and conduction, and accounts for 5 units. Both of these processes are shown on the far right side of the diagram. This now clarifies the importance of these heat transfer processes to the temperature and moisture structure of the atmosphere for if it were not for conduction and convection, the atmosphere would be cooling everywhere. Thus the ground acts as a source of both moisture and heat and this explains why, in general, moisture and temperature decrease upward in the atmosphere.

Let us now turn our attention to the average conditions for each latitude belt. Instead of looking at the

---

* This estimate is based on the latest satellite measurements of the albedo.
  H. Jacobowitz, W. L. Smith, H. B. Howell, and F. W. Nagle, The First 18 Months of Planetary Radiation Budget Measurements from the Nimbus—6 ERB Experiment. *J. Atmos. Sci.*, Vol. 36 (1979), 501–507.

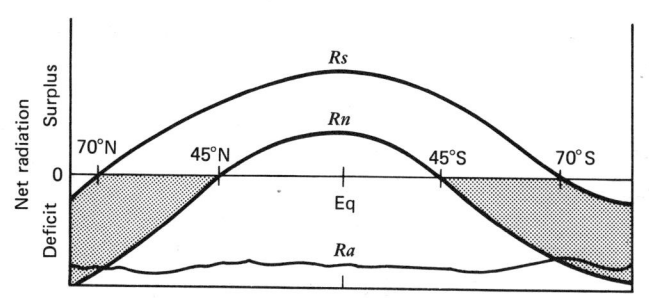

Figure 3-12 A schematic representation of the variation of net radiation (incoming short wave minus outgoing longwave energy) with latitude. Rs is the net radiation of the earth's surface, Ra the net radiation of the atmosphere, and Rn the net radiation of the combined earth atmosphere system. Note that $Rn = Rs + Ra$.

long and shortwave portions separately, let us examine the radiation surplus or deficit, known as the *net radiation*. We obtain the net radiation simply by subtracting the outgoing radiation of the system from the radiation that has been absorbed. Positive values indicate a surplus and negative values a deficit. This is shown in Figure 3-12 by the curve labeled Rn for the combined earth–atmosphere, Ra for the atmosphere, and Rs for the earth's surface. We see immediately from the diagram that the atmosphere always exhibits a radiation deficit, that is, it emits more longwave radiation than it absorbs shortwave radiation. The surface however, exhibits a radiation surplus, absorbing more energy than it emits, except for the polar regions. The combined system exhibits a radiation surplus between about 45°N and 45°S and a radiation deficit poleward from those latitudes.

If this was the only process operating, we would expect warming in the surplus regions and cooling in the areas of radiation deficit. Temperature observations, however, do not show any evidence of warming or cooling. This leads us to conclude that in order to keep the poles from getting colder and tropical regions from getting warmer heat must be transported poleward from areas of surplus to areas of deficit. This horizontal heat transport is carried out by the atmospheric circulation and ocean currents. Recent estimates indicate that the atmosphere contributes about 50% and the oceans about 50% to this transport. Thus, based on radiation balance considerations we see how the ocean and atmospheric circulations are developed and that heat must be transferred from lower latitudes to the polar regions.

Despite the simplified explanations presented, the heat balance is quite complex. Any change in incoming or outgoing radiation may very well result in a much different climatic regime than the one in which we now live. This will be further discussed in Chapter 13.

4
TEMPERATURE OF THE AIR

Galileo Galilei (1564–1642) was an Italian astronomer and physicist who invented the first thermometer. Using the theory that fluids expand and contract with the addition or depletion of heat, Galileo constructed a water-in-glass instrument that indicated relative degrees of hotness and coolness.

LEARNING OBJECTIVES

After reading this chapter, you should be able to:

1. Describe how a standard, maximum, or minimum thermometer operates.
2. Convert a temperature in one scale to a temperature in another scale.
3. List the freezing and boiling points of each of the three major temperature scales.
4. Explain the major controls of the earth's surface temperature.
5. Compare vertical and horizontal temperature variations.
6. Define *positive*, *negative*, and *isothermal lapse rates*.
7. List and describe the methods by which air is forced aloft.
8. Describe the adiabatic process.
9. Explain stability, instability, and neutrality of air.

In the previous chapter we described not only the difference between heat and temperature but also the close relationship between them. Since heat energy, combined with moisture, is a major component of all weather phenomena, temperature is one of the most important variables meteorologists measure in order to understand the atmosphere. In this chapter we shall discuss some of the more common temperature measuring devices, consider the various temperature scales, and conclude by indicating what factors control the temperature both in the horizontal and vertical. New concepts will be introduced that should give you a better appreciation of why there are temperature variations around the world.

TEMPERATURE MEASURING INSTRUMENTS

In meteorology the measurement of temperature can be considered as two separate problems: the measurement of air temperature at the surface and that of altitudes above the surface.*

Surface Measurements

The most commonly used instrument is the *liquid-in-glass thermometer*. Mercury or alcohol, the measuring medium, expands or contracts easily with a change in temperature. This movement is made within the thin long tube from the bulb end. The tube itself is etched with a graduated temperature scale and then often mounted on a metal backing for support. Although the liquid-in-glass thermometer can be quite accurate for instantaneous readings, it is quite fragile and easily broken (Fig. 4-1).

Since the standard air thermometer yields only instantaneous readings, a person would have to continuously observe it for a 24-hour time period in order to record the highest temperature for the day. This, of course, would be most impractical and a waste of manpower. Therefore, a self-registering liquid-in-glass thermometer called a *maximum thermometer* was developed. It is constructed similar to the clinical thermometer used to take a person's temperature. At the end of the bore, close to the bulb, there is a small constriction. This constriction impedes the flow of mercury but does not restrict its rise or fall. In order for the mercury to flow past this point, a force such as that produced by expansion is needed. Thus, as the temperature of the air increases, the mercury expands and the force of expansion is sufficient to push past the constriction (Fig. 4-2). However, when the temperature starts to fall, the contraction of the mercury does not have the force capable of retreating past the constriction and the thin column of mercury is broken off from the reservoir. Thus, the mercury remains at its highest level and the maximum temperature is recorded. In order to reset this thermometer, a spinning or shaking motion is needed. This provides the necessary force to allow the mercury to flow through the narrow constriction.

The lowest temperature is just as important as the highest temperature of any time period to many of our daily activities. This is especially true in agriculture and in the determination of water runoff from snowmelt in the spring. Therefore, a similar specialized thermometer for minimum temperatures was constructed to record these data. Resembling the maximum thermome-

---

*In meteorology the term *surface* indicates measurements at approximately eye level.

38 TEMPERATURE OF THE AIR

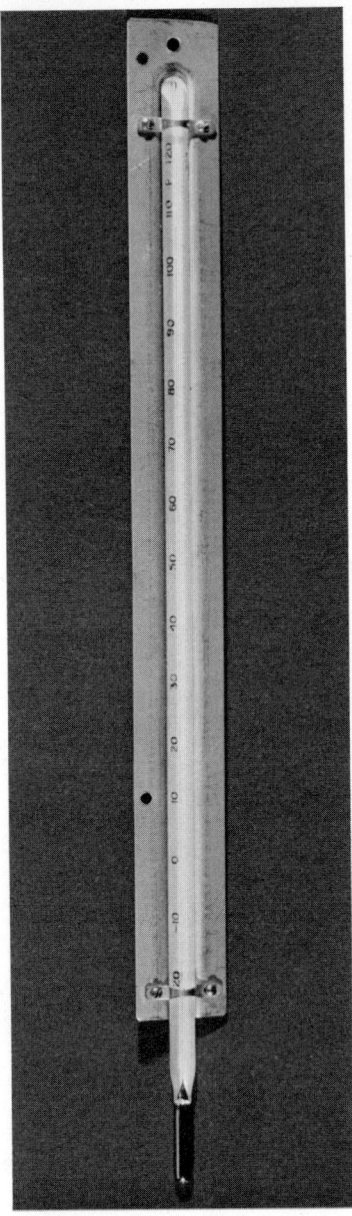

Figure 4-1 A standard air thermometer for measuring instantaneous temperatures. (Courtesy Weather Measure, Division of Systron-Donner, Sacramento, California).

ter in physical appearance only, the minimum thermometer uses an activating element of alcohol or some other low density liquid. Immersed within this liquid is a small *glass dumbbell-like weight*, which can move easily through the liquid (Fig. 4-3).

The thermometer is placed in a horizontal position with the weight resting against the edge of the alcohol. As the alcohol retreats because of lowering temperature, the dumbbell is dragged down with the alcohol edge (*meniscus*) for it cannot break through this point without extreme force. When the temperature rises, the alcohol flows around this index leaving it at its lowest point. Therefore, the right side of the dumbbell is the lowest temperature reached while the alcohol edge is always at the present air temperature. Tilting the thermometer easily resets it for a future time period.

A third type of specialized temperature measuring device was invented utilizing the expansion method. There was a definite need to observe the continuous fluctuation of temperature and the time period that elapsed between increases and decreases. Therefore, a *self-recording thermograph* with an activating element

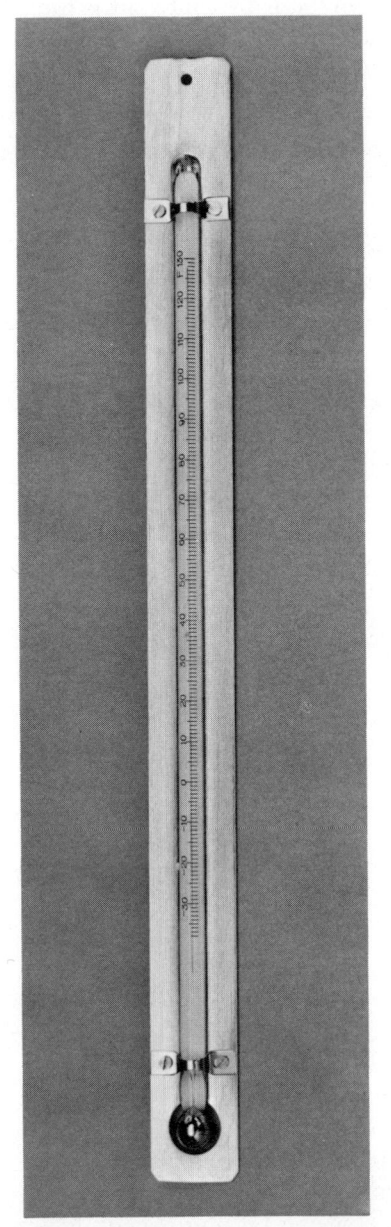

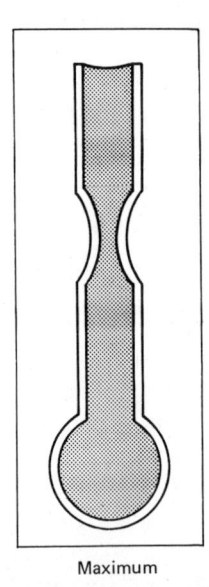

Figure 4-2 The maximum thermometer was developed for manufacturing processes and the recording of climatological data. (Courtesy Weather Measure, Division of Systron-Donner, Sacramento, California)

TEMPERATURE MEASURING INSTRUMENTS 39

similar fashion is connected to a pen arm. Although it reproduces a temperature trace, it is not as accurate as the liquid-in-glass thermometer (Fig. 4-4).

Temperature measuring devices have also been developed to measure energy flow under odd types of conditions or circumstances. How would you measure the temperature of the soil 6 feet below the surface? Certainly not with a liquid-in-glass thermometer. How could you measure the temperature at the moon's surface if you were on the earth, the temperature of the tops of clouds, or the energy in the air one mile ahead of you? These problems gave rise to instruments such as the *spectrophotometer*, which uses the color of visible light as a method of determining temperature, the *electrical resistance thermometer*, whereby the flow of energy varies with the resistance of a ceramic resistor, which is in turn influenced by temperature, and *infrared radiometer* used to measure temperature differences remotely as in determining clear air turbulence or snow cover.

Upper Air Measurements

What about temperatures at 10,000 feet or 15,000 feet above the earth's surface? These measurements are

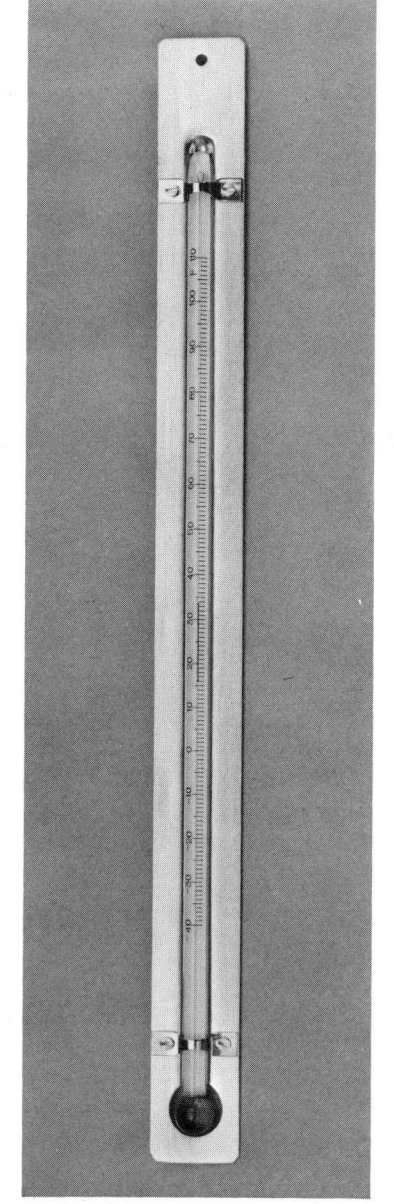

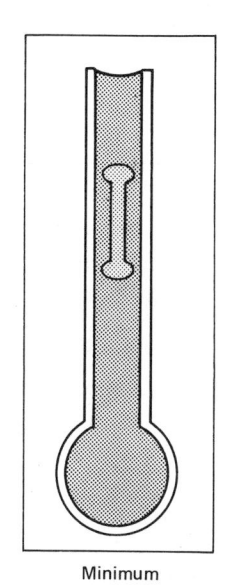

Figure 4-3 Principal of the minimum thermometer. (Courtesy Weather Measure, Division of Systron-Donner, Sacramento, California)

consisting of either a sensitive bimetallic strip or a curved tube filled with liquid was developed. The bimetallic strip is similar to our simple thermostats in that it consists of two metals having different expansion rates. They are welded together back to back, and the different expansion rates cause a curvature in the strip as the temperature changes. The bimetallic strip is attached to a pen arm by a series of linkages, which in turn records temperature information on a revolving chart. The *Bourdon tube* (curved tube filled with liquid) thermometer bends as the volume of liquid inside expands or contracts with temperature changes and in a

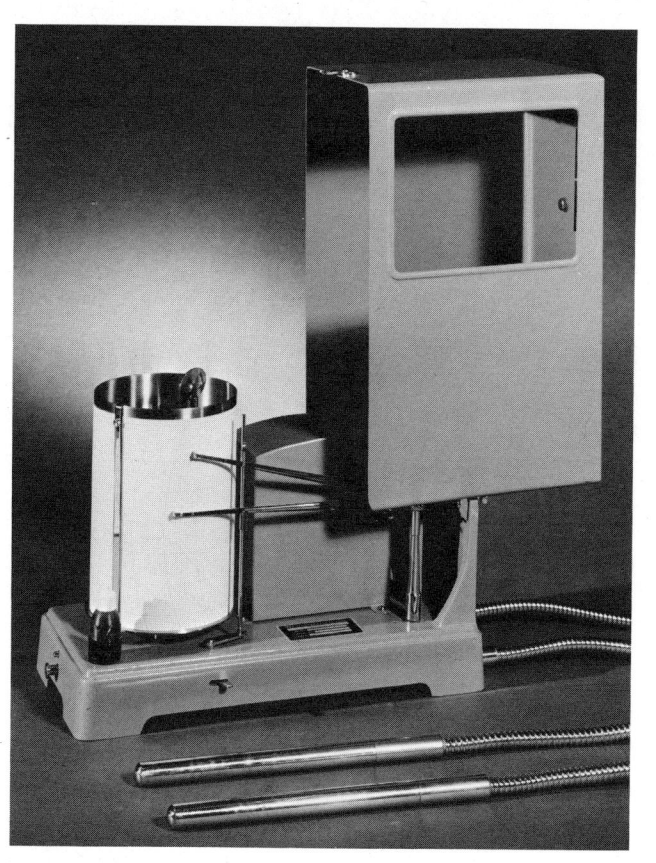

Figure 4-4 A recording thermograph for observing temperature traces. (Courtesy Weather Measure, Division of Systron-Donner, Sacramento, California)

40 TEMPERATURE OF THE AIR

needed quite frequently so the meteorologist can obtain a three-dimensional view of the changes in the atmosphere. Without this ability the forecasting of weather and air pollution episodes would be quite difficult. As discussed in Chapter 2, upper air temperatures are routinely measured by balloon-borne instruments called *radiosondes* or by *meteorological satellites*. While both are designed to measure temperature of the air, these instruments work by different physical principles.

The radiosonde system depends on a *thermocouple*, which is a device that converts heat energy to electrical energy. Since heat energy of the atmosphere is closely related to the air temperature, the thermocouple is actually converting air temperature to electrical energy. The electrical energy is then transmitted in the form of radio signals to a ground-based receiver where the signals are plotted as temperatures. The heights at which the temperatures occur are determined from other *transducers* (devices converting energy from one form to another) that are also received by the ground stations.

The determination of temperatures from satellites depends on the radiative characteristic of the atmosphere. Recall from Chapter 3 that the infrared radiation of the atmosphere depends mainly on the temperature. Scientists have discovered that radiation in a narrow band of wavelengths centered at about 15 μm are sensitive indicators of the vertical temperature structure of the atmosphere. By measuring radiation in this band as well as several other bands and by applying sophisticated mathematical techniques, it is possible to determine the vertical temperature structure of the atmosphere. This type of measurement is known as remote sensing, since the satellite is at altitudes of about 1500 km (900 miles) and senses radiation from the troposphere and stratosphere. In contrast, a radiosonde system measures temperature only at the place where it is located.

TEMPERATURE SCALES

There are currently three temperature scales used throughout the world. They are the *Fahrenheit*, *Celsius*, (formally called Centigrade), and *Kelvin* scales. Table 4-1 summarizes the known freezing and boiling points of water of each scale.

TABLE 4-1 **Comparison of Temperature Scales**

	Freezing Point	Boiling Point	Differences Between Fixed Points
Fahrenheit (F)	32	212	180
Celsius (C)	0	100	100
Kelvin or Absolute (K)	273.16	373.16	100

Fahrenheit Scale

Devised by a German scientist, Gabriel Fahrenheit (1686–1736), in the early 1700s, this scale is only used in the United States. However, this scale along with the entire English system of measurement will be replaced by the easier to use metric form of measurement. The Fahrenheit temperature scale utilizes fixed freezing and boiling points of water, which are +32°F for freezing (or melting point for ice), and +212°F for boiling (or condensation point for gaseous water).

Celsius Scale

Invented by the Swedish astronomer, Anders Celsius (1701–1744), only two years before his death, this scale is used by the entire scientific community and generally by all other countries of the world. Grudgingly, the United States is beginning to use this system of measurement. The reference points for Celsius are 100° for boiling and 0° for freezing of water.

Kelvin (K) or Absolute (A) Scale

Neither one of the previous scales satisfied the requirements for scientific investigation. Therefore, a third scale was developed by Lord Kelvin of England (1824–1907), which utilized a starting point at which no molecular activity existed. This point is designated as 0°K. Thus, the scale could only go up and, therefore, have no negative numbers. This point called *absolute zero* is equivalent to −273.16°C (for practical applications the number is rounded to −273°C) and represents that point at which all molecular activity ceases to exist. The freezing point of water is represented by 273.16°K (0°C) and the boiling point by 373.16°K (100°C). Notice, like the Celsius scale there are only 100 divisions separating the two points, which makes conversion of the two scales quite simple.

TEMPERATURE CONVERSIONS

Celsius to Kelvin

The easiest conversion is between Celsius and Kelvin, or vice versa. The difference between the two temperatures is a constant 273.16°, with the Kelvin temperature always the highest. Therefore,

$$K = C + 273.16°$$

and

$$C = K - 273.16°$$

Examples
1. What are the Kelvin temperature equivalents for 10°C and −10°C?

$$K = 10° + 273.16° = 283.16°$$
$$K = -10° + 273.16° = 263.16°$$

2. What is the Celsius temperature equivalent for 298°K?

C = 298° − 273.16 = 24.84°

Celsius to Fahrenheit

The conversion from Celsius to Fahrenheit is somewhat more involved but still rather easy. It can be done by formula:

$$F = \frac{9}{5}C + 32°$$

or by the *forty method* which is a three-step nonformula method. It involves the following three steps:

(a) Celsius temperature + 40.
(b) Multiply by $\frac{9}{5}$.
(c) Subtract the 40 you started with.

Examples
What is the Fahrenheit temperature for 20°C?

Formula Method

$$F = \frac{9}{5}(20) + 32 = \frac{180}{5} + 32 = 68°F$$

Forty Method
(a) 20° + 40° = 60°
(b) $(60°)(\frac{9}{5}) = 108°$
(c) 108° − 40° = 68°F

Fahrenheit to Celsius

Once again either the formula method

$$C = \frac{5}{9}(F - 32)$$

or the *forty method* can be used.

(a) Fahrenheit temperature + 40.
(b) Multiply by $\frac{5}{9}$.
(c) Subtract the 40 you started with.

Examples
What is the Celsius equivalent for 77°F?

Formula Method

$$C = \frac{5}{9}(77 - 32) = \frac{5(45)}{9} = 25°C$$

Forty Method
(a) 77°F + 40° = 117°
(b) $(117°)(\frac{5}{9}) = 65°$
(c) 65° − 40° = 25°

Table 4-2 summarizes conversions between the three systems for every 5°C. It is done in this way to provide some reference for making subjective evaluations of the Celsius temperature range for human comfort, particularly for people accustomed to using Fahrenheit temperatures. As can be seen from the table, temperatures lower than 0°C are dangerously cold for humans without protection. The temperature range 0 to 10°C is cold to cool, 10 to 20°C is cool to pleasant, 20 to 30°C is pleasant to warm, and above 30°C the temperatures can be considered hot.

TABLE 4-2 Temperature Conversions Between the Celsius, Fahrenheit, and Kelvin Scales

Celsius	Fahrenheit	Kelvin[a]
0	32	273
5	41	278
10	50	283
15	59	288
20	68	293
25	77	298
30	86	303
35	95	308
40	104	313
45	113	318
50	122	323
55	131	328
60	140	333
65	149	338
70	158	343
75	167	348
80	176	353
85	185	358
90	194	363
95	203	368
100	212	373

[a] Rounded to nearest degree.

TEMPERATURE VARIATIONS

Variations in air temperature are usually broken down into the heat or energy distribution from place to place over the globe. However, even if the amount of solar energy reaching the earth's surface was uniform, the resulting temperatures would not be. It is, therefore, quite obvious that other factors must contribute to the irregular temperature patterns found on earth. The main controls of these temperature changes are:

1. Latitude
2. Season of the year
3. Continentality versus oceanity
4. Horizontal movement of warm and cold air (temperature advection)
5. Irregularities of the earth's elevation
6. Ocean currents

Latitude and Season

These influences on temperature are related to the angle of incidence of incoming energy, which was discussed in Chapter 3. That is, as the latitude increases away from the equatorial regions where the sun's rays are more direct, the sun's energy hits the earth at a more oblique angle, resulting in less heating. During the summer the sun's rays are more nearly perpendicular to the earth's surface than during the winter, resulting in more energy per unit area and generally higher temperature. This is illustrated in Figure 4-5, which is a graph of the north-south temperature variation along the 95°W longitude line. This meridional graph, which runs roughly through the center of the United States, stretches from 29°N to 48°N latitude or from Houston, Texas to International Falls, Minnesota. Notice that the temperatures in both seasons of the year decrease from south to north but during January decrease much more rapidly. Also, during the summer, the seasonal effect influences the magnitude of the temperature in that it is much higher for every latitude. Lastly, the seasonal *range* of temperature, (difference between temperatures at the same location) is much smaller at lower latitudes than those closer to the poles. This is because in the southern regions the change of angle of incident solar radiation is much smaller through the year than at higher latitudes.

Continentality versus Oceanity

Large bodies of water are heated or cooled much more slowly than large land areas. There are two principal reasons. The first was mentioned in Chapter 3 and is due to the *specific heat* capacity of the land and water. Water has a specific heat capacity 2 to 3 times larger than land; therefore, it requires 2 to 3 times more energy to experience the same temperature change. Second, because it is continuously moving, water is much more efficient in carrying heat away from the place where it is applied. Since this heat will spread over a much larger mass, the temperature rise experienced will be much smaller than the land areas. Thus the combination of high specific heats and the ability to conduct large amounts of heat results in smaller temperature variations for large water bodies as compared to land surfaces. This of course has very practical implications. The temperatures of cities located at approximately the same latitude (which eliminates latitude and seasonal effect) will show a marked temperature difference depending on how close they are to large bodies of water. Cities near the ocean will experience milder winters and cooler summers than those situated in the interior of continents. For example, New York City, which is located on the coast of the Atlantic Ocean, has an average January temperature of 1.1°C (34°F) and July temperature of 25°C (77°F). Compare this with Lincoln, Nebraska, a continental location at the same latitude, that has an average January temperature of −3.9°C (25°F) and July temperature of 26.7°C (80°F).

Horizontal Movement of Air

A large body of air moving over an area from another region usually has a dominating influence on temperature. As this invading mass replaces the present air, it is accompanied by its own unique physical characteristics, and a temperature change, often extreme and sudden, can occur. This condition is most noticeable with the passage of weather systems. For example, it is common in the mid-latitudes to experience a wintertime situation in which below freezing temperatures are followed by 24 to 48 hours of much higher readings only to be followed by frigid arctic air once again. The weather situations creating these conditions are discussed in Chapter 12.

Irregularities of the Earth's Surface

Aside from the ocean-continent irregularity already discussed, the nonhomogeneous composition of the earth's surface creates distinct temperature variations.

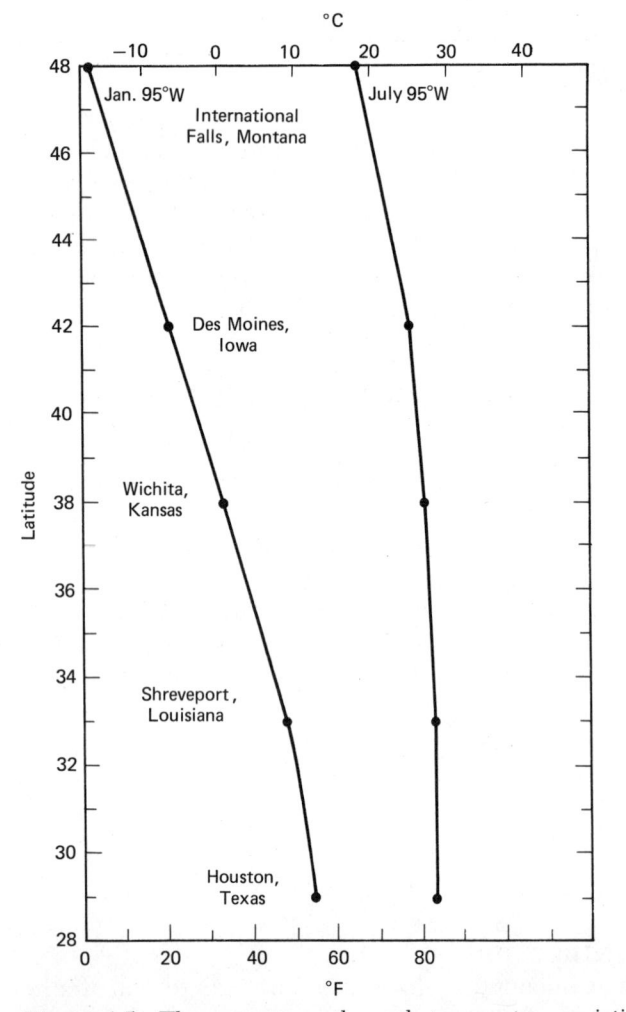

Figure 4-5 The average north-south temperature variation along 95°W longitude for January (left) and July (right).

These variations are due to elevation differences, type of soil, air drainage, and the slope of the land.

Since the earth's surface acts as a heat source for the atmosphere, temperatures decrease with altitude. This is similar to moving a finger away from a lighted match. This decrease equals, on the average, 6.5°C/km (3.5°F/1000 ft), and it can now be seen why millions of Americans seek summer vacations in the mountains. A mountain 5000 ft high could be 19.5°F cooler than a city at sea level.

Mountain ranges represent one of the most significant surface irregularities found on earth. Their major effect on weather in general and temperature in particular is that they are an effective physical barrier to the movement of large air masses. Thus, it is possible for mountain ranges to prevent the movement of cold or warm air and, in this way, exert a major influence on the temperatures of large areas. For example, in the continental United States, the major mountain ranges are located near the coastal areas and are essentially oriented north–south. They not only block the flow of moist ocean air from reaching the interior, but they allow the free passage of cold arctic or warm topical air into the mid-section of the continent. In Europe, mountain ranges such as the Alps, are oriented east–west. Thus they effectively block the invasion of cold air into the southern latitudes, and a comparison of locations at the same latitude and elevation from one continent to another would indicate higher winter temperatures in Europe where mountain chains influence weather conditions.

Because of the daily path of the sun, land sloping toward the west will be cool in the morning and warm in the afternoon, while the opposite is true for eastward-sloping land. Southern slopes will tend to remain warm all day while northern slopes that get little sun will remain relatively cool. The slope of the land also influence the drainage of air and subsequent temperature variations. On calm clear nights the air close to the ground tends to cool rapidly. On sloping surfaces the air at higher elevations often cools faster than air at lower levels. Because cold air is more dense, it tends to sink, resulting in the coldest air collecting at the bottoms of valleys and depressions.

Bare, dry soil heats and cools more rapidly than a clayey, wet soil or soil covered with vegetation. Light-colored surfaces reflect a large portion of the incident solar energy; therefore they tend to stay somewhat cooler than dark surfaces, which absorb much of the solar radiation.

Ocean Currents

Ocean currents can have a strong impact on the temperature of coastal areas. One well-known ocean current is the Gulf Stream, which carries warm water from the Florida Straits northward along the Atlantic coast of North America and then northeastward and eastward toward the British Isles. This meandering flow of warm water heats up the air in contact with it and where there are onshore winds this heated air is carried inland. The same process occurs with cold currents such as the California current, which flows north to south. Here cold moist air is carried in by *prevailing westerly winds* and locations such as San Francisco are much cooler throughout the year than similar cities not influenced by ocean currents. In general, coastal regions are much more moderate in temperature than those situated inland because of the smaller seasonal temperature differences of the water.

Vertical Temperature Variations

Vertical temperature variations are much greater than the horizontal variations just discussed. In a vertical distance of 10 km (6 miles) the temperature may vary as much as 100°C (180°F). This situation is encountered in the horizontal only when considering temperature differences between the hottest and coldest locations on the surface of the earth with distances being on the order of thousands of kilometers.

In determining vertical temperature variations we must distinguish between *environmental air* and *rising air*. Environmental air can be thought of as air that is always at a constant level. It may change its temperature through any of the processes previously described, but it can be identified as the normal air mass that envelopes us. In contrast, *rising air*, as its name implies, rises through the environmental air, much as a helium-filled balloon does. We distinguish rising air from the environment because it undergoes vertical temperature changes that are uniquely different from environmental air.

Environmental Air

As we discussed in Chapter 2, the temperature of the air decreases with height from the surface to the tropopause. This represents average conditions and was determined from measurements taken at many places and over long periods of time. The variation of temperature with height is known as the lapse rate and for average conditions is often called the *normal lapse rate* and is 6.5°C/km (3.5°F/100 ft). However, at any one time the vertical temperature profile of the lower atmosphere can exhibit large variations from average conditions. Figure 4-6 is an example of a temperature sounding that could be measured by a *radiosonde*. It indicates three different kinds of lapse rates. The *positive lapse rate* (located at points 2, 5, and 6) indicate a decrease of temperature with height. A lapse rate of zero, one that has

44 TEMPERATURE OF THE AIR

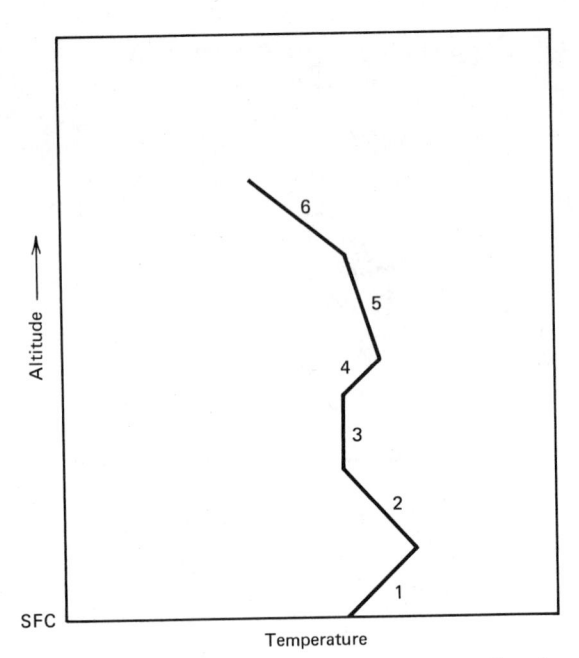

Figure 4-6 A schematic temperature trace of environmental air commonly found in the atmosphere.

no temperature change with height, is located at position 3 and is termed *isothermal*. Lastly, the temperature profile located at 1 and 4 are *negative lapse rates*, which imply an increase of temperature with height. This condition is given the special name of *temperature inversion* and plays a very important part in human comfort.

Inversions
Surface inversions (point 1, Fig. 4-6) result from air near the ground cooling faster than the air above. They often develop on clear, cool, calm nights when the ground can radiate energy away quite easily, resulting in low ground temperatures. Thus the air in contact with the ground cools by conduction while the air above this lowest layer changes very little. These inversions tend to break up when the sun's radiation increases the temperature of the lower air layers. Upper air inversions, (point 4, Fig. 4-6), usually form when warm air slides over a colder layer or when cold air sinks and warms. Whatever the cause, inversions retard the upward movement of air, thus trapping all the noxious gases and smoke in the inversion layer. Close to the earth this trapping may result in devastating air pollution episodes while aloft visibility for pilots may be reduced drastically. Surface inversions such as that which took place in Donora, Pennsylvania in October, 1948 produced a heavy dense *smog*, which is a term coined to show the combination of smoke and fog. Similar to that pictured in Figure 4-7, the Donora episode resulted in 20 deaths and close to 6000 people becoming ill. It took air pollution incidents at Donora, Pennsylvania or London, England in 1952, with 5000 deaths, to make people realize that while we cannot control the formation or dissipation of inversions, we should control the contaminants that we put into the air we breathe. Technology has been applied to this problem and many improvements have been made.

Rising Air
Vertical motions in the atmosphere result from a variety of physical processes. These include air flow over a moutain, the collision of two air masses of different densities, heating the air, and the movement of air into or out of a common area (Fig. 4-8).

Orographic lifting is the name given to the mechanism that causes air to rise up and over a mountain barrier (Fig. 4-8a). This is quite important in the United

Figure 4-7 Typical inversion fog. (Courtesy H. Michael Mogil)

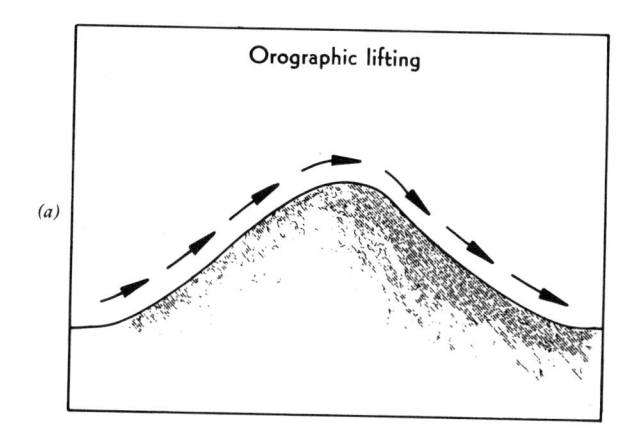

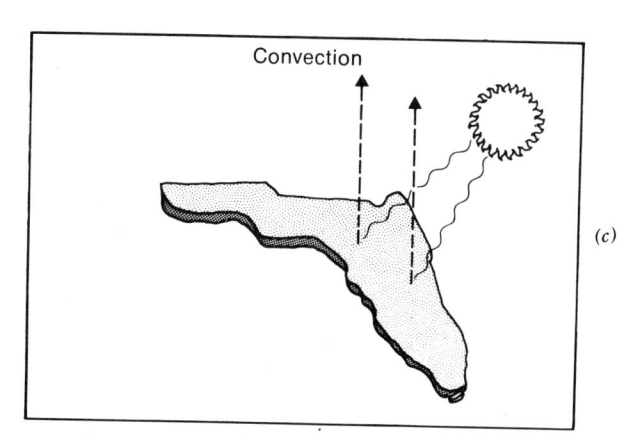

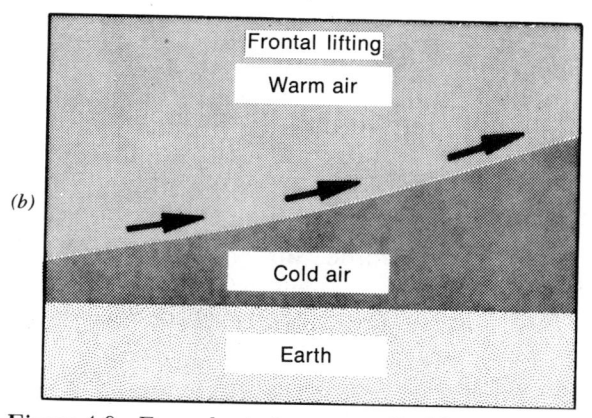

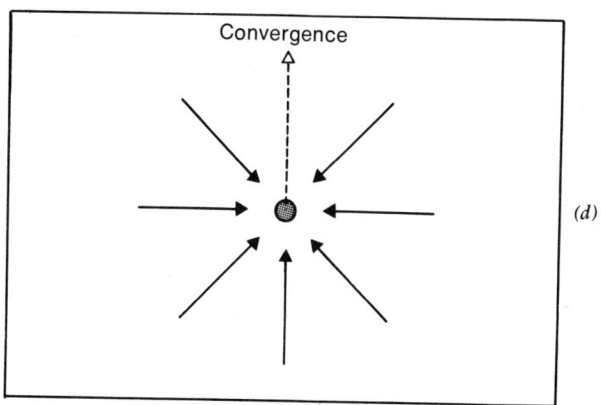

Figure 4-8 Four physical processes for lifting air: (a) orographic lifting, (b) frontal lifting, (c) convection, (d) convergence/divergence. (From Thomas L. Burrus, and Herbert J. Spiegel, *Earth in Crisis*, 2nd Ed., St. Louis, 1980, The C. V. Mosby Co.)

States because as previously discussed the north-south alignment of the two major mountain chains is perpendicular to the flow of the prevailing westerly winds.

Frontal lifting, the collision of two air masses, results in the warmer or less dense air rising or being pushed up over the colder more dense medium. The rate of lifting depends upon the steepness of the boundary between the cold and warm air. Warm fronts such as that pictured in Figure 4-8b usually produce a gentle upward gliding of warm air over the cold air mass. On the other hand cold fronts mostly generate a violent uprising of the warm air as the colder air mass pushes in under the warmer air.

The third process, *convection*, generates vertical air motions simply by heating the air. Upon being heated, the air becomes less dense than its surroundings and tends to rise much like a balloon filled with hot air or helium (Figure 4-8d). Conversely, air that is colder than its surroundings will tend to sink. Thus, uneven heating of the earth's surface can lead to air that is unevenly heated, with the result that pockets of air can be rising and falling. The rate at which the air moves depends on the intensity of the heating. Experiencing bumpiness in an aircraft is nothing more than flying horizontally through these vertically moving air currents.

Convergence and *divergence* takes place when air moves into the same area (converges) or is transported out of a region (diverges). Since converging air cannot occupy the same space, it follows, that the air must move vertically. If this air is close to the ground it will move up, since the surface acts as a physical barrier to downward motion. When air diverges the opposite occurs and air is transported away from a place. Because air is physically constrained from forming a vacuum the diverging air near the surface leads to downward motion. The wind flow associated with pressure systems result in upward motions near the center of lows and downward motions with highs. In other words, low pressure areas and the convergence process are related (see Fig. 4-8c) while divergence is associated with high pressure systems.

Temperature Changes of Rising Air

To understand the way in which the temperature changes in vertically moving air, let us consider a parcel of air at the earth's surface that is forced to rise. Let us consider this parcel as rising through the environmental air, which remains essentially stationary. As the parcel rises, it encounters lower pressures, which causes it to expand. The air parcel, in order to expand, pushes on its surroundings and, therefore, does work on its environment. Since work is a form of energy and there are no external sources of this energy for the air parcel to use, it uses its own heat energy. By using its own energy to do this work, the temperature of the air parcel drops as it expands. A simple analogy is the rapid escape of air from a bicycle or automobile tire. As the air escapes, it expands rapidly with a rapid decrease of temperature. The air feels cool and the metal part of the tire stem is quite cold.

Conversely, when the parcel of air sinks, it encounters higher pressure from the environment. In this case, work is done not by the parcel but on it, and the air compresses with a resulting temperature rise. This is like squeezing air in a tire pump, resulting in the heating of the pump.

This process of temperature change as a result of pressure change that involves no external energy sources is called an *adiabatic process*. The rate at which the temperature of a rising parcel of air changes while undergoing the adiabatic process is called the *adiabatic lapse rate*. For dry air this rate is a constant 10°C. If it sinks 1 km, its temperature will increase by 10°C.

Moist Adiabatic Lapse Rate

Notice that the previous lapse rates were specified for dry air. Recall that air contains water vapor, which can affect the lapse rate of rising air. As an air parcel rises, expands, and cools, the moisture content approaches the saturated value, which is the maximum amount of water vapor the air can hold. As the temperature lowers past this point, liquid water starts to condense out of the parcel. In Chapter 3 we saw that as water vapor condensed heat was given off, and returned to the air. The amount of heat given off depends on the amount of condensation, which varies according to the moisture content of the atmosphere. For the average atmosphere near the surface the heat liberated in rising air due to condensation results in a warming of about 4.2°C/km (+2.3°F/1000 ft). Thus, if air is saturated the adiabatic lapse rate is reduced from 10°C/km to about 5.8°C/km. This reduced lapse rate is known as the *moist adiabatic lapse rate* and exhibits a slower decrease of temperature than the dry adiabatic lapse rate. Note also that as the moisture content of the rising air changes, so will the moist adiabatic lapse rate. When there is no more moisture available for condensation, then the moist and dry adiabatic lapse rates are identical. Figure 4-9 shows the environmental, dry adiabatic, and moist adiabatic lapse rates for the standard atmosphere.

IMPORTANCE OF VARIOUS LAPSE RATES AND ATMOSPHERIC STABILITY

Consider a parcel of air located at the surface of the earth that is forced to rise. Suppose that the initial force that caused it to rise was removed. What would happen to that parcel? Would it continue to rise, sink back to the surface, or simply remain where it was? The answers to these questions are related to the atmospheric stability. If the parcel of air continues to rise, the atmosphere is considered *unstable*. Unstable air then is characterized by a parcel that will continue to move away from its location when the initial mechanism which forced it aloft is no longer operating. Conversely, if the parcel sinks the atmosphere is considered *stable*; that is, once the forcing mechanism is removed, the parcel will sink back to its original position. If the parcel neither rises nor sinks, the atmosphere is considered neutral. Let us illustrate this somewhat complicated process through the use of a ball and a bowl. If the ball is placed inside the bowl and pushed upward it will eventually come back to the bottom, as long as it is not pushed out of the bowl. This is similar to stable conditions. If the bowl is inverted and the ball pushed off the top, the ball will continue to move away from its starting point; in other words, it will be unstable.

Neutral conditions can be described in another way: If a ball rolls along a level surface, it eventually stops at another location. It doesn't continue to move forward or come back to its original position.

The determination of whether the atmosphere is considered stable, neutral, or unstable depends on whether the air parcel is colder than, equal to, or warmer than the environmental air. Air that is colder than the environmental air will be denser or heavier and thus sink to its original location while warmer air will be lighter and thus continue to rise. Parcels with temperature equal to the environmental temperature will neither rise nor fall.

The temperature of the rising air parcel is determined by the adiabatic lapse rate (moist or dry). The surrounding air temperature is determined by the environmental lapse rate. Therefore, atmospheric stability is governed by how these lapse rates are related to each other. This is best explained by referring to Figure 4-10, which illustrates the relationship between the lapse rates and atmospheric stability. First, consider Figure 4-10a. Here the dry adiabatic lapse rate is always greater than the environmental lapse rate. Thus a rising air parcel will always be colder than its environment and would be stable. Figure 4-10b shows the case where the adiabatic lapse rate is less than that of the environment. A lifted parcel would then be warmer than its environment; it

IMPORTANCE OF VARIOUS LAPSE RATES AND ATMOSPHERIC STABILITY 47

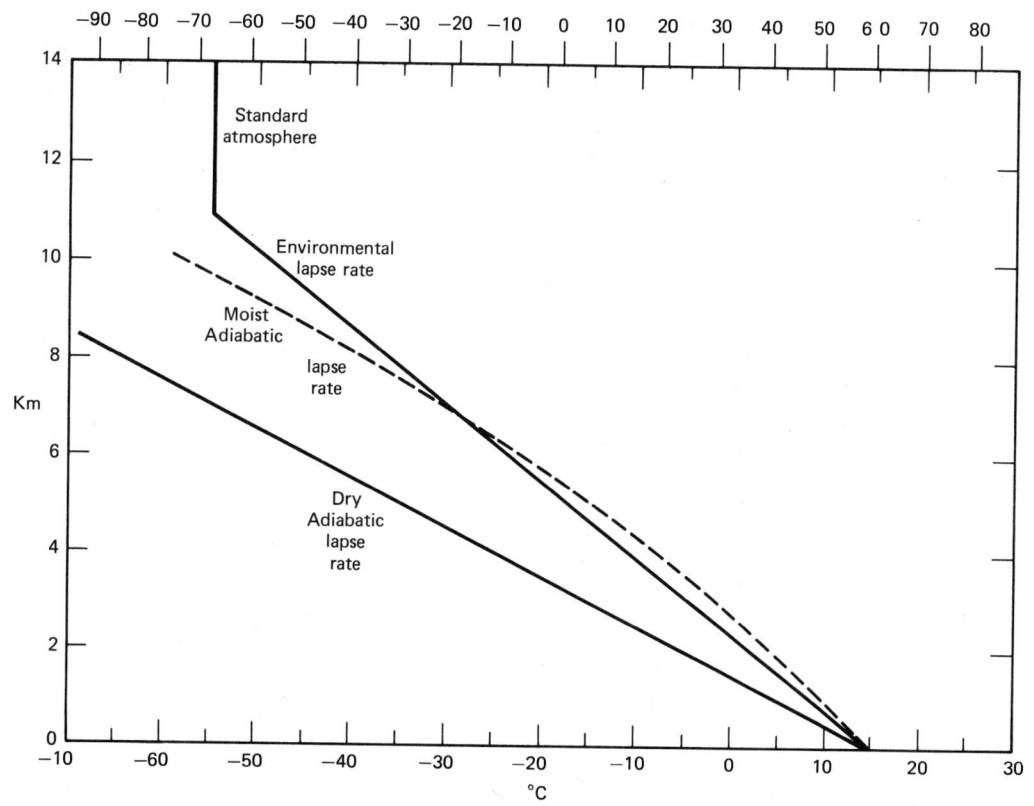

Figure 4-9 Dry and moist adiabatic lapse rates for rising air compared to the normal lapse rate of environmental air when both are computed using the standard atmosphere. Notice that the moist adiabatic lapse rate is less than the normal and dry adiabatic lapse rates and also that at 8 km (approximately 5 miles) both the dry and moist adiabatic lapse rates are parallel to one another, indicating the lack of moisture available for condensation.

would be less dense and would continue to rise. This is an instabile condition. Finally, Figure 4-10c illustrates the case where the environment and parcel have the same lapse rate. A rising parcel would then have the same temperature of the environment and be considered neutral.

The conditions for stability can be summarized as follows:

Instability. Environmental lapse rate greater (faster temperature decrease) than the dry adiabatic lapse rate.
Neutrality. Environmental lapse rate equal to the dry adiabatic lapse rate (same temperature decrease).
Stability. Environmental lapse rate less (slower temperature decrease) than the dry adiabatic lapse rate.

If the atmosphere is saturated, the same conditions

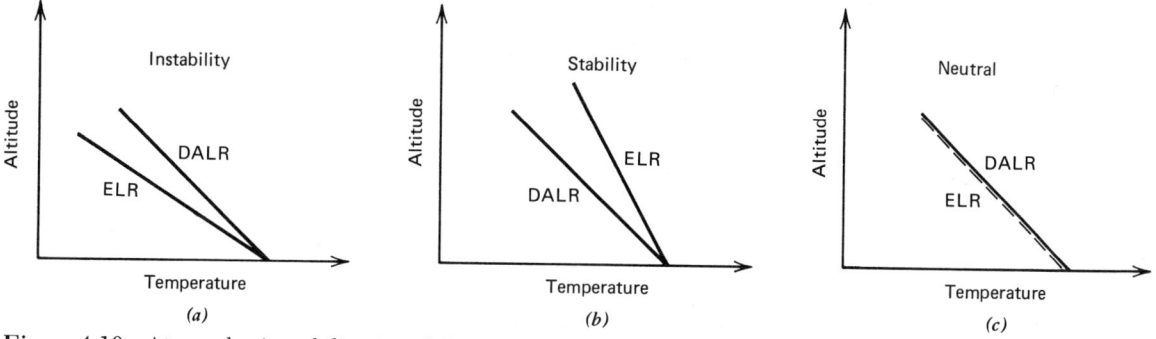

Figure 4-10 Atmospheric stability, instability, and neutrality. DALR is *d*ry *a*diabatic *l*apse *r*ate and ELR stands for *e*nvironmental *l*apse *r*ate.

48 TEMPERATURE OF THE AIR

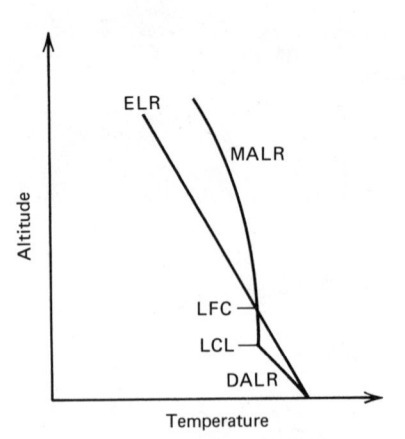

Figure 4-11 A schematic illustration of conditional instability. DALR and ELR are as defined in Figure 4-10. MALR is the *m*oist *a*diabatic *l*apse *r*ate, LCL is the *l*ifting *c*ondensation *l*evel and LFC is the *l*evel of *f*ree *c*onvection.

would hold except that the dry adiabatic lapse rate would be replaced by the moist adiabatic lapse rate.

In most cases air parcels start their ascent as unsaturated air and follow the dry adiabatic lapse rate. However, if the water vapor content is high enough, they reach condensation shortly above the surface and then follow a moist adiabatic lapse rate. Under those circumstances it is possible for a parcel to be stable with respect to the dry and unstable with respect to the moist adiabatic process. This situation is shown in Figure 4-11. The point where the parcel first becomes saturated is called the *lifting condensation level* (LCL). From that point onward the parcel will follow the moist adiabatic lapse rate. The point marked LFC is called the *level of free convection*. Above this point the parcel is unstable, that is, it will accelerate upward when the external forcing mechanism is removed. This situation is known as *conditional instability*; here *conditional* indicates that if a parcel could get to the LFC, instability would result.

Knowledge of the equilibrium state of the atmosphere is a valuable tool for the forecasting of severe weather and thunderstorm activities. Therefore, twice a day radiosondes are sent aloft at many weather stations around the world. These instruments transmit back to earth data that allows the meteorologist to calculate the stability of the atmosphere.

5
WATER IN THE ATMOSPHERE

Rudolf Clausius (1822–1888), a Prussian mathematical physicist, is credited with having made thermodynamics a science. He formulated the theoretical relationship between saturated vapor pressure and temperature that was first discovered by B. P. Clapeyron. The equation as it now exists is called the Clausius-Clapeyron equation.

LEARNING OBJECTIVES

After reading this chapter, you should be able to:

1. Draw the hydrologic cycle of water.
2. List the types of liquid, gaseous, and solid forms of water found in the atmosphere and on earth.
3. Describe the changes of state that take place in the atmosphere.
4. Define *vapor pressure*, *dew point temperature*, *humidity*, and *saturation*.
5. Compare relative, absolute, and specific humidity.
6. Explain how a sling psychrometer operates.
7. Use a sling psychrometer and calculate relative humidity.
8. List the major sources of atmospheric moisture.

Although the amount of water vapor present in the atmosphere is quite small when compared to the total mass of the atmosphere, it is an extremely important atmospheric variable. Its importance results from its role in atmospheric physics and because it is a principal ingredient in most aspects of society's well-being. The latter comes about through the *hydrologic cycle* (the cycle of evaporation, condensation, and precipitation) in which the atmosphere represents a major branch. For example, the importance of precipitation in agricultural use, domestic animal production, power supply, and human consumption needs no further explanation. Too little or too much precipitation can obviously have grave consequences to society's well-being.

Water is a unique substance. Under the normal range of temperatures and pressures found on earth it can exist as a liquid, solid, or gas. In pure form liquid water is an oderless, colorless, tasteless substance. However, because of organic and mineral impurities, the color and taste of water can vary a great deal. In the atmosphere or at ground level, liquid water is found as *drizzle*, *cloud*, *rain*, *fog*, and *dew*.

Ice is the solid state of water, and is found in the atmosphere in a variety of forms such as *snow*, *hail*, *sleet*, *ice crystal clouds*, and *snow pellets*.

The gaseous form of water is called water vapor. It mixes easily with the other atmospheric gases, and because it is capable of changing form, it is an important weather variable. Although it contributes to the total mass of the atmosphere, the actual amount of vapor in the atmosphere is quite variable, unlike most other gases. Consequently, its contribution to the total atmospheric pressure is also variable. The pressure exerted by water vapor in the atmosphere is called *vapor pressure*.

CHANGES OF STATE

Water is transformed from one state to another by the addition or subtraction of energy, which effectively controls the activity of the water molecules.

Fusion

The transformation from solid to liquid is called fusion or melting. Under normal atmospheric pressure this process takes place at 0°C (32°F) and depends upon the amount of impurities in the water. Pure water at normal atmospheric pressure solidifies exactly at 0°C. When heat is absorbed by the ice, the water molecules move more rapidly. The increase of motion causes molecules to slip more freely over one another, and the solid liquefies. The amount of heat required to melt enough ice to yield 1 gram of water is 80 calories,* and is known as the *heat of fusion*.

Solidification

The process by which a liquid changes to a solid is called solidification, or freezing. It is the reverse of fusion and occurs at the same temperature. Therefore, the same amount of heat that is absorbed to melt ice (approximately 80 cal/g at 0°C) is released when freezing takes place. When water freezes it expands slightly rather than contracts as do most substances. This unique behavior is significant, since it causes ice to be less dense than liquid water. This is why ice floats, and why water does not sink when it freezes at the surface of a lake.

---

* For practical purposes a calorie is the energy needed to raise the temperature of 1 g of water 1°C. Note however, that in a change of state the heat energy is used to accomplish the change and not to raise the temperature of the water.

The release of heat when freezing occurs is the principle that allows farmers to use the sprinkling method for frost protection. When the fine mistlike water droplets freeze, heat is released into the air, lessening both the drop in temperature and the danger of severe frost.

Evaporation

As energy is added to liquid water, molecules attain speeds that allow them to escape into the air as water vapor. This process requires a large amount of energy (approximately 600 cal/g at 0°C) which is commonly called the *latent heat of vaporization*. If this energy is not supplied by an external source such as the sun, it is taken from an evaporating surface. This surface could be a water body, soil, vegetative cover, or the surface of a person's skin, and the energy transfer needed to accomplish this evaporation process results in a temperature drop. This is the reason for the cool feeling one has when water evaporates from the human body. In fact, the evaporation of perspiration from the skin is an important mechanism for keeping the body cool and helps regulate a person's internal temperature.

Condensation

Condensation is the reverse of evaporation. That is, it is the process through which vapor is transformed to a liquid. Since in this process the water molecules tend to slow down as compared to the gaseous state, heat is released when condensation occurs. The amount of heat is exactly equal to the heat that is required to evaporate moisture.

Sublimation

The process of a solid being transformed directly to a gas or vice versa is called *sublimation*. This means that ice and snow can change to a gaseous state without going through the intermediate step of melting. Two common substances that also *sublime* are solid carbon dioxide (CO_2), which is frequently called dry ice because it changes from solid to gaseous without melting, and naphthene, which is the main ingredient of moth balls. Naphthene sublimes at room temperatures, and CO_2 sublimes at −40°C.

For water, the heat necessary to accomplish this change is the sum of the latent heats of fusion and evaporation and is about 680 cal/g. Thus, going from solid to vapor requires 680 cal/g while the change from gas to solid releases 680 cal/g. As one might have guessed, this carries the name *latent heat of sublimation*. The most common sublimation process is the formation of ice crystals at very high levels in the atmosphere where the air temperature is quite low.

By plotting points corresponding to a change in the state of water at different temperatures and vapor pressure, we can gain a better understanding of how the phase of water is affected by changes in temperature and pressure. Such a plot is called a phase diagram and is shown in Figure 5-1 for saturated conditions with temperature as abscissa and pressure as ordinate.

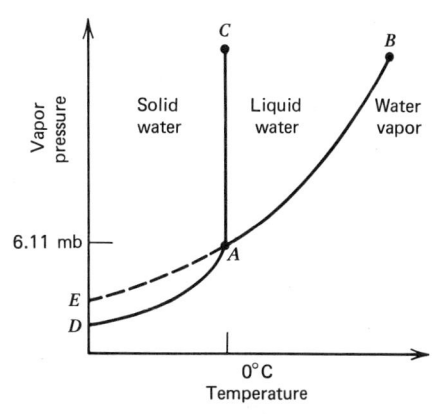

Figure 5-1 The phase of water depends on pressure and temperature under saturated conditions.

Along the line AB both liquid and vapor exist in equilibrium. To the right of AB only vapor can exist and to the left only liquid can exist. The boiling point of water is reached when the vapor pressure is equal to the atmospheric pressure. That is why at high altitudes (low atmospheric pressure), water boils at lower temperatures. Automobile radiators are put under pressure because the higher pressure raises the boiling point of water and thus allows cars to operate at temperatures that exceed the normal boiling point of water.

The line AC represents the conditions under which liquid and solid water can exist, while AD represents the equilibrium conditions between the solid and gaseous phases of water.

At point A all three phases of water exist. This is called the triple point and is located at a vapor pressure of 6.11 mb and a temperature of 0°C. In some clouds, ice and liquid water and vapor can coexist.

It is well known that water can exist in liquid form at temperatures far below 0°C. Liquid water at temperatures below 0°C is called *supercooled water*. The relationship between supercooled water and water vapor is given by the dashed curve AE, which is seen to be a smooth extension of AB. Notice that for a given temperature the vapor pressure over supercooled water is larger than that over ice. This has important consequences for the growth of ice crystals and precipitation, and will be discussed later.

MOISTURE PARAMETERS

There are several parameters used to describe the quantity of water vapor in the air at any given instant.

Vapor Pressure

As we have already mentioned, the pressure exerted by the gaseous form of water is known as vapor pressure. For any one temperature and pressure, the amount of vapor the air can hold is limited. When that limit is reached the air is said to be *saturated*, and the vapor pressure is at a maximum. When this maximum point is reached the resulting pressure is called the *saturation vapor pressure*. The saturation vapor pressure is dependent on temperature only; the higher the temperature, the higher the saturation vapor pressure. Since vapor pressure depends on the amount of vapor present, it is clear that air is capable of holding more water vapor when the temperature is high than when the temperature is low.

Saturation

Saturation is the term given to air that is holding the maximum amount of water vapor it is capable of at a given temperature. Air can become saturated in two ways: by adding moisture with no change of temperature or by lowering the temperature until the maximum amount of vapor the air can hold is equal to the amount of moisture actually present. Reaching saturation through these processes is shown in Figure 5-2. At 20°C (68°F) a 1 cubic meter (m³) volume of air is capable of holding 17.7 g of water vapor. If the air contains only 9.5 g, it is said to be undersaturated. However, if the temperature is held constant and water is added by evaporation, the parcel will eventually reach 17.7 g, its maximum, and the air will become saturated. This means of saturation is important over bodies of water where large amounts of evaporation can produce saturated air.

The second method of reaching saturation is shown in Figure 5-2b. As the temperature drops to 10°C (50°F), the maximum amount of moisture the air can hold is reduced until it reaches 9.5 g. At this point, although there is less moisture available, saturation is reached. The formation of dew is basically produced by this mechanism. The air near the surface cools by radiational heat loss and when it reaches the saturation temperature for the original vapor content of the air, condensation forms and dew is deposited. If the cooling occurs through a deep enough layer, fog will form.

Dew Point Temperature

This is the temperature to which air must be cooled, at fixed water vapor content and pressure, in order to achieve saturation. When temperatures are below

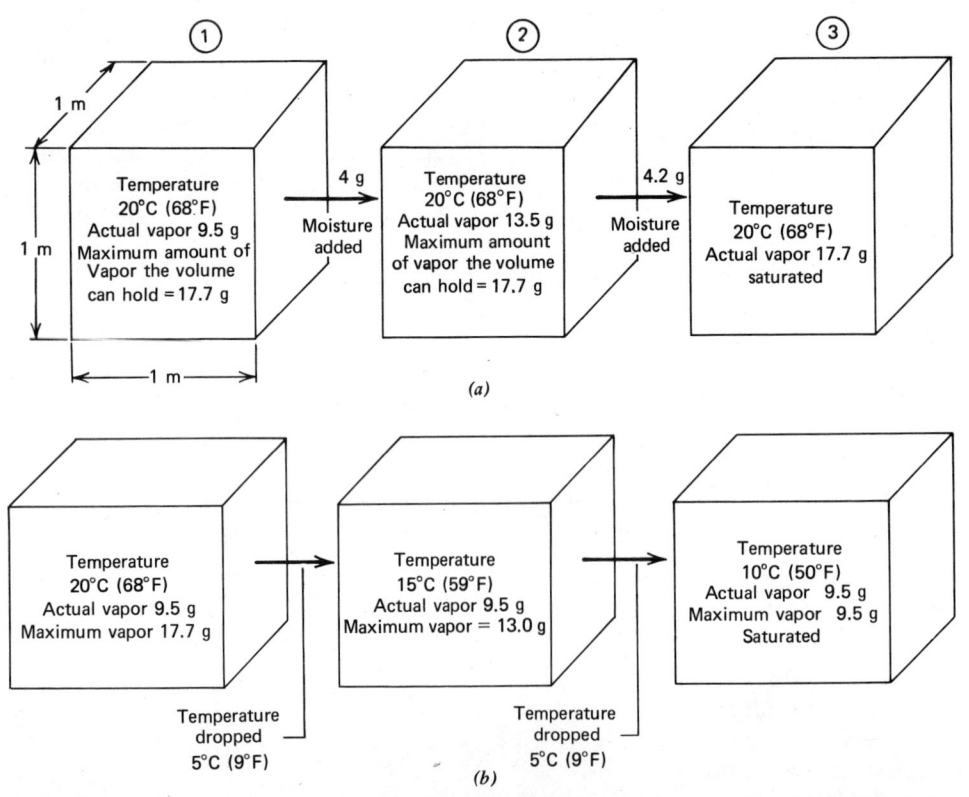

Figure 5-2 Two mechanisms for bringing about saturation. (*a*) Increasing the amount of water vapor at constant temperature until the capacity of the air is reached. (*b*) Decreasing the temperature that reduces the maximum amount of water vapor that the air can hold.

freezing it is sometimes called the *frost point*. By using the example of Figure 5-2b we find that the dew point temperature is 10°C (50°F). The dew point can never be higher than the air temperature, and is equal to the air temperature when the air is saturated.

Absolute Humidity

Absolute humidity is the density of water vapor. It is the mass of water vapor per unit volume of air, and expressed as grams per cubic centimeter (g/cm³). Absolute humidity is affected by a change in moisture content or by a change in volume. The latter can be brought about through variations in air temperature or pressure. In Figure 5-2a the absolute humidity is 9.5 g/m³ or 0.0000095 g/cm³.

Relative Humidity

Relative humidity is one of the most familiar terms used to describe moisture in the atmosphere. It is an easy concept to understand for it is the ratio of the actual amount of moisture in the air to the maximum amount of moisture the air can hold at that same pressure and temperature, expressed as a percent. For the example of Figure 5-2a the relative humidity is 9.5/17.7 × 100 = 53.6%.

Relative humidity is an important parameter in determining human comfort. For any given temperature a person's comfort is very dependent on the relative humidity of the air. This fact leads to a practical application in residential heating systems. During the winter months when both the temperature and moisture content of the air is low, heated air becomes extremely dry because of low relative humidity. By adding moisture to the heated air human comfort is greatly increased.

Specific Humidity

Specific humidity is defined as the mass of water vapor in a unit mass of moist air, and usually is expressed as grams of water vapor per kilogram of moist air. For the example figure 5-2a with a pressure of 1000 mb the volume of air would have a total mass of 1183 g and the resulting specific humidity would be 8.0 g water/kg air.

WATER VAPOR MEASURING INSTRUMENTS

Just as with temperature, the measurement of atmospheric water vapor can be separated into surface and upper air determinations.

Surface measurements are usually quite accurate because sensitive, well-calibrated instruments are used. One of the most accurate instruments is known as the *sling psychrometer* (Fig. 5-3). This instrument consists of two identical mercury thermometers attached to a strong backing. One of the thermometer bulbs is encased in a wick, which is saturated with water. The instrument is whirled, allowing water to evaporate from the wick. Since the thermometer supplies the latent heat required for evaporation, the temperature around the water saturated wick (called the wet bulb) cools. On a dry day, more evaporation takes place and the bulb reading is lower. The second thermometer measures the air temperature. The reading of the wet bulb at the lowest point (i.e., when it is not changing) is called the wet bulb temperature. The difference between that and the air temperature (called the dry bulb temperature) is related to the relative humidity of the air and is generally determined from precalculated tabulations as shown in Table 5-1.

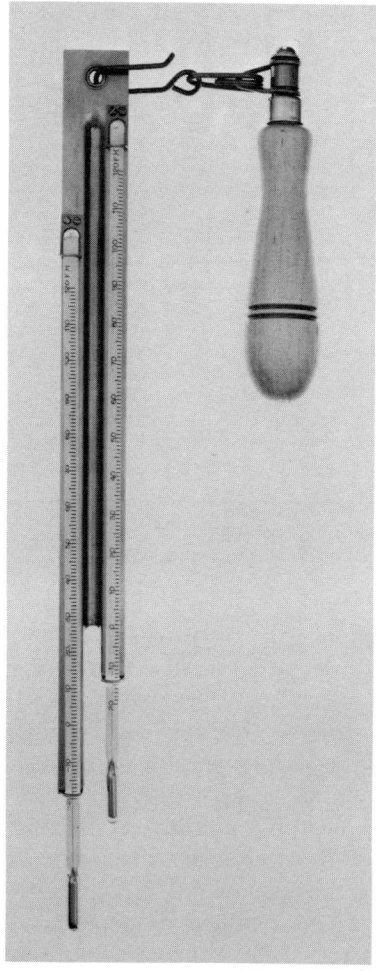

Figure 5-3 Sling psychrometers are commonly used to obtain accurate measures of the moisture content of the air. (Courtesy Science Associates, Princeton, N. J.)

For recording the relative humidity on a continuous basis a *hygrograph* is used. This instrument depends on sensing elements, which contract or expand because of the absorption of water vapor. Human hair is frequently used, although there are other sensing elements that will work. When the sensing element

54 WATER IN THE ATMOSPHERE

TABLE 5-1 Relative Humidity in Percent for Combination of Air Temperatures and Wet Bulb Temperature Depression[a]

Air Temperature °C	(°F)	\multicolumn{8}{c}{Wet Bulb Temperature Depression °C ($T_{air} - T_{wet\ bulb}$)[b]}
		0	2.5(4.5)	5.0(9.0)	7.5(13.5)	10.0(18.0)	12.5(22.5)	15(27)	17.5(31.5)
0	(32)	100	54	11					
5	(41)	100	65	31					
10	(50)	100	71	43	18				
15	(59)	100	75	52	31	13			
20	(68)	100	78	58	40	24	9		
25	(77)	100	81	63	47	33	21	7	
30	(86)	100	83	66	52	39	27	18	6
35	(95)	100	84	69	55	43	32	23	13
40	(104)	100	85	71	59	48	39	29	21
45	(113)	100	86	73	62	52	41	33	26
50	(122)	100	87	77	64	54	45	37	30

[a] *Note:* A much larger table could be generated for each 1°C.
[b] Values given are in °C with °F values in parentheses.

changes in size, it activates a series of levers and springs that move a pen across a rotating drum, inscribing a humidity trace (Fig. 5-4).

Measurement of moisture in the upper air is routinely accomplished through the use of radiosondes. These instruments contain an element with an electrical resistance that depends on the moisture content of the air. These resistance changes cause variations in electrical signals, which are telemetered at a ground-based receiving station. The principle and operation of this device is analogous to the way temperature in the upper air is measured, as described in Chapter 4.

Remote sensing devices are also used to obtain moisture parameters in the upper air. Similar to the way temperature is measured, the measuring process of these instruments depends on the absorption of radiation by atmospheric water vapor. By applying sophisticated mathematical techniques to the radiation that is measured by the sensors, it is possible to calculate atmospheric moisture content. Such instruments are routinely flown aboard meteorological satellites and complement the temperature measurements described in Chapter 4.

EVAPORATION, CONDENSATION, AND PRECIPITATION PROCESSES

Evaporation Processes

Simply stated, evaporation is the transformation of a liquid into a gas. Physically, it represents the process by which water molecules escape from the liquid to the air and become vapor. Two conditions must be met to evaporate water into the atmosphere from a source such as a lake or ocean. First, there must be an energy source to accomplish the transformation from liquid to

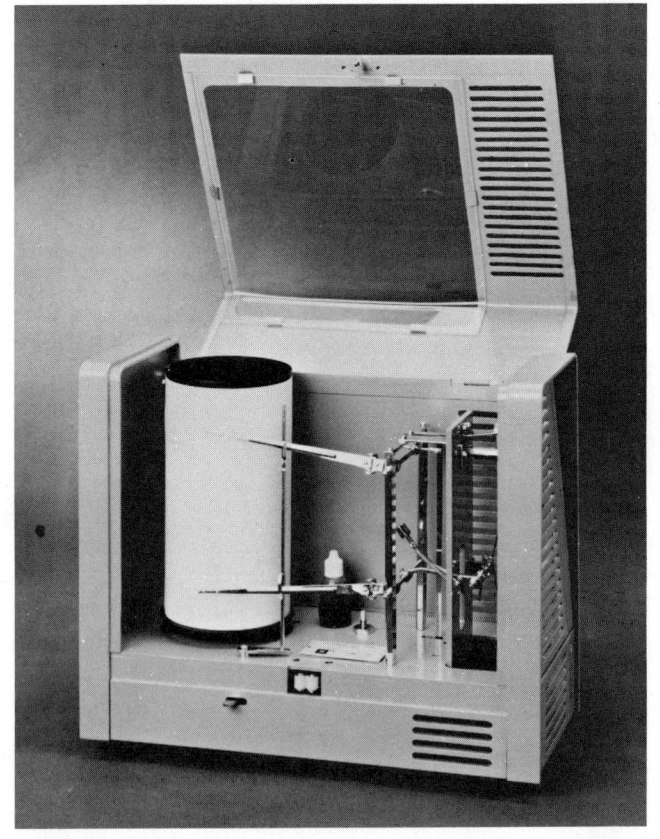

Figure 5-4 Recording instruments allow for continuous monitoring of weather variables. Pictured here is hygrothermograph, which simultaneously records temperature and water vapor. (Courtesy Weather Measure, Division of Systron-Donner, Sacramento, California)

gas; second, the air must be able to hold the vapor, or be unsaturated, as has been described previously. If neither of these conditions are met, evaporation will

EVAPORATION, CONDENSATION, AND PRECIPITATION PROCESSES 55

not take place. The energy received from the sun is the primary energy source for evaporation, although evaporation still can take place when the sun is not shining. When the sun is not shining the energy stored in the water itself is used for the process. This is an important factor in the cooling of water bodies.

There are large amounts of evaporation when the energy source is plentiful and the air is dry. However, when the air is humid and when the energy source is not plentiful, only small amounts of evaporation take place. This is consistent with our experience in drying clothes outdoors. On sunny dry days clothes dry quite rapidly, while on overcast humid days clothes take much longer to dry.

Condensation Processes

As previously stated, condensation is the change from a gaseous to a liquid state. The change from a gaseous to a solid state is called sublimation. The determining factor in whether sublimation or condensation takes place is the temperature of the air at the time the transformation occurs.

For condensation to occur the air must become saturated through the cooling or moisture adding process, and *condensation nuclei* must be present in the atmosphere. Condensation nuclei are microscopic particles such as salt, smoke, or dust, which are suspended in air. These particles are important because they make it easier for condensation to take place, sometimes at relative

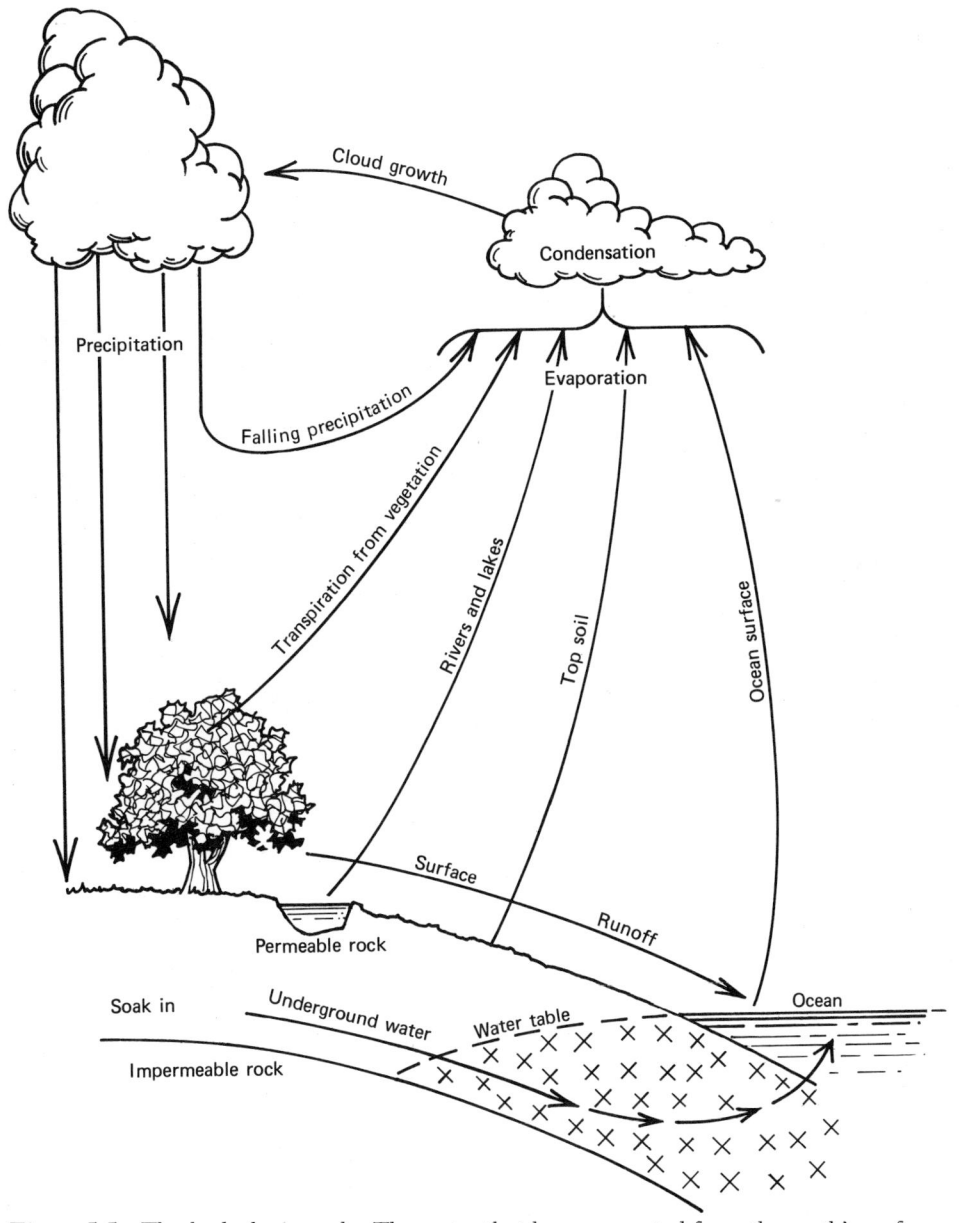

Figure 5-5 The hydrologic cycle. The water that has evaporated from the earth's surface may fall as precipitation many thousands of miles from where it entered the atmosphere.

humidities of less than 100%. Conversely, there are cases when there are few condensation nuclei present and the moisture supply is plentiful. Under such conditions it is possible for the air to hold enough vapor so the relative humidity is greater than 100%. This is called *super saturation*, and is a short-lived phenomena.

Precipitation Processes

The condensation of water vapor and the formation of clouds do not automatically result in precipitation. In order for rain to occur cloud droplets must grow to sufficient size so that gravitational force exceeds the upward forces that keep them suspended. There are two processes for drop growth. One is the growth of water drops through coalescence of smaller drops within clouds. The second mechanism occurs when supercooled water and ice particles exist in the same cloud. The ice crystals then grow at the expense of the water drops, capture more droplets as they fall, and if the air temperature is above freezing in the lower layers of the atmosphere, they melt and appear as rain. If the air temperature is below freezing snow or some other form of solid precipitation will occur.

These processes and the different forms of precipitation are discussed in greater detail in Chapter 6.

SOURCES OF ATMOSPHERIC MOISTURE

The most obvious and common source of atmospheric moisture is the *hydrosphere*, which consists of all bodies of water that reside on the earth's surface. However, two other sources of moisture can be significant. The first is the amount of moisture given off by vegetation, known as *transpiration*. Since this represents a source of moisture at the earth's surface, it is frequently combined with evaporation and called evapotranspiration. Another source appears through volcanic activity. When they erupt, volcanoes spew forth enormous quantities of moisture. The importance of this process depends on the extent of vulcanism and is believed to have been of primary importance in the formation of the atmosphere though it is of minor importance today.

HYDROLOGIC CYCLE

The process whereby liquid or solid water is evaporated into the atmosphere and then returned to the earth is known as the *hydrologic cycle*. Figure 5-5 shows how the cycle operates. Evaporation from land and water surfaces provides vapor to the atmosphere. Over land areas this process includes transpiration. Condensation of the vapor occurs in the atmosphere, and subsequent precipitation returns the vapor to the earth's surface in either liquid or solid form. When averaged over a long period of time and over the entire globe, the total evaporation equals or balances the total precipitation. For shorter time periods a state of balance may not exist, and it is possible that evaporation may exceed precipitation. In that case the atmosphere will gain vapor while the earth's surface loses liquid water, and the atmosphere is considered as storing water vapor. In contrast, when precipitation exceeds evaporation, the earth's surface stores liquid water while the atmosphere has a deficit of water vapor.

6
FOG, DEW, CLOUDS, AND PRECIPITATION

Luke Howard (1772–1864), an English chemist, in 1803 introduced a cloud classification that is the basis of our present day system. His scheme included three types of cloud descriptions: hair-like (cirrus), heaped (cumulus), and layered (stratus). Today the cloud names have been expanded to include their genus and species, the variety, the supplementary features, and the mother cloud.

58 FOG, DEW, CLOUDS, AND PRECIPITATION

LEARNING OBJECTIVES

After reading this chapter, you should be able to:

1. List and describe the products of the condensation and sublimation processes.
2. Explain how water droplets grow and the factors that limit their size.
3. Describe the formation and major features of cooling and evaporation fogs.
4. Explain the main methods of fog dispersal.
5. Describe a cloud from its name.
6. Predict the general location of a cloud from its name.
7. Visually identify clouds in the sky.
8. Predict, in a general way, upcoming weather from observing clouds in a time sequence.
9. Explain the precipitation process.
10. List precipitation form by categories.
11. Describe the three main precipitation-measuring devices.
12. Define *hygroscopic*, *supercooled water droplets*, *ground fog*, *virga*, and the prefixes *strato*, *cumulo*, *nimbo*, *cirro*, *alto*, and *fracto*.
13. Explain how lenticular clouds are formed and what their formation means.

As previously mentioned in Chapter 5, three ingredients, water vapor, condensation nuclei, and a cooling process, must be present for water droplets to form. If any of these are absent, visible liquid water will not result. If conditions exist such that condensation or sublimation can take place, the earth-atmosphere system will produce one or more of the following four end products: *dew*, *frost*, *fog*, or *clouds*. However, it must be kept in mind that this initial process can produce only very small particles, and will not result in precipitation.

WATER DROPLET SIZE

Water droplets grow on minute particles called condensation nuclei. Their growth depends on the type and size of a particle, the particle's affinity for water, and the shape of the droplet.

Particle Size

Condensation nuclei are called microscopic because they cannot be seen by the human eye. When the very smallest nuclei are present, the water droplets that begin to form are so small that evaporation immediately exceeds condensation, and the droplet never really materializes. In this case it is possible to have 100% relative humidity and not have visible moisture, simply because humidities much greater than 100% are needed to form and sustain these droplets. This same situation can occur in relatively clean air when there are very few nuclei to form water droplets.

Type of Particle and Its Affinity for Water

There are many types and sources of condensation nuclei. In nature, volcanic eruptions, forest fire debris, duststorms, and ocean salt from evaporation put billions of particles into the atmosphere every day. To this total are added the man-made nuclei from automobile exhaust and industrial complexes, which in many instances exceed the natural infusion of solid particles. Figure 6-1 shows man's neglect of the environment when the air is polluted with an overabundance of condensation nuclei.

Many of these particles are termed *hygroscopic*, which means that they have an affinity for moisture. Such particles can speed up the condensation process by absorbing water molecules even when relative humidities are well below 100%. A good example of a hygroscopic nuclei is salt. When the moisture content of the air begins to rise, the free salt particle begins to absorb moisture. In large crystals of table salt, this absorption results in an expanded crystal, which is then too large to be shaken out of the top of a salt shaker. But in the atmosphere, the absorption of water molecules by the salt results in a salt solution that becomes weaker and weaker as a greater number of water molecules adhere to the nuclei. The weaker the solution, the less affinity the nuclei has for additional molecules; eventually a point is reached at which the salt no longer has an attraction for water molecules. At this point the droplet has grown to its maximum size. However, its size is too

Figure 6-1 Although individual nuclei cannot be seen because of their size, an accumulation of these nuclei result in the ever familiar air pollution problem. (Courtesy NOAA)

small to be pulled earthward by gravity and classified as precipitation. This process is called the *solute effect* with the nuclei being the *solute* and the water droplet the *solution*. It should be noted that all nuclei are not hygroscopic and some nuclei resist the condensation of water molecules on them.

In an environment of below-freezing temperatures, certain nuclei can form ice crystals rather than water droplets. These particles, called *ice nuclei*, promote the sublimation process, which produces ice crystals that can make up *clouds* and *fog* as well as the more familiar *snow* and *frost*. In certain instances it is possible for water droplets to remain liquid at temperatures below 0°C (32°F). These water droplets are called *supercooled* and, when they are transformed into a solid, they resemble an ice crystal such as that pictured in Figure 6-2.

Droplet Shape

Water droplets are formed in a spherical shape. Because of this formation, the curvature of the drop and the water's surface tension also affect its size. This factor is much more important in the development of raindrops than it is in droplet growth.

FOG

One of the four products of the condensation–sublimation process is fog. It is formed by the accumulation of water droplets, which are small enough to be suspended in the air close to the earth's surface. The National Weather Service defines fog as visible moisture that begins at a height of lower than 15.3 m (50 ft). If the visible moisture begins at or above 15.3 m, it is called a cloud.

In the previous chapter it was determined that saturation could be reached by two primary methods: cooling and evaporation. Since fog occurs only when the air is saturated, there can be only two major categories of fog: *cooling* and *evaporation fogs*.

Cooling Fogs

Cooling of the air near the ground to its dewpoint temperature can produce fog. This cooling can occur in one of four ways:

1. Air in contact with a colder surface.
2. Radiational cooling of the air itself.
3. Collision of two air masses of different temperatures.
4. Air cooling as it rises.

Some of the more common cooling fogs include *radiation fog*, *advection fog*, and *upslope fog*. Radiation fog forms during the nighttime hours and lingers into the early morning. This type of fog is formed in the following way. As sundown approaches, the earth's surface begins to cool radiatively and the shallow layer of air close to the surface cools by conduction. If this air is moist enough, the small amount of cooling will enable it to reach its saturation point, and visible water drop-

Figure 6-2 Contrails from an airplane flying well above the freezing level produce a cloud composed entirely of ice crystals. These contrails were produced over Braunschweig, Germany, March 6, 1940. (Courtesy H. K. Weikmann)

Figure 6-3 Morning ground fog caused by radiational cooling. This type of fog forms over land during the night. (Courtesy H. Michael Mogil)

lets (or ice crystals, depending upon temperature) will form. In many instances these droplets will be thick and will restrict visibility so much that they are called fog. *Ground fog*, as radiation fog is sometimes called, seldom lasts past noon, because of the increase of radiation during the daylight hours. In many instances the fog is patchy and is found only in low-lying areas, where the cold air sinks because it is heavy. Conditions favorable for ground fog such as that shown in Figure 6-3 are clear skies, light winds, cool nights, and moist air. These conditions are found frequently over land during the autumn and winter months.

Advection Fogs These fogs are common over both land and water. They form when warm moist air moves over a colder surface; hence the name advection, which means to move in the horizontal. As the warm moist air moves over a colder surface, the air is cooled to its saturation point, and thick layers of fog are formed. The west coasts of continents, especially if a cold ocean current is present, are especially vulnerable to advection fog. In the United States, the California coast fits the criteria exactly. Prevailing westerly winds move air from over a warm ocean area to over the colder waters off the California coast. Fog forms over this current and is carried inland by the prevailing westerly wind flow. When this fog combines with pollutants, *smog* can result. (Originally, the term smog was coined to describe the combination of smoke and fog. Now the word is used to describe any combination of particulates and gaseous pollution, regardless of water content.) Since there is currently no way to keep the fog away from California, smogs can only be prevented by reducing the pollutant level. The mandatory use of catalytic converters in all automobiles produced in the United States after 1975 has helped to reduce significantly the automobiles' contribution to this smog problem.

Upslope Fog This fog is associated with any rising terrain, such as mountain areas. It results from the *orographic lifting* of moist air on the windward side of mountains. Because of the adiabatic cooling process, the air that is forced aloft cools to its saturation point, and visible moisture appears. Remembering the height rule for fog, a person standing on the mountain would be in fog, but another person standing at the base of the mountain would be looking at a cloud.

Evaporation Fog

When evaporation of liquid water occurs to such an extent that the air becomes saturated, *evaporation fogs* occur. In some instances this moist air is also cooled, so that the saturation point is reached much more quickly than usual. A good example is the smoking-steam effect when one breaths air on a very cold morning. As it leaves the mouth, the moisture-laden air that a person exhales cools quickly, resulting in condensation and the formation of a miniature "cloud." Two of the more important evaporation type fogs are *prefrontal* and *steam fog*.

Prefrontal Fog Prefrontal fog is sometimes called precipitation fog, because it is formed from the precipitation that falls from clouds ahead of a slowly advancing mass of warm moist air. The triggering mechanism comes from warm moist air being forced to rise over colder air. As in the warm front shown in Figure 6-4, the warm air cools adiabatically, produces clouds, and eventually produces precipitation. This precipitation falls through a colder drier air mass and starts to evaporate. If enough evaporation takes place in the lower layers, the air will become saturated and fog will form. It does not take too much evaporation to reach saturation because the air is cold and its capacity to hold water is quite small. Fogs that are formed in this manner usually cover very large areas and are sometimes called *continental fogs*.

Steam Fog Many a hunter or fisherman has seen ringlets of steam rising from a lake, pond, or canal early in the morning. Swimming pool owners have experienced the sight of their pools becoming almost invisible from the columns of moisture rising from the warmer water. What they see is steam fog, which forms when warm water evaporated into a thin overlying area

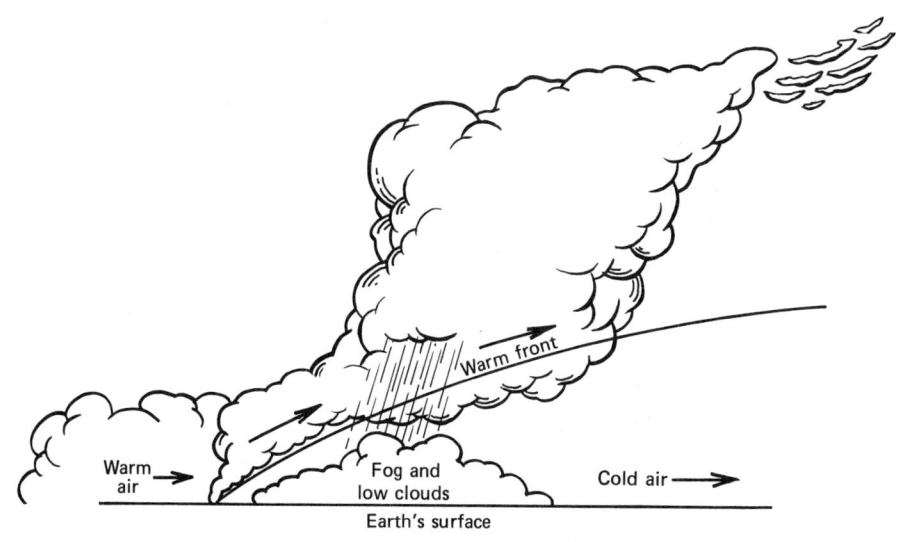

Figure 6-4 Diagram of prefrontal fog with an approaching warm front.

(Fig. 6-5). This moist air becomes buoyant and rises into cooler air above, which quickly becomes saturated and forms visible water droplets. By looking closely, you can see a small clean area above the water surface (invisible water vapor); above that area the ringlets of steam begin. Steam fog gets its name because its appearance is similar to a steam kettle, which has a clear space just above its lip.

Fog Dispersal

If fog were categorized according to temperature, three categories would prevail: *warm fog*, in which the temperature of the water droplets were above freezing; *cold fog*, made up of supercooled water droplets; and *ice fog*, composed of ice crystals with temperatures below −30°C (−22°F). Since there are three types of fog, specific methods must be used to disperse each one.

Figure 6-5 Steam rising off a roof after a rainstorm. When similar occurrences take place over ponds and lakes it is called steam fog. (Courtesy H. Michael Mogil)

Since fog is dangerous to both air and land transportation, methods have been devised to modify atmospheric conditions that cause fog formation. Led by the U. S. Air Force and commercial aviation interests, the following methods have been used successfully.

1. **Dry Ice Seeding.** Aircraft disperse crushed, dry ice over a cold-fog area. This allows ice crystals to grow as they rob the fog of its supercooled water droplets. The ice crystals grow large enough to fall and a large clear area is formed over an airport. The seeding must be done upwind of a runway so that the wind can carry the dry ice to the right location (Fig. 6-6).
2. **Liquid Propane System.** This method, pioneered at Orly Airport in Paris, France, has been successfully used to disperse cold fog within a small area. A ground-based network of dispensers emit tiny droplets of propane into the fog, which immediately evaporates and produces ice crystals. The crystals grow at the expense of the droplets, and the area becomes free of fog. This method has its limitations, for during light winds the propane gas will not drift through the fog and there is insufficient time for the ice crystals to form. However, the advantage to this method is that fog dissipation can be carried on by ground-based equipment that surrounds an airport.
3. **Glycerine Seeding.** Used for warm-fog dispersal, glycerine is sprayed on the top of the fog layer. As it falls through the fog, the glycerine's high affinity for water allows it to capture some of the available moisture in the fog. If enough is captured, the fog will dissipate by precipitation. This is a relatively new technique that shows much promise.
4. **Air Mixing.** The U. S. Air Force has conducted tests in which helicopters are flown over shallow fog areas. It was hoped that the air currents set up by the rotor blades would allow drier air to subside into the warm fog and disperse it. Recent tests have met with limited success.

62 FOG, DEW, CLOUDS, AND PRECIPITATION

Figure 6-6 The effects of seeding a cloud with dry ice produces a hole in the thick cloud layer. (Courtesy NOAA)

5. **Evaporative Technique.** Strictly used in warm fog, this technique uses heat to cause evaporation of water droplets. At airports where this method is carried out, jet engines are placed at strategic locations underground. The blast from these engines heats up the air and produces turbulence. Both the heat and the turbulence tend to dissipate the fog. This method has been used at Orly Airport in Paris since 1970.

At present, there is no known method for dispersing ice fog. Recent tests in Alaska, where ice fog predominates, have proven ineffective.

DEW AND FROST

Two additional outcomes of the condensation/sublimation process are *dew* and *frost*. Both take place during the night or early morning hours when calm conditions and clear skies are present. During this time the earth's surface cools by radiative cooling, and the very thin layer in contact with it cools by conduction. Just as with fog, if this thin layer cools sufficiently to reach its *dew point temperature*, the saturated air will produce small water droplets, which will then be deposited as dew on all exposed objects. When the air temperature is below freezing, ice crystals will form instead of dew, and frost will appear.

The process by which dew forms can readily be seen by filling up a glass with a mixture of ice and water. As time passes, the ice cools the water, which in turn chills the glass. The cooler glass lowers the temperature of the air that comes in contact with it, and water droplets appear on the outside surface of the glass. These water droplets, similar to dew, are produced by the condensation of water vapor from the surrounding air as it cools to its dew point temperature.

Farmers and growers are frequently faced with the problem of protecting their crops and trees from frost.

As was briefly explained in Chapter 1, methods have been developed in different areas to counteract the formation of frost. In the citrus groves and low-lying areas of Florida and California, elevated blower or wind machines (Fig. 6-7) are used. Both are intended to stir the cold air near the surface with the warmer air above in the hope of keeping the ground air temperature above freezing. A second method is to use heaters that directly warm the air. Many years ago, piles of old tires were burned and commercial smudge pots were used to produce smoke. It was thought that the smoke particles would absorb the outgoing long-wave radiation and hold in the heat. However, the inefficiency and pollution problems of smoke caused growers to turn to the more effective and cleaner methods of frost protection.

CLOUDS

Similar to fog, dew, and frost, clouds result from the condensation/sublimation process. The only difference between clouds and the other forms of visible moisture is the location of formation.

Clouds can form anywhere in the troposphere. These visible aggregates of condensed liquid or ice crystals are light enough to float in the air and move from place to place by the wind currents (Fig. 6-8).

Because so many different cloud forms can be seen in the sky, it would appear difficult to devise an acceptable system in which they could be identified and classified. However, by observing their characteristics, meteorologists have been able to classify clouds according to their *appearance* and *height*. The most widely used classification scheme was devised by Luke Howard in 1803, and adopted by the World Meteorological Organization.

Figure 6-7 Wind machines are a common sight in the citrus groves of central Florida. As long as the temperature drop is not accompanied by a strong northerly wind, the wind machines provide adequate protection against freezing temperatures and frost. (Courtesy Florida Citrus Council)

Figure 6-8 Cloud cover over the Pacific Ocean and adjacent land masses as seen by the GOES satellite. Notice the hurricane with its distinctive eye in the lower right. (Courtesy NOAA)

Based on appearance only, clouds can be separated into two major types:

Cumuliform. Clouds of vertical development formed by the condensation of rising air.
Stratiform. Layered clouds formed by the condensation of air that is mostly devoid of vertical currents.

Structured according to the height of the cloud bases, four divisions within each major type can be identified. These are *high*, *middle*, *low* and *vertical development* clouds.

In order to better understand the meaning of cloud names, the following root words should be defined:

Stratus, strato	Layered or sheetlike
Cumulus, cumulo	Puffy, heaped (prefix for a vertical cloud)
Nimbus, nimbo	Dark and rainy
Cirrus, cirro	Curly, featherlike (prefix for a high cloud)
Alto	High (prefix for a middle cloud)
Fracto	Broken

Using the two classification categories and the root words, all cloud forms can be identified and described (see Table 6-1).

High Clouds

The upper and lower limits of high clouds vary with latitude. Close to the polar region high clouds can be found as low as 3500 m (approximately 10,000 ft), while in the tropics, they can form as high as 18,500 m (60,000 ft). For classification purposes only, the average value of 6000 m (20,000 ft) is used. Clouds in the high category are always composed of ice crystals or frozen water droplets in a featherlike or thin tufted transparent form. The three types of high clouds all have the word *cirrus* or prefix *cirro* incorporated into their names.

Cirrus (Fig. 6-9). A white-streaked, delicately fibrous band of thin clouds, which stretches many miles across

64 FOG, DEW, CLOUDS, AND PRECIPITATION

TABLE 6-1 **Clouds and Their Descriptions**

Classification	Cloud Name	Composition	Average Height of Bases
High clouds	Cirrus, cirrocumulus, cirrostratus	Frozen water droplets or ice crystals	6 km (20,000 ft.)
Middle clouds	Altostratus, altocumulus	Ice crystals and/or water droplets	2–6 km (6500–20,000 ft.)
Low clouds	Nimbostratus, stratus, stratocumulus, fractostratus, fractocumulus	Water droplets (ice crystals in winter)	15m–2 km (50–6500 ft.)
Vertical development clouds	Cumulus, cumulonimbus	Water droplets at lower levels and ice crystals at upper levels	In low cloud range

the sky. Because of height considerations, moisture availability is small, and the resultant cloud is both thin and without shadows. A special type of cirrus known as *cirrus mares tails* is sometimes seen. It is thought that this special form results from the falling of ice crystals and frozen water droplets into areas of different wind directions. This *wind shear* results in the particles being carried along with the wind, and it gives the appearance of large hooks or mares' tails. Even though they are moved horizontally by the wind, the clouds still fall and eventually reach a level at which they evaporate. Figure 6-10 is a good example of mares' tail cirrus.

Cirrus clouds usually indicate an advancing weather front, but can also be generated by jet contrails (Fig. 6-11) and the sweeping off of the tops of thunderstorm clouds.

Cirrocumulus (Fig. 6-12). Resulting from high level turbulence, cirrocumulus clouds are shaped into tiny whitish globular rolls, which have a ripplelike appearance that is similar to a sandbar at the beach. In some instances, this appearance has been identified with the skin of a mackerel fish, and when these clouds develop they are given the name "mackerel sky." Although cirrocumulus are quite rare and generally occur only in northern latitudes, they are usually associated with the approach of a cold front.

Cirrostratus (Fig. 6-13). Formed as a thin whitish veil with a milky texture, cirrostratus are most famous for their ability to produce a halo around the sun or moon. As with all high clouds, cirrostratus are composed of ice crystals or frozen water, and usually cover the entire sky. These clouds are so thin that they allow the sun or moon to be seen quite clearly, and on a moonless night it is sometimes difficult to identify their presence. When cirrostratus are observed, they may very well indicate that a warm front is approaching and that precipitation could occur within 24 to 48 hours.

Middle Clouds

The lower and upper limits of middle clouds also vary with latitude. In the polar region a range of 2100 to 4000 m (6500 to 13,000 ft) is common, while at the equator the range is 2100 to 7500 m (6500 to 25,000 ft). Middle clouds are identified by the prefix *alto*. (*Note.* Alto means "high," but it identifies a "middle" cloud.)

Altostratus (Fig. 6-14). Formed by warm moist air rising slowly over a colder denser layer of air, a relatively uniform flat cloud takes shape. These clouds have a bluish or bluish-grey color and are dense enough to hide the sun or moon. On occasion, the layer will show a light spot, as if the sun is shining through a translucent screen. Its appearance would be similar to shining

Figure 6-9 Cirrus clouds. (Courtesy H. Michael Mogil)

Figure 6-10 Mare's tail cirrus. (Courtesy Norman Mendelson)

Figure 6-12 Cirrocumulus clouds. (Courtesy NOAA)

a flashlight through ground glass. When altostratus clouds are seen, it usually means that a warm front is close and that precipitation will occur within a short (12 to 24 hours) period of time.

Altocumulus (Fig. 6-15). Generally composed of water droplets except when above the freezing level, these clouds result from warm air rising slowly over colder air. Composed of white or gray layers of globular rolls, altocumulus clouds are quite familiar in mountainous areas. Very reminiscent of wool on a sheep's back, they are commonly called "sheepback clouds." In mountainous areas a special form, altocumulus lenticularis or *lenticular* clouds are occasionally seen. These almost stationary clouds have a smooth edge and are shaped like a giant lens. Lenticular clouds form on the lee side of a mountain as air flows over the mountain ridge in a wavelike pattern (Fig. 6-16). Cloud droplets form in the rising air, while in the descending air the droplets evaporate. Because of formation on one side and evaporation on the other, the cloud appears to be standing still. The presence of lenticular clouds, such as those shown in Figure 6-17 means that extremely turbulent air is prevalent in the area.

Low Clouds

Unlike the high and middle clouds, the upper and lower limits of low clouds do not vary with latitude. They range from a few meters to 2100 meters (6500 feet). For the most part, low clouds are more capable than middle clouds of producing a larger quantity of precipitation, simply because there is more moisture available. With their lead gray dense appearance, low clouds are also thicker and more ominous looking.

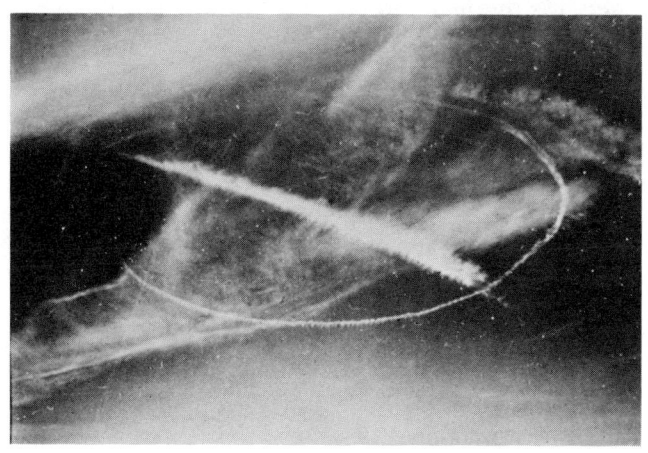

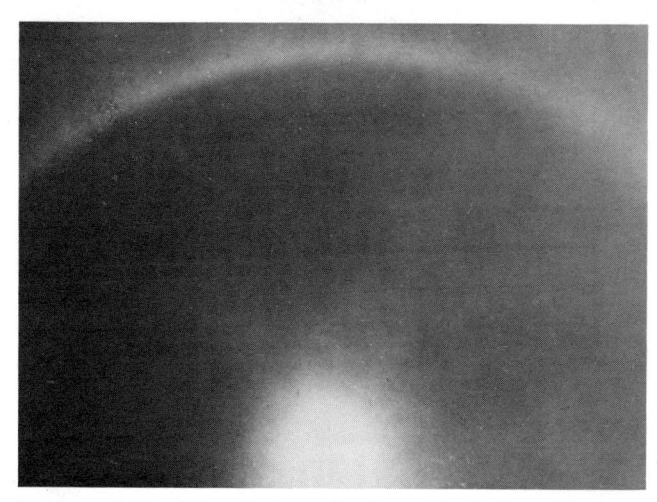

Figure 6-11 Transverse cirrus clouds filaments formed by jet aircraft contrails. (Courtesy H. K. Weikmann)

Figure 6-13 Cirrostratus clouds with halo. (Courtesy NOAA)

Figure 6-14 Altostratus clouds; note the contrasting light spot caused by the sun. (Courtesy H. Michael Mogil)

Stratus (Fig. 6-18). The lowest of all the clouds found in the sky, these clouds are formed whenever moist air becomes saturated at lower altitudes. Stratus can be described as a thick gray uniform layer resembling fog, and looks to the observer as though the skies are about to release all the moisture they are holding. Commonly found in low-lying areas such as valleys, mountainous areas on the windward side, lands near the seashore, and in regions with the passage of a warm front, stratus clouds are only capable of producing drizzle or light rain.

A special form of stratus cloud is called *fractostratus*. These clouds, which are torn shreds of the mother cloud, skid along the sky at the same speed as the wind.

Figure 6-15 Altocumulus clouds.

They are called "scud clouds," and usually are widespread after the passage of a warm front.

Nimbostratus (Fig. 6-19). Better known as the *dark rainy layer*, these thick ominous-looking clouds stretch from horizon to horizon. Because they are uniform and shapeless, an observer has a difficult time determining where this diffuse layer starts and ends. Associated with this overcast layer are steady rain or snow, which are heavy at times. It is so thick that it blots out the sun easily, and in the winter nimbostratus clouds can be composed of water droplets, ice crystals, and snowflakes. This cloud will often produce precipitation that evaporates before it reaches the ground, a situation called *virga*.

Stratocumulus (Fig. 6-20). Often occurring as a continuous layer of parallel rolls, these dark gray clouds appear as if they will produce extremely heavy precipitation. However, when precipitation does fall, it is usually only of light intensity. These clouds form when unstable air exists below the cloud base with stable air above the tops. At other times stratocumulus exists as the result of the lifting of stratus clouds or the lowering of altostratus clouds.

Clouds of Vertical Development

Vertical development clouds are a special type of cloud family. They form from convective currents and have their bases anywhere between the low and middle cloud category. Their tops may go as high as 20,000 m (65,000 ft) in the tropics, and the cloud itself builds through all three of the other layers. Convection currents can result from any process that starts air moving in the vertical with unstable air taking over after the mechanical lifting force ends. The four methods of lifting as described in Chapter 4 are *convection*, *frontal lifting*, *convergence*, and *orographic lifting*. (The first method is thermally forced lifting; the other three are mechanically forced lifting.)

Cumulus (Fig. 6-21). Cumulus clouds form as lifted air reaches its saturation point. These clouds typically form over a heated surface during the summer. Around 10:00 to 11:00 a.m., the ground has heated up sufficiently to raise the temperature of the air above it. This air becomes buoyant and unstable. As the air rises, small white detached clouds with cauliflowerlike tops begin to form at the level of saturation. These clouds are composed mainly of water droplets except when instability forces the cloud to grow past the freezing level. The resulting tall thin cloud is called a *towering cumulus* and is composed of water droplets, super-cooled water droplets, and ice crystals. Towering cumulus clouds may produce showery precipitation, which can

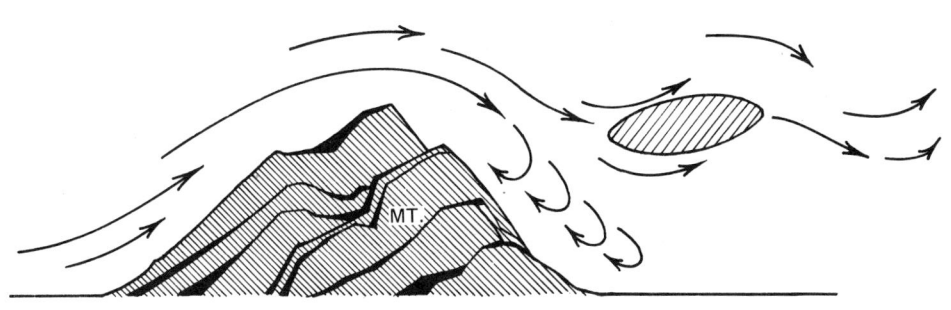

Figure 6-16 Formation of lenticularis clouds occurs when strong winds blow across a mountain ridge, which sets up standing waves of up and downdrafts. The clouds form on the updraft side of the standing wave.

be quite heavy at times. As these clouds become well developed, they line themselves with the wind flow and appear to form into *cloud streets*, with alternating clear spaces between each row of clouds. The clear spaces result from air that subsides between two rising air columns.

Cumulonimbus (Fig. 6-22). Better known as the *thunderstorm cloud*, the cumulonimbus is nothing more than an extension of the towering cumulus. After the formation of the towering cumulus, the air may still rise because of its instability, and the cloud will become very dense and massive, almost resembling a mountain. As it reaches the higher levels of the troposphere it hits the tropopause, which acts as a lid to further growth. The top may then be swept off by the high winds, and an anvil top will appear. Because of its thickness, the base of the cumulonimbus cloud will be very dark. Its makeup is complex, with water droplets forming at its lower levels, supercooled water and ice crystals above its freezing level, and ice crystals and snowflakes at its very top.

A special cloud feature that often accompanies a thunderstorm cloud is the *mammatus*. These mammatus are heavy pouches of large globular rolls dangling from the mother cloud especially from under the anvil. Their appearance indicates an extremely well-developed thunderstorm and turbulent conditions.

Since the entire topic of thunderstorms and their development is extremely important, it will be covered in greater detail in Chapter 9.

PRECIPITATION PROCESS

As explained earlier in this chapter, the condensation process is only capable of producing small water droplets or ice crystals, which are easily held aloft by air currents. Therefore, there must be other physical methods by which the ice crystals or water droplets can grow so that they may be pulled earthward by gravity as *precipitation*. There are two principal processes that result in water particles large enough to be classified as precipitation: the *ice crystal* and *coalescence processes*.

Figure 6-17 Lenticularis clouds. (Courtesy H. Michael Mogil)

Figure 6-18 Mountain stratus clouds. (Courtesy Norman Mendelson)

Figure 6-19 Nimbostratus clouds.

Ice Crystal Process (Cold Clouds)

In the early 1900s, W. Findeisen, a German meteorologist, discovered that water droplets and ice crystals could exist side by side at temperatures below freezing. In the free atmosphere, clouds have a composition such that water droplets and ice crystals do coexist. It is well known that the small pressure exerted by the vapor around a water droplet (saturation vapor pressure over water) is much greater than the small pressure exerted by the vapor around the ice crystal (saturation vapor pressure over ice) at the same temperature. When ice and water coexist, there is an imbalance of vapor pressure forces. This imbalance of forces results in water vapor molecules being transferred from the water droplet (high pressure) to the ice crystal (low pressure). Thus, the droplet starts evaporating while the ice crystal grows larger.

Eventually, the crystal grows large enough to fall through the cloud, where it continues to grow through contact with other crystals and with supercooled water droplets. As the precipitation in the form of snow begins to fall, it may encounter a layer of air whose temperature is above freezing. The snowflake melts and a cold rain results. If, on the other hand, a cold layer lies below the upper warm layer, the snowflake will melt and then freeze to become *sleet*. A third possibility exists in which temperatures all the way to the ground level may be below freezing. In this case snow will pile up on the ground. A fourth and rarer possibility is to have rain freeze on objects at the earth's surface as in an ice storm.

Coalescence Process (Warm Clouds)

Early experiments by scientists indicated that water droplets carried electric charges. One such investigation was the *Rayleigh Fountain Experiment*, in which water was forced through a very small opening in a glass tube. The water came out in a fountain spray except when a comb charged with static electricity was placed close to it. The water was repelled into a large mass, and it was concluded that this repulsion could take place only if the water had the same charge as the comb, since only unlike charges repel.

Figure 6-20 Stratocumulus clouds. (Courtesy Dick Jaffe)

Figure 6-21 Fair weather cumulus.

Figure 6-22 Cumulonimbus with the familiar anvil top. (Courtesy National Severe Storms Laboratory, NOAA)

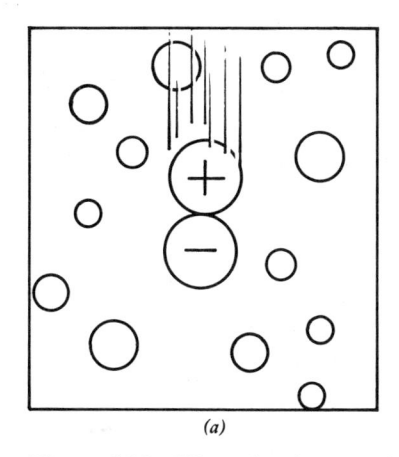

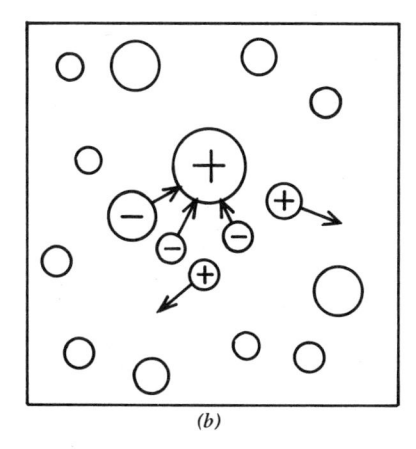

Figure 6-23 Water droplet growth by coalescence; (a) wake capture and (b) direct capture.

In subsequent experiments, it was found that droplets encountering an electric field caused the electrical charge on the droplet to separate into a positive and a negative side. It was also discovered that when two droplets or ice crystals collided under extremely turbulent conditions, an electron was rubbed away from one particle, which became positively charged, and then transferred to the second particle, which took on a negative charge. It was here that *coalescence* (the merging of two particles) was found to be important. Building upon the previous experiments, it was established that the collision of two water droplets with like charges would not result in droplet growth. In experiments, it was proved that the droplets with like charges bounced off one another. However, if the droplets were oppositely charged, they were attracted to one another and adhered on contact. This method of droplet growth is quite important in the tropics, where large amounts of droplets are readily available, because of high humidities and an abundance of salt nuclei.

Coalescence is sometimes called the *capture process* because one particle always captures another. Depending on their size, particles will fall at different velocities. When a larger droplet falls, it sweeps aside all smaller falling particles, picking up those with opposite charges (Fig. 6-23). This is called the *direct capture process*. When two droplets fall through the same path, the wake left by the first allows the second to travel faster. This is called *wake capture* and is similar to a race car trailing very close in the wake of another car in order to reduce friction caused by air molecules.

PRECIPITATION FORMS

There are many types of precipitation, and the form a falling water particle takes depends upon the air temperature at the time of formation and the layers of air it must fall through.

Meteorologists use the term "hydrometeor" to define, in a general way, all forms of precipitation, whether falling, suspended in the air, blown by the wind, or deposited on the ground. Although there are 50 specific types included within seven major categories, only the most commonly observed forms will be discussed.

Liquid Precipitation

Drizzle and *rain* are the two major components of this category. Drizzle consists of numerous water droplets less than 0.5 mm (0.02 in.) in diameter. Since the droplets are so small, they fall very slowly and appear to be floating. In some instances, drizzle takes so long to reach ground level that it evaporates before hitting the ground. When this occurs, the drizzle is called *mist*. Both drizzle and mist are formed from low stratus clouds where their closeness to the earth's surface do not give the droplets an opportunity to grow either by coalescence or the ice crystal process.

Rain, the most common type of precipitation, is composed of droplets greater than 0.5 mm (0.02 in.) in diameter. Because they are larger, raindrops are not as numerous as drizzle droplets. Rain normally falls from higher clouds such as nimbostratus, altostratus, and cumulonimbus where the two growth processes can readily take place.

Freezing Precipitation

This hydrometeor must not be confused with *frozen precipitation*, the third category. Freezing indicates that the droplet was supercooled liquid while it fell and froze upon impact with a colder surface. Frozen implies that the falling particle was solid before it hit the ground. *Freezing drizzle* and *freezing rain* are the two precipitation forms making up this category. Freezing drizzle creates a light coating of ice on any exposed sur-

Figure 6-24 Vegetation covered with a thick coating of ice after an ice storm. (Courtesy H. Michael Mogil)

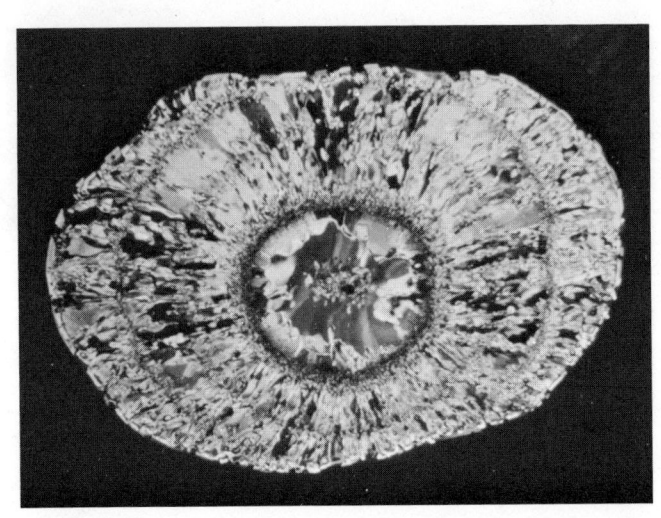

Figure 6-25 A cutaway view of the internal structure of a hailstone. The concentric layers resemble an onion cut in half. (Courtesy C. and N. Knight)

face and is called *light glaze*. It creates hidden hazards for pedestrians and motorists because they do not know glaze is present until it is too late. Freezing rain ends up as a thick coating of clear ice and is called *heavy glaze*. A more common name is an *ice storm*. When heavy accumulations of ice occur because of freezing rain, tree limbs snap, power lines break, and food crops are destroyed (Fig. 6-24).

Frozen Precipitation

Snow, *sleet*, and *hail* comprise the major elements of frozen precipitation. Although they are all solid prior to hitting earth, they differ in formation, makeup, and effect.

Snow results from the accumulation of ice crystals because of the growth processes. Since an ice crystal is hexagonal (six sided) in shape as a result of the structure of the water molecules, snowflakes are also hexagonal. The size of the flake depends to a large extent upon how many collisions occur during its journey earthward and upon the amount of water vapor available at the time of crystal formation. In extremely low temperatures, the amount of water vapor available is very small, and as a result the size and amount of snowflakes will be very small.

As water droplets or melted snowflakes fall, they may encounter an air layer with temperatures below freezing. The liquid particles freeze into solid transparent or translucent grains of ice, which are called sleet. Sleet falls into a broader category called *ice pellets*, which also includes *small hail*.

A third representative of this group is hail, a hard pellet of ice that is formed by concentric layers of frozen water (Fig. 6-25). Hail is generated only in thunderstorm clouds as water droplets from the lower levels of the cloud are carried aloft by the strong vertical convective currents. After the droplets freeze they fall through the cloud, picking up additional water droplets. Before they can fall out of the cloud, the updrafts catch them and carry the ice and water combination aloft once again. This process continues, with successive rings of ice being deposited one on top of another. Eventually the hailstones become so heavy that the updrafts can no longer carry them to the top of the cloud, and gravity pulls them earthward.

Hailstones are made up of both clear and milky textured ice, depending on their rate of freezing. The size of the stone is related to the velocity of the updrafts, with the strongest vertical currents yielding the larger hailstones. Some hailstones in the midwestern part of the United States have grown to the size of oranges.

PRECIPITATION MEASURING DEVICES

One of the most common precipitation measuring devices is called the *standard 8-inch rain gauge*. (Fig. 6-26). Its use is widespread because of its simple construction, low cost, and easy use. The rain gauge is made up of an 8-inch diameter collector top, which is funnel-shaped at the bottom. This allows the water to enter a narrow 2.53-inch diameter measuring tube. The area of the receiver is 10 times larger than the area of the measuring tube, which magnifies the amount collected by 10 times. Therefore, the meteorologist can accurately measure rainfall to the nearest one hundredth of an inch. The measuring tube is housed in an overflow cas-

PRECIPITATION MEASURING DEVICES 71

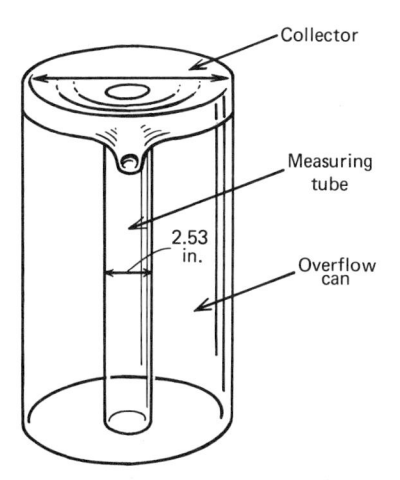

Figure 6-26 The standard 8 inch rain gauge.

ing to allow for spillover in case of excessive rainfall. The measurement is done by a precalibrated dip stick, which is ruled so that 10 inches of length equals 1 inch of water.

A second type of measuring device is the *tipping bucket rain gauge*. Like the standard rain gauge, it has a receiver top that is funneled at the bottom. The water is conducted through this funnel to a seesaw type arrangement with catch buckets at either side of a fulcrum (Fig. 6-27). As water collects in one bucket an imbalance occurs and the mechanism tilts, which activates a mercury switch. The mercury switch energizes a recording pen arm or digital readout to register rainfall to the nearest one hundredth of an inch. After tipping, the bucket empties out and the other side receives water until it becomes unbalanced and the whole process begins again.

A third method of recording rainfall is by the use of a *weighing rain gauge* (Fig. 6-28). This instrument is constructed to measure the weight of water. It has the same receiver top and funnel as in the previous two methods, and the water funnels into a large container that is set on a spring. The weight of the water is transformed into rainfall amount by utilizing the known density of water.

Of the three gauges, the standard is the most accurate. With excessive rainfall rates, the tipping bucket cannot keep up, and the weighing rain gauge loses some of its accuracy simply because the spring is not a good water-measuring device. However, the weighing and tipping bucket gauges are used when permanent records of rainfall are required.

Figure 6-27 The tipping bucket rain gauge. (Courtesy Weather Measure, Division of Systron-Donner, Sacramento, California)

Figure 6-28 A weighing rain gauge. (Courtesy Science Associates, Princeton, N. J.)

A new method of rainfall measurement has recently been developed. It is termed *remote sensing*, and radar is the primary remote sensing technique. Radar emits microwave radiation in the range of 0.1 to 10 cm of wavelength. At these wavelengths the microwave energy is reflected by only the larger water droplets or snowflakes. Therefore, the radar can determine the area of precipitation, track its movement, and determine rainfall rates within reasonable bounds of accuracy.

CLOUD SEEDING

Fresh water is the very lifeblood of our existence on earth. Therefore, any possible improvement to increase our fresh water supply is worth pursuing. Meteorologists have been experimenting with cloud-seeding techniques in order to enhance the precipitation process.

The two most common substances used in cloud seeding are *silver iodide* and *dry ice*. Both have a crystal structure that closely resembles the natural ice crystal. When these chemicals are introduced into a cloud containing supercooled water droplets, the ice crystal process begins immediately. As in the natural process, the crystals grow at the expense of the water droplets until their size causes them to fall. Cloud-seeding experiments have been carried out with limited success in many parts of the world, but it has proven most efficient on windward sides of mountains.

In warm cumuliform clouds, aircraft spread crushed salt particles in the hope of introducing more condensation nuclei for increased rainfall.

To date, evidence of cloud-seeding experiments has indicated that only under ideal circumstances can success be guaranteed. Until such time as meteorologists have a better understanding of the artificial nucleation process, other means of fresh water replenishment must be found.

7
ATMOSPHERIC PRESSURE

Evangelista Torricelli (1608–1647) is accredited with the invention of the barometer. This Italian mathematician used the ideas of his predecessors Galileo and Berti to construct a water instrument that could measure the weight of air. In 1644, he converted his 60-foot tube of water into a 32-inch glass tube of mercury. The mercurial barometer is still in use, and current pressure instruments are called Torricellian barometers in honor of his discovery.

LEARNING OBJECTIVES

After reading this chapter, you should be able to:

1. Define and explain atmospheric pressure.
2. Describe how a mercurial and aneroid barometer operate.
3. Explain how temperature affects pressure.
4. Define *isobar*, *pressure gradient*, *trough*, *ridge*, and *millibar*.
5. Given a pressure reading in one system of measurement, calculate its equivalent in another system.
6. Given a pressure system and central temperature at the surface of the earth, predict the type of pressure system aloft.
7. Explain the ideal model of the earth's semipermanent pressure systems.

One of the basic elements used by meteorologists in producing forecasts is atmospheric pressure. Defined as a force per unit area, atmospheric pressure is the result of the gravitational attraction acting on a column of air that extends from the surface of the earth up to the point where outer space begins. This force is produced by the acceleration downward of gravity (g) on the mass of air (M); the product of the two is called weight. Therefore, atmospheric pressure at any point is the total weight of the air above that point per unit area.

$$\text{Force} = \frac{\text{Weight}}{\text{Area}} = \frac{M \times g}{\text{Area}}$$

Because air is a gas, the pressure exerted by the atmosphere can also be related to the number and speed with which molecules bombard a surface. It is possible to relate pressure to both temperature and density* changes since these factors affect the number of molecules within a given volume and the speed at which they move. A parcel of air is said to have a low density when it has only a few molecules for its volume; it therefore has low pressure. Density can be changed by removing mass from a volume or by expanding or contracting the volume while the mass remains unchanged. The speed at which molecules move is related to the temperature: the higher the temperature, the faster the molecules move. Therefore, for a fixed volume the pressure gets higher as the temperature rises.

One of the most common examples of the interrelationship of pressure, temperature, and density is the behavior of an automobile tire. The density of an automobile tire is essentially fixed, so that when it gets hot from use, the air pressure increases. A simple experiment to prove this is to measure the air pressure using a tire gauge first when the tire is cold and then after having ridden on it for several miles. The second measurement will show that the increase in temperature increased the air pressure significantly. This is an example of the relationship of increased pressure occurring with increased temperature, when the density of the air is held constant.

The relationship between pressure and density can also be illustrated with an automobile tire. When air is added to a tire, it is essentially added at a constant temperature. In this case, the air pressure is increased because mass is added to the volume, increasing its density.

The above examples depended on either the density or the temperature being constant. In general, the pressure temperature and density are interrelated, and the relationship between them is called the *equation of state*. This equation, as well as other gas laws, are explained in more detail in Appendix 2.

PRESSURE-MEASURING INSTRUMENTS

There are three basic instruments for measuring pressure: the mercurial barometer, the aneroid barometer, and the barograph. As previously discussed in Chapter 1, the first mercury barometer was credited to Evangelista Torricelli in 1644. It is based on the principle of balancing a column of mercury whose weight is equal to the weight of the atmosphere. Figure 7-1 illustrates this principle. The air in a long hollow glass tube is evacuated and the tube is placed with its open end into a cistern, which is filled with mercury. The weight of the outside atmosphere forces the mercury from the cistern into the vacuum tube. Since the space inside the tube is a vacuum, it has no air pressure, and the col-

---

* Density is defined as mass (M) per unit volume (V). That is, M/V.

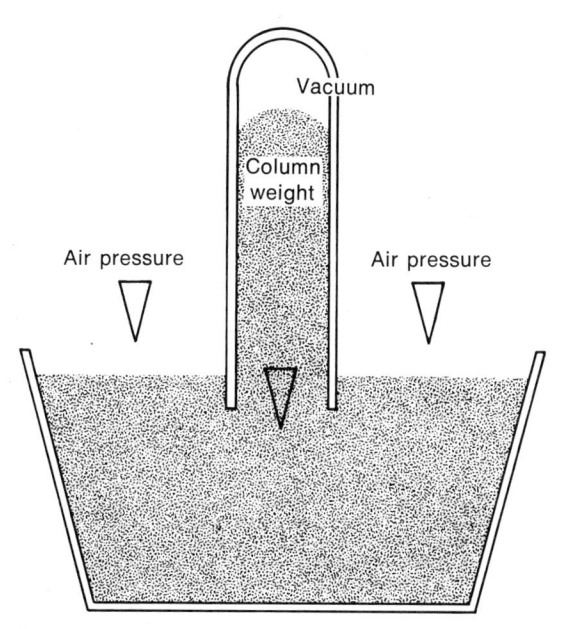

Figure 7-1 The principle of the mercury barometer is the balance of the weight of mercury column by the atmospheric pressure acting on the vessel of mercury. (From Thomas L. Burrus and Herbert J. Spiegel, *Earth in Crisis*, 2nd Ed., St. Louis, 1980, The C. V. Mosby Co.)

umn of mercury will rise to just the level that equals the outside air pressure. Under normal or average pressure conditions the column of mercury will be 760 mm (29.92 in.) high and is called *normal atmospheric pressure*, at sea level. Mercurial barometers are very accurate, however they require small corrections for air temperature, latitude, and elevation as well as for possible manufacturing errors.

Aneroid Barometer

The aneroid barometer (Fig. 7-2) is a mechanical device used for measuring pressure. This device uses a series of partially evacuated chambers, which alter their shape as the surrounding air pressure changes. This change is magnified and trasmitted by a series of levers and pulleys to a pointer attached to a dial. The dial is scaled in the same units as the mercury barometer. Although not as accurate as a mercurial barometer, it is widely used since it is not as fragile and does not require a latitude or temperature correction. However, like the mercurical barometer, it does require an instrument and altitude correction.

The barograph is a constant recording aneroid barometer. Other than its external design the only difference between the two instruments is that the barograph uses a recording pen instead of a pointer. This pen traces the pressure variations on a graph attached to a rotating drum (Fig. 7-3), allowing a permanent record to be kept.

PRESSURE UNITS

As stated earlier, pressure represents a force per unit area. There are a variety of systems expressing this force: metric units, English units, and cgs (for centimeter, gram, seconds) systems as well as the measure of length of a mercury column. In the United States the commonly used unit to express atmospheric pressure is in inches or mercury (the length of a mercury column). However, in scientific usage the commonly used unit of pressure is the *millibar* (mb). A millibar is a cgs units and is 1000 dynes per square centimeter. A dyne is the cgs units of force. Therefore, the millibar is a measure of the force per unit area. Under average conditions, the pressure is 1013.2 mb, which is equivalent to 760 mm (29.92 in.) of mercury.

VERTICAL PRESSURE VARIATIONS

As has already been pointed out, the pressure at the surface of the earth is the weight of the overlying air column per unit area. Above the earth's surface, the pressure would decrease, since there would be less atmosphere.

The change of pressure with height can be calculated from a formula known as the *hydrostatic equation*. This

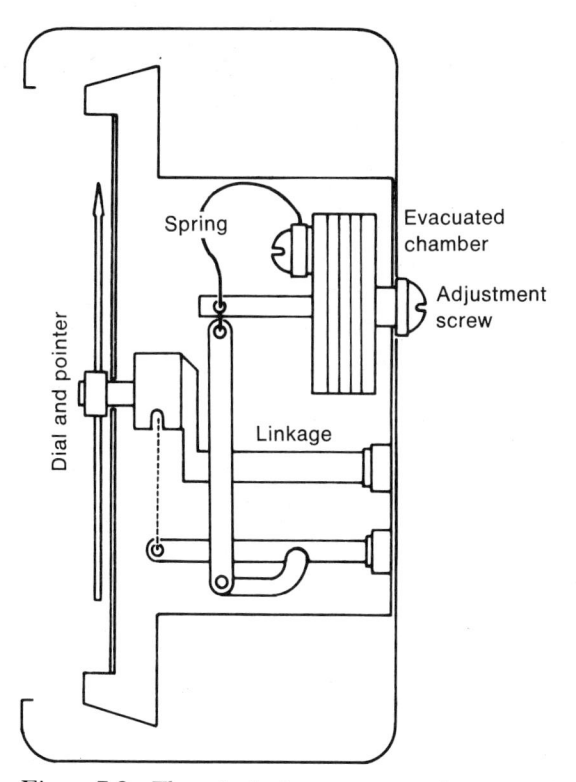

Figure 7-2 The principal components of an aneroid barometer. (From Thomas L. Burrus and Herbert J. Spiegel, *Earth in Crisis*, 2nd Ed., St. Louis, 1980, The C. V. Mosby Co.)

76 ATMOSPHERIC PRESSURE

Figure 7-3 The barograph enables pressure observations to be recorded over a period of time on a paper-covered rotating drum. (Courtesy Weather Measure, Division of Systron-Donner, Sacramento, California)

equation is developed and explained in Appendix II. In this section we shall discuss the main feature of vertical variations of pressure.

It is known that the change of pressure with height depends on the density of air. Pressure changes rapidly with height if the air density is high and more slowly if the air density is low. Figure 7-4 shows a plot of pressure against height. Notice that pressure varies most rapidly at the low levels, where the density is high. For example, pressure changes of 200 mb occur with a height change of 2 km in the lowest layer, while a 2 km change in the upper troposphere accounts for only about a 90 mb pressure change.

Because the air density is related to temperature as well as pressure, the variation of pressure with height can be calculated if we know how the temperature in an air column behaves. This is important because temperature is easily and routinely measured. For a given pressure, the air density is low at high temperatures and high at low temperatures, meaning that a given volume of cold air weighs more than the same volume of warm air. This is significant because for the same height change, air whose average temperature is high will have a smaller change in pressure in the vertical than air whose average temperature is low.

Figure 7-5 shows a situation where the surface pressure is uniform and where the average air temperature on the left-hand side is higher than that on the right-hand side. The lines drawn on the diagram are lines of constant pressure, called *isobars*. As we ascend in

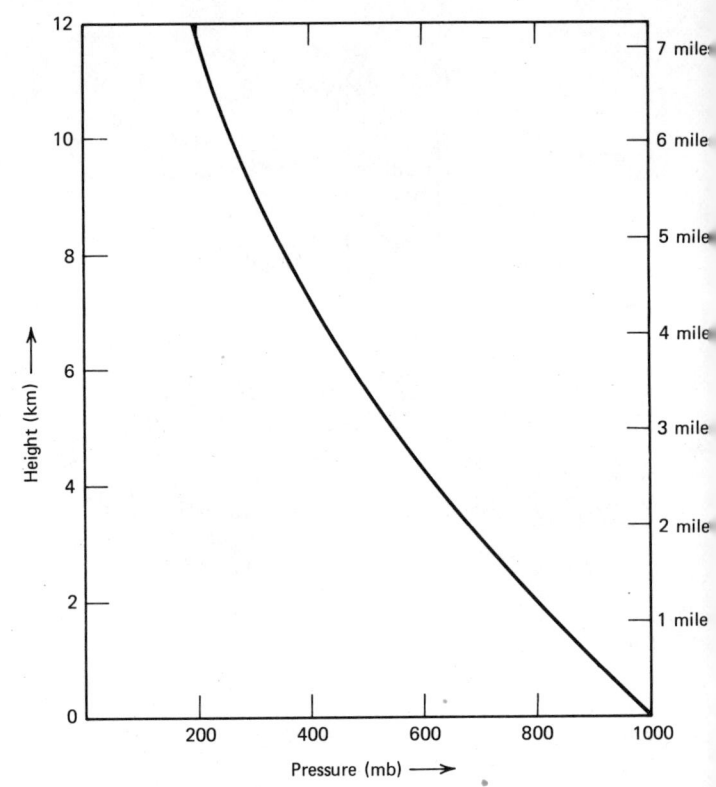

Figure 7-4 The variation of pressure with height in the standard atmosphere.

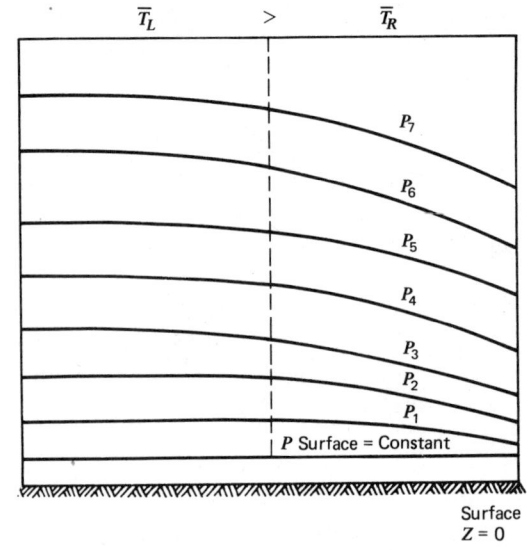

Figure 7-5 The vertical variation of pressure depends on the average temperature in the column. $\overline{T}_L$ is the average temperature on the left hand side of the diagram and $\overline{T}_R$ is the average temperature on the right-hand side of the diagram. The isobars above the surface are labeled P_1, P_2, etc.

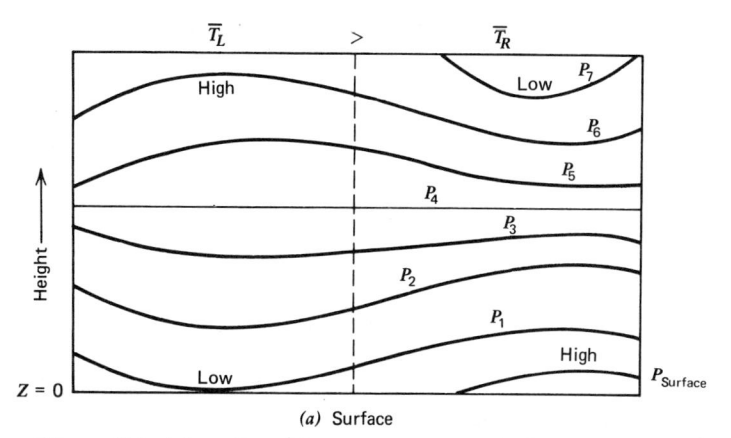

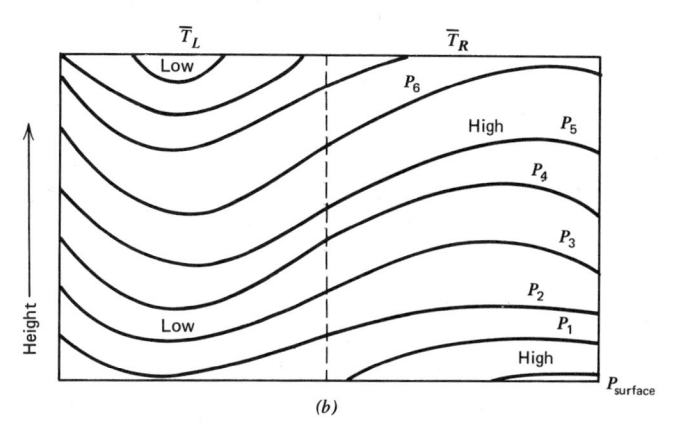

Figure 7-6 The effect of mean temperature of an air column on the vertical and consequent horizontal distribution (a) case of a warm low and cold high, (b) case of a cold low and warm high. Symbols have the same meaning as in Figure 7-5.

height, we notice that the pressure is no longer uniform across the diagram but is higher on the left than on the right. This is so because at higher temperatures, the left side shows a smaller decrease in pressure than is shown on the right. Thus, the vertical temperature structure has led to horizontal pressure variations. These horizontal pressure variations are especially important, since they are the principal forces that produce winds in the atmosphere.

Another example of how the average temperature in a column of air can affect the vertical and horizontal pressure distribution is shown in Figures 7-6a and 7-6b. Figure 7-6a shows a situation with low surface pressure on the left and high surface pressure on the right. In this diagram, the left side has a mean temperature in the air column that is higher than the mean temperature shown on the right. Thus, the right side will exhibit a larger pressure change in the vertical than will the left side. As a result, the pressure on the right as we ascend decreases more rapidly than it does on the left, and at the top the horizontal pressure distribution has been reversed so that the high pressure is now on the left and the low pressure is on the right. This situation is commonly known as warm lows and cold highs.

Finally, Figure 7-6b shows on the left-hand side the case of low surface pressure and low mean temperature; on the right-hand side it shows high surface pressure and high mean temperature. In this case, pressure will decrease more rapidly on the left than on the right. In other words, the low and high pressure areas increase their intensity with height.

This section can best be summarized as follows: cold lows and warm highs will increase with height, warm lows and cold highs will decrease with height, and constant pressure at the surface will develop high pressure aloft when the air column is warm and low pressures when the air column is cold. This concept is used in forecasting the intensification or dissipation of high and low pressure areas.

ALTIMETRY

An airplane altimeter is an important application of the hydrostatic principles. The altimeter is an aneroid barometer, which while measuring pressure is scaled in altitude through use of the hydrostatic equation. Because the altitude scale is set to zero for the surface pressure at the ground, the altitude at any pressure P is obtained by assuming an average atmosphere (see Chapter 2). If the actual atmosphere is colder than the average value used, the altimeter will read too high, and conversely, if the atmosphere is warmer than the average value used, the altimeter will read too low. Clearly, with poor visibility, flying with a pressure altimeter can be hazardous, unless corrections are made to account for variations in the temperature of the air column between the aircraft and the ground.

HORIZONTAL PRESSURE VARIATIONS

Pressure variations in the horizontal are much smaller than pressure variations in the vertical. However, horizontal pressure variations are extremely important, for they are the forces that drive the atmospheric winds. Before describing the horizontal pressure differences, some commonly used pressure terms need to be explained.

PRESSURE TERMS

A number of terms are used in describing pressure patterns. They are *isobar, pressure gradient, trough,* and *ridge*.

78 ATMOSPHERIC PRESSURE

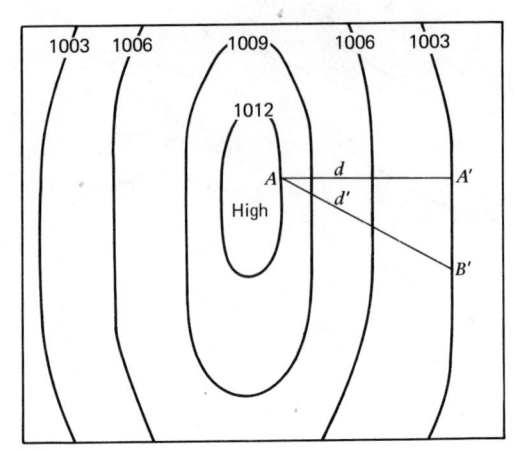

Figure 7-7 An idealized example of isobars at the surface. Each isobar is drawn for pressure intervals of 3 mb. The line AA' is the path along which the pressure gradient is measured. See text for explanation.

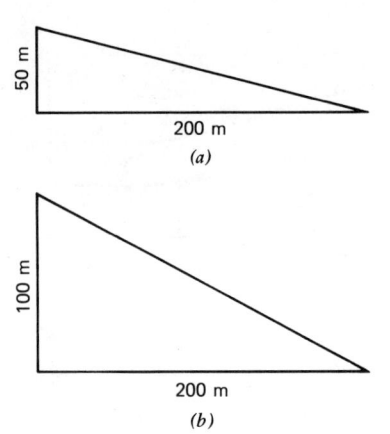

Figure 7-8 The magnitude of the pressure gradient force by analogy to an inclined surface. (a) An elevation change of 50 m (164 ft) over 200 m (656 ft) and (b) an elevation change of 100 m (328 ft) over 200 m (656 ft). The slope of (b) is much greater than (a).

Isobar

We have already been introduced to the concept of an isobar in the discussion of vertical pressure variations. An isobar is a line that connects points of equal pressure. For surface pressure, isobars are usually drawn at pressure intervals of 3 mb. An idealized example of isobars is shown in Figure 7-7.

Pressure Gradient

The pressure gradient is the change in pressure with horizontal distance. It is measured from high to low values, perpendicular to adjacent isobars, and is the shortest distance between them. It represents the greatest change of pressure in the shortest distance. In addition to having a direction, the pressure gradient also provides information on the magnitude of this change with distance, and is often called pressure steepness.

The direction of the pressure gradient is illustrated by referring to Figure 7-7. If we measure the change in pressure over the distance d along the line AA' we see that the pressure change is 9 mb. If we now measure the pressure change along the line AB', distance d' we see that it has also changed 9 mb but over a larger distance, since d' is larger than d. Obviously the change is greatest over distance AA', which is perpendicular to the isobars.

The steepness of the gradient is established by comparing it to an inclined surface as shown in Figure 7-8. In Figure 7-8a the surface has a slope such that the elevation change is 50 meters over a horizontal distance of 200 meters. In Figure 7-8b the elevation change is 100 meters over the same horizontal distance. The magnitude of the gradient is larger in the second case than in the first and obviously steeper. Again in the case of atmospheric pressure, the greater the change in the horizontal over a fixed distance, the steeper the gradient.

Troughs and Ridges

When examining pressure maps the patterns often appear elongated and wavy, with areas of low pressures sandwiched between areas of high pressure or vice-versa, as shown in Figure 7-9. The elongated area of low pressure is called a trough and the elongated area of high pressure is known as a ridge. These terms are analogous to surface features such as mountain ridges and associated elongated valleys or troughs.

SEMIPERMANENT PRESSURE PATTERNS

There are two major pressure patterns that are related to weather changes. The first is a nearly stationary or

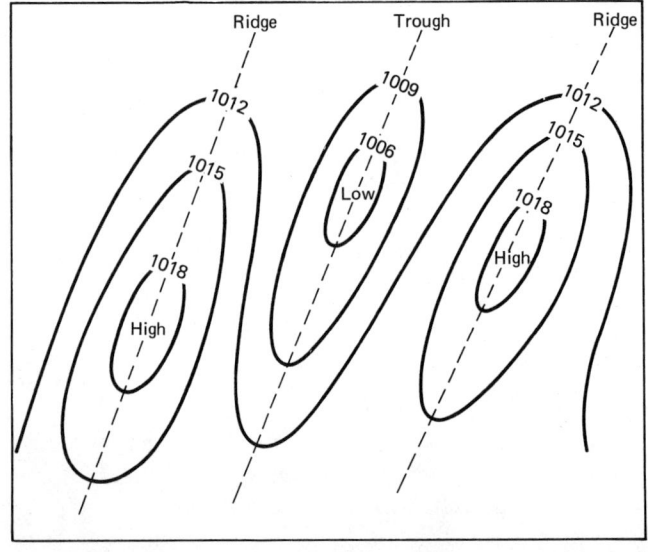

Figure 7-9 Wavy pressure patterns and the associated axes of the ridges and trough.

semipermanent pressure pattern related to the average heat distributed over the earth's surface. The second pattern is related to the dynamics of air flow associated with daily weather. These systems have relatively short time spans that change from day to day, and they meander over the face of the earth. We will call them the *transient pressure patterns*. In this section we shall consider semipermanent pressure patterns, leaving the transient ones for discussion in Chapter 10.

If the earth's surface were perfectly smooth and of uniform composition the sea level pressure pattern would assume a zonal arrangement, that is, they would be arranged along latitude circles. Figure 7-10 shows the pressure distribution to be expected under such conditions. The low pressure area at the equator is in response to solar heating that takes place at the equatorial region. This heating results in less dense air and consequently lower pressures. This zone of low pressure that has been created is known as the *equatorial trough*. Conversely, high pressure at the poles is in response to the dense air resulting from low temperatures. This region is known as the polar high pressure zone. In between, the pressure patterns are determined by a combination of solar heating and large scale dynamics of a rotating atmosphere. These mechanisms produce an irregular belt of low pressures at 60° north and south, which are termed subpolar lows. At 30° north and south there is an irregular high pressure zone called the subtropical high. Each of these pressure patterns have wind circulations associated with them (see Chapter 8).

However, this drawing is an idealized one and has to be modified when we consider the influence of continents and oceans. The major effect is manifested on how the heat is shared between them, is discussed in Chapter 3. Briefly, the land heats and cools much more rapidly than the oceans, with the result that, on an annual and seasonal basis, variations in temperature are much greater over land than over water. Figure 7-11a, which depicts the average worldwide sea level pressure in July, also shows these influences on the pressure distribution. In July we see that the subtropical belt of high pressure is well developed over the cool oceans at 30°N but is broken by low pressure areas over the heated land, most noticeably over North Africa, South Asia, and Mexico. The region of low pressure at 60°N, while evident, is not as well established as the high pressure zones over the oceans. In the southern hemisphere the zone of high pressure between 20°S and 40°S is nearly continuous. It is winter season there, and the cool land areas exhibit high pressure. Minor disturbances in the high pressure occur over South America and Africa.

During the northern hemisphere winter, represented in Figure 7-11b by the average pressure for January, the contrast between ocean and land is greater. The low pressure areas at 60°N in the North Pacific and North Atlantic are now well developed; they represent warm areas relative to the cold land surrounding them. These lows are known as the Aleutian and Icelandic lows, respectively. The cold land areas at 60°N exhibit very high pressure, as we should expect. The subtropical highs over the oceans are still evident, but somewhat weaker than in the summer. Minor perturbations in this belt are caused by heated land masses near Central America, South Asia, and the Middle East. This zone of high pressure has shifted southward from its July position. In the southern hemisphere, January represents the summer season. We observe low pressure over the continental areas and well-defined high pressure over the oceans. Note that the position of the high pressure belt has shifted further southward as compared to July.

The equatorial trough, or the low pressure zone that is located near the equator, also shifts with the seasons. This zone is much farther south in January than in July.

Finally, it is seen from Figures 7-11a and 7-11b that the seasonal changes in the pressure pattern are much more pronounced in the northern hemisphere than in the southern hemisphere. This is because the southern hemisphere has large expanses of ocean and little land mass, while the northern hemisphere has much more continental area.

These patterns of high and low pressure have a strong

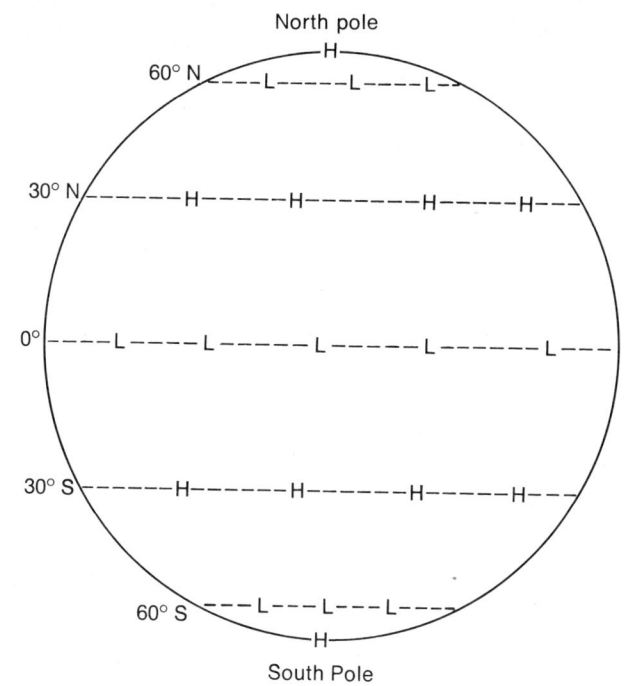

Figure 7-10 Sea level pressure distribution expected if the earth's surface was perfectly smooth and of uniform composition.

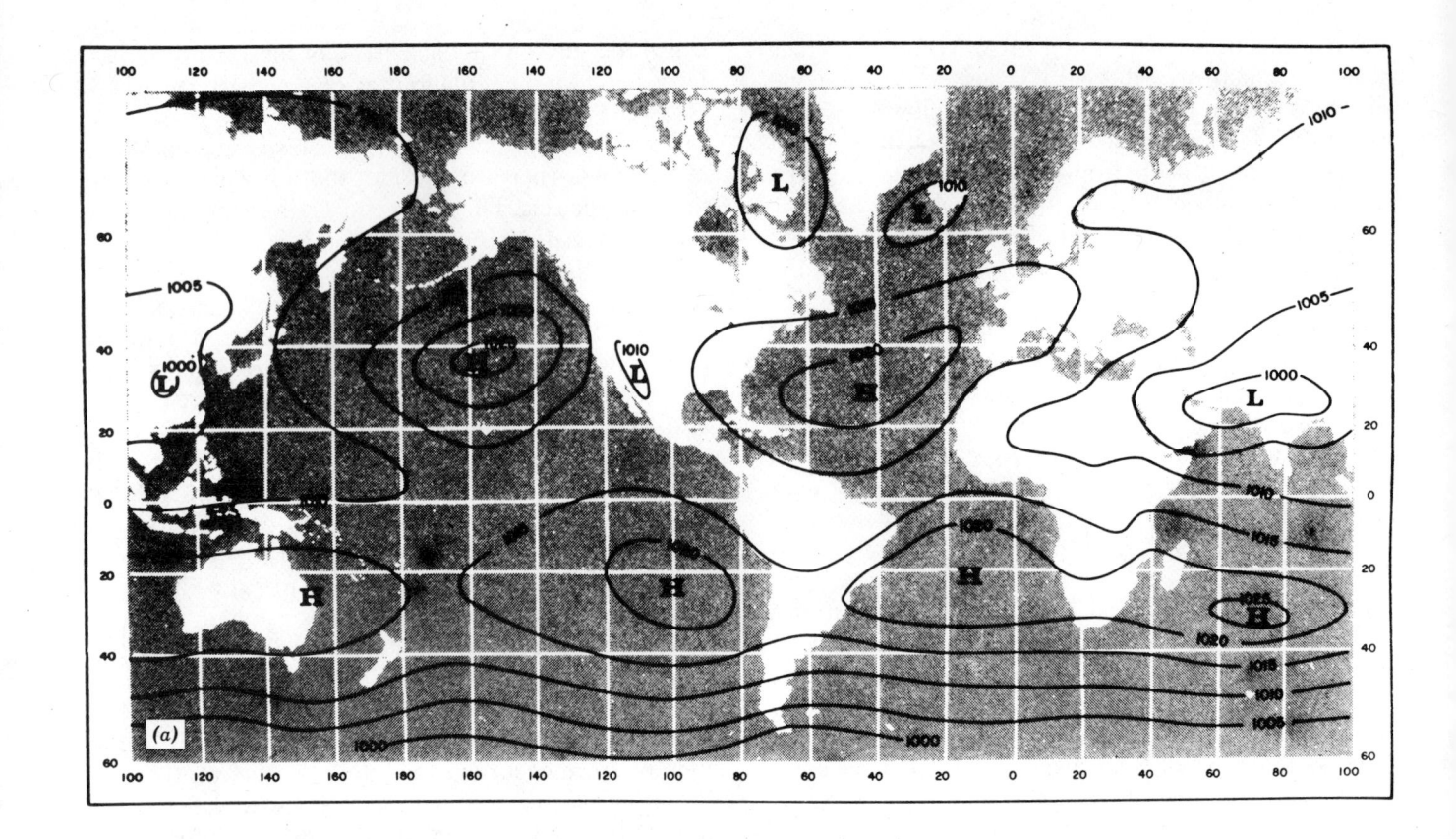

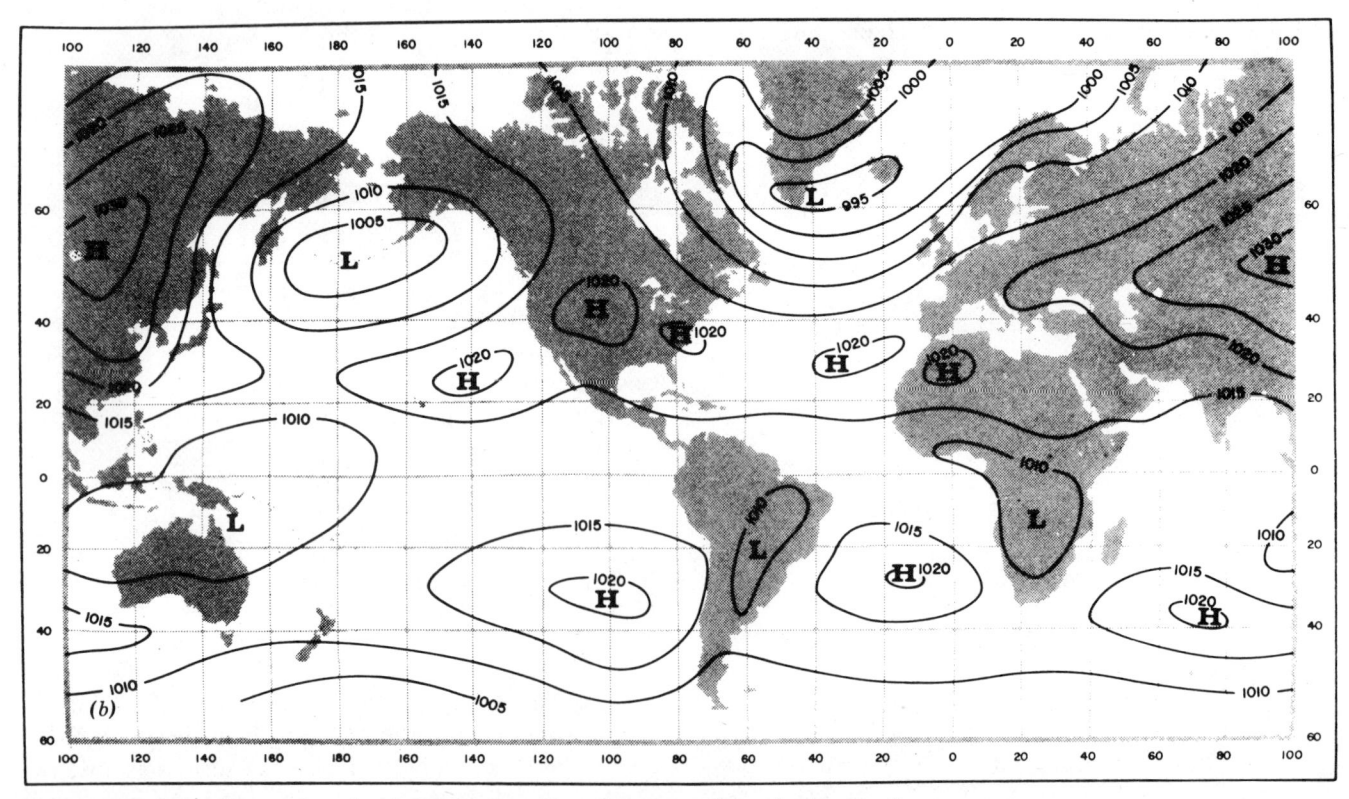

Figure 7-11 The observed sea level distribution. Note departures from the idealized pattern caused by the difference in heating between land and ocean. (*a*) July and, (*b*) January. (Courtesy NOAA)

influence on the weather of a region. This comes about because high pressure is generally associated with sinking motions, which result in clear skies and no precipitation, while low pressure areas are associated with rising motions and precipitation.

Thus, the seasonal shifts in the major pressure patterns are associated with shifts in weather disturbances. For example, the north-south shifting of the equatorial trough region causes a seasonal shift in precipitation in that area of the world. Storm tracks shift northward during summer as the zone of high pressure moves northward, and southward in winter as both the low pressure zone at 60°N and the high pressure belt at 30°N push toward the equator.

The actual day-to-day variations in the weather are more closely related to the transient pressure patterns. However, the interaction between the semipermanent and transient pressure patterns is what shapes the weather in a given area.

The semipermanent large scale pressure distribution also has a pronounced effect on the climate of a region. This will be discussed further in Chapter 8.

8
WINDS OF THE WORLD

Christoph Buys-Ballot (1817–1890), a former director of the Netherlands Meteorological Institute, formulated the *baric wind law*. This law, known more commonly as the Buys-Ballot's law, states that in the northern hemisphere, when a person stands with his back to the wind, the lowest pressure is always toward the left. In the southern hemisphere just the opposite occurs.

LEARNING OBJECTIVES

After reading this chapter, you, should be able to:

1. Briefly explain Newton's "Three Laws of Motion."
2. List and describe the major and secondary forces affecting air movement.
3. Explain the balance of wind forces around high and low pressure areas and relate them to the resultant wind flow.
4. Develop a step process to describe the general wind pattern of the world.
5. Draw the six major wind belts of the earth.
6. Explain the jet stream.
7. Describe the mechanisms responsible for land and sea breezes, monsoon winds, mountain and valley breezes, and the foehn (chinook) winds.
8. Given the air temperature and wind speed, calculate the wind chill factor.
9. Define *turbulence, pressure gradient force, coriolis force, centrifugal and centripetal forces, frictional force, geostrophic* and *gradient wind, macroscale, mesoscale, microscale, cyclone,* and *anticyclone.*

Everything that happens in nature is the result of an imbalance. Lightning is caused by the buildup of negative charges, clouds by the overabundance of water vapor, and an avalanche by the gravitational force that overcomes the frictional force of material on the side of a mountain. These occurrences are just nature's way of obtaining a balance of natural forces.

In the atmosphere, imbalance constantly takes place because of many separate interacting forces. In order to compensate for these forces the air moves in order to achieve a balance once again.

All physical laws governing moving particles can be related to the atmosphere. Most important are the three laws of motion attributed to *Sir Isaac Newton*, a noted eighteenth century physicist. Although it is impossible to discuss these laws in detail here because of the nature of this text, a brief summary of them will offer an excellent introduction into the reasons for atmospheric behavior.

THE THREE LAWS OF MOTION

The Law of Inertia

The first law of motion states that a body at rest or in motion will remain that way unless an outside force acts upon it. Therefore, calm winds will change to windy conditions only if a force is exerted within the atmosphere to cause an imbalance.

The Law of Acceleration

The second law of motion asserts that a change in motion of a body relates directly to the force trying to move it. Once set in motion, a body will continue to move in a straight path and at the same speed unless an outside force acts on that body. This movement will be in the same direction as the force responsible for its movement. In the atmosphere, air particles will move in a straight line if only one force produces the motion. This is strictly a short-lived phenomenon, because many forces eventually react with the moving air.

The Action-Reaction Law

The third law of motion implies that whenever a force is exerted on a body, an equal and opposite force is exerted by the second body on the first. Simply stated, for every action there is an equal and opposite reaction. In the atmosphere there cannot exist just one force or just one body and, if air is to move, at least two forces must be present.

PRIMARY FORCE AFFECTING THE WIND

In its simplest form, the definition of wind is "the horizontal movement of air relative to the earth's surface." This definition eliminates all vertically moving air, which meteorologists term *convection currents*, or *turbulence*. Although convective air is important in the formation of clouds and eventually precipitation, horizontally moving air is much more significant to the weather process. Many theories and models have been proposed to explain atmospheric motion, but a complete understanding is still in the future because of the complex forces responsible for this phenomenon.

84 WINDS OF THE WORLD

The major force responsible for initiating horizontal air flow is the *pressure gradient force*. This force results from a difference of pressure, which in turn comes about from a difference in air temperature. As land and water bodies, multicolored soils, and forested versus desert areas receive energy from the sun, they heat up at different rates. This differential in heating is mirrored in the air directly above the heated surface, and the atmospheric temperature imbalance provides the ingredients for pressure differences. (Since warmer air weighs less than cooler air, a low pressure area is created within the warmer air.) These pressure differences create a pressure gradient that drives the wind because of nature's desire to achieve an equilibrium of pressure.

Two simple experiments can be performed to show how pressure differences relate to air movements. The first is to open a vacuum packed can and listen to the air rush into the can from the outside higher pressure. The second is to fill up a balloon with air and puncture it. The air inside the balloon, which has a higher pressure due to its entrapment, rushes to the outside to achieve a balance in pressure. In both cases, it should be realized that the air is moving from higher to lower pressure, and that the greater the pressure difference the faster the air traveled.

Just as in the two examples, the atmosphere attempts to achieve an equilibrium in the horizontal pressure patterns such as that shown in Figure 8-1. On weather maps a pressure gradient, which is the rate of change of pressure from one place to another, can be determined by the spacing of the isobars around the pressure system. The closer the isobars are to one another, the steeper the gradient and the faster the wind speed.

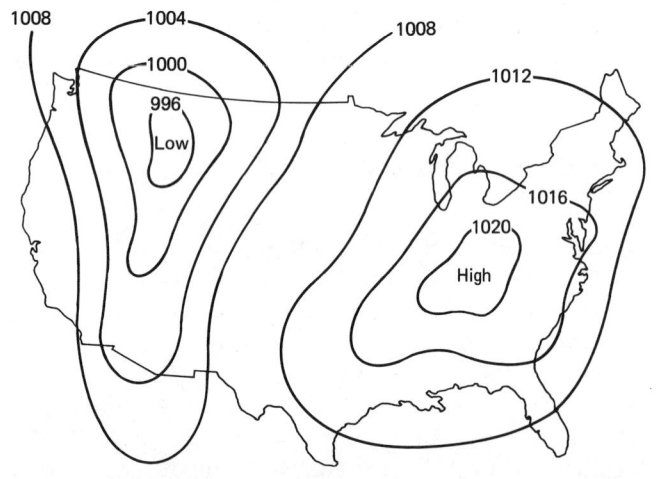

Figure 8-1 Isobars on a weather map depict high and low pressure areas. If only the pressure gradient force is considered it would be easy to predict the wind flow. Notice that the wind speed would be higher around the low pressure area.

SECONDARY FORCES AFFECTING THE WIND

Secondary forces are those forces that react with the air only after the air begins to move. There are three important secondary forces, which when applied to the moving gas mixture, cause a curved path flow in many different directions. These forces are *coriolis force*, *centrifugal force*, and *friction*.

Coriolis Force

This force arises solely because of the rotation of the earth and is actually an apparent force. To an observer on the earth's surface the effect of this force in the northern hemisphere is to deflect moving air to the right of its direction of motion and to the left in the southern hemisphere. To an observer in a fixed position in space, looking down on the earth, the air would be observed to move in a straight line with the earth's surface moving away; thus, the term apparent force. A good example would be an attempt to draw a straight line with a piece of chalk on a rotating phonograph record. Even though your hand moved in a straight path, the chalk trace on the record would be curved (Fig. 8-2).

Our earth rotates on its axis at the rate of 1674 km (1041 mph) at the equator. The speed decreases with increasing latitude until it is essentially zero at the poles. This reduction in speed results from the smaller latitude circles poleward of the equator (Fig. 8-3). An object at the equator will make one complete circle of approximately 40,126 km (24,000 miles) in a 24 hour period and therefore, will have a speed of 1041 mph. The same object at 60°N would travel 20,112 km/hr (12,504 miles) in the 24 hours and have a speed of 838 km/hr (521 mph). At the north pole the linear speed would be zero because there would be no distance traveled.

As an object such as a parcel of air or a rocket starts to move in a straight path from the equator to the north pole, its eastward speed will be 1674 km/hr (1041 mph). As the object travels northward, its eastward movement will be faster than the eastward movement of the surface of the earth at higher latitudes. Therefore, it will run ahead of any object at higher latitudes, and appear to an earth based observer to be curving to the right (Fig. 8-4). If the path were from north pole to equator, the object would have a zero eastward movement, and it would lag behind a lower latitude object whose eastward movement would be faster. To an observer on the earth this motion would appear to be to the right of the direction of motion.

While this force is called an apparent force, its effects are very real and must be taken into account. For example, in order to be accurate, long range artillery must account for the effects of the coriolis force.

SECONDARY FORCES AFFECTING THE WIND 85

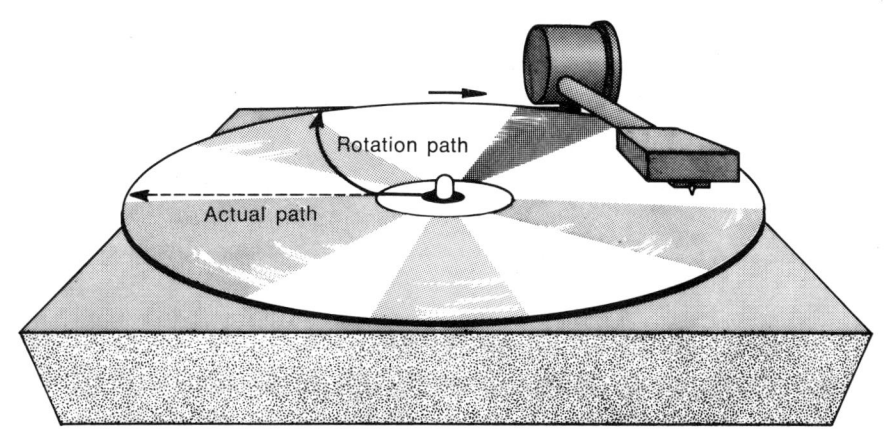

Figure 8-2 Because of the rotational effect of the phonograph record, a curved path results. If the rotation is clockwise the path is to the right of the direction of motion (northern hemisphere) and counterclockwise to the left (southern hemisphere). (From Thomas L. Burrus and Herbert J. Spiegel, *Earth in Crisis*, 2nd Ed., St. Louis, 1980, The C. V. Mosby Co.)

An obvious question should come to mind. Why to the right in the northern hemisphere and to the left in the southern hemisphere? In the southern hemisphere the sense of rotation is opposite to that in the northern hemisphere. This can be visualized by the following thought experiment (Fig. 8-5). Imagine yourself looking down on the north pole. The sense of rotation would be counterclockwise. Now imagine yourself looking down at the south pole. The sense of rotation is now clockwise. Thus deflection occurs to the right in the northern hemisphere and to the left in the southern hemisphere. Since there is no rotation about the vertical if you are standing at the equator, there is no deflection and thus no coriolis force. The coriolis forces varies with latitude, taking on values proportional to the sine of the latitude. It is also proportional to wind speed, being larger at greater wind speeds. However, maximum deflection occurs at the poles resulting in maximum coriolis force.

Centrifugal Force

If a rock tied to a string were swung, its path would be a circle with constant radius. At any point along that circle, the rock's preferred path would be a straight line, as

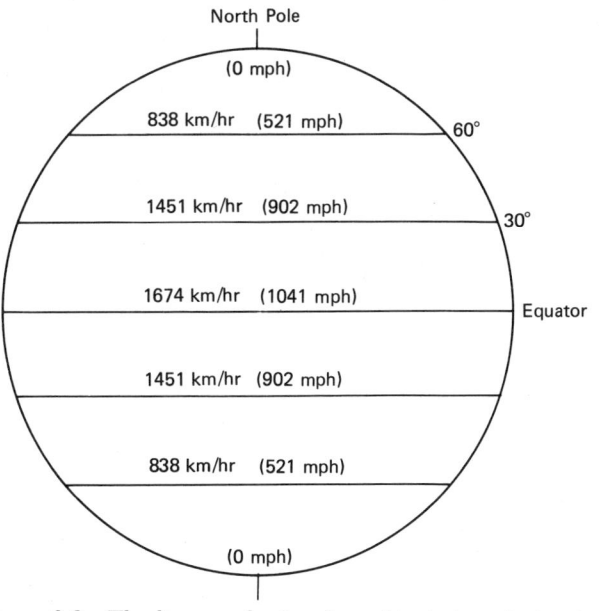

Figure 8-3 The linear velocity of an object at predesignated latitude lines.

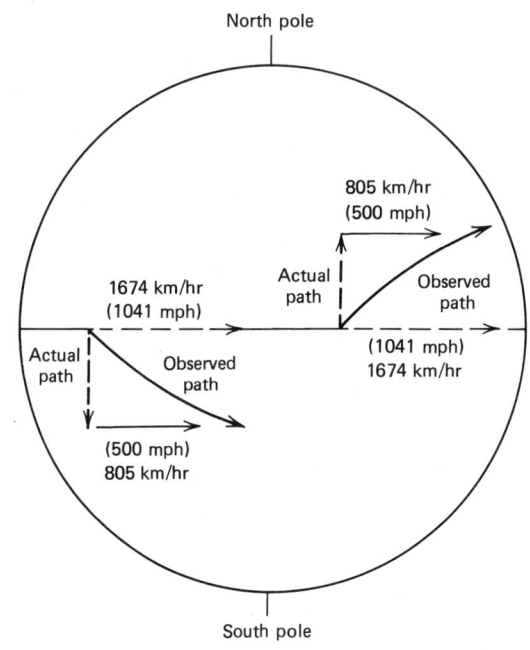

Figure 8-4 A diagrammatic view of the coriolis effect.

86 WINDS OF THE WORLD

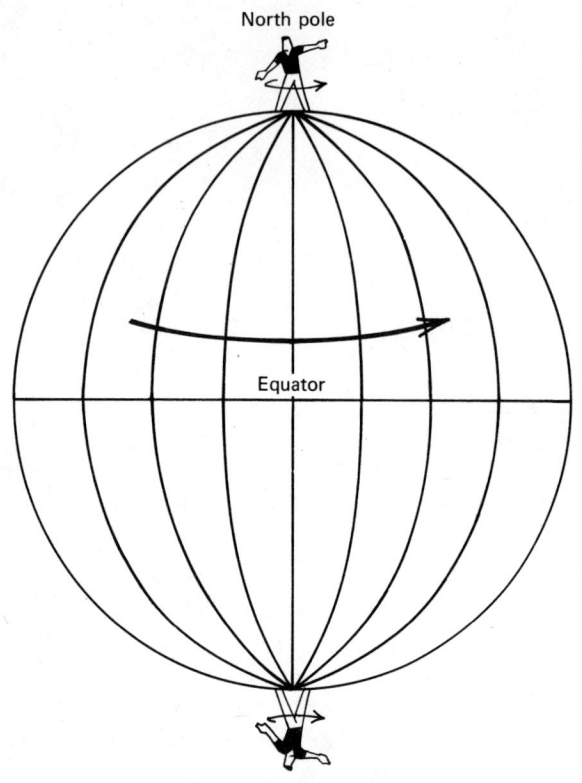

Figure 8-5 Although the earth always rotates west to east, the spin of a person standing in free space is opposite in each hemisphere.

explained by Newton's first law of motion. However, the curved path is maintained by the inward pull of the swinger's arm, and this pull inward is called *centripetal force* (Fig. 8-6). If this were the only force acting on the rock, the rock would accelerate inward and hit the person on the head. Since this does not happen, and since the object stays at a constant radius, there must be an equal and opposite force termed *centrifugal* in accordance with Newton's third law of motion. Centrifugal force is one of the reasons why wind circulation differs around high and low pressure areas.

Frictional Force

Any moving object close to the surface of the earth will be affected by friction produced by the interaction of the moving body on an uneven terrain. As altitude increases, the frictional effect of the earth decreases, until for all practical purposes, it is essentially zero at 610 m (2000 ft) above the earth's surface. Friction tends to slow down the flow of air, for the force acts in a direction opposite to the direction of movement. Since roughness of a surface is the most important ingredient in frictional effect, wind speeds over larger water bodies are on the average greater than wind speeds over land. When the frictional effect is considered, it is easy to understand why wind speeds at higher altitudes are much greater than winds close to the earth's surface.

As the air slows down through friction the coriolis effect is reduced, and the air deviates from its original path. This deviation can range from 10° over smooth water surfaces to 45° over rough terrain. This effect will be discussed later in the chapter.

THE BALANCE OF WIND FORCES

As was stated previously, nature will not tolerate an imbalance. When an imbalance does occur, nature will take action to counter any force that has created it. It is apparent that eventually, there must be a balance among all four forces previously discussed. In order to understand this continual drive for balance and the resultant air flow around pressure systems, we will go through a four-step process.

Step 1: Geostrophic Wind (Wind Moving Parallel to Straight Isobars). In the initial stage, let's assume that we are dealing with a straight line wind in the northern hemisphere above the *friction level* at 610 m (2000 ft). Because of the limitations we have placed on location, both friction and centrifugal force are eliminated from consideration, and only pressure gradient force and the coriolis force must be dealt with. If pressure gradient force were the only force, air would flow from high to

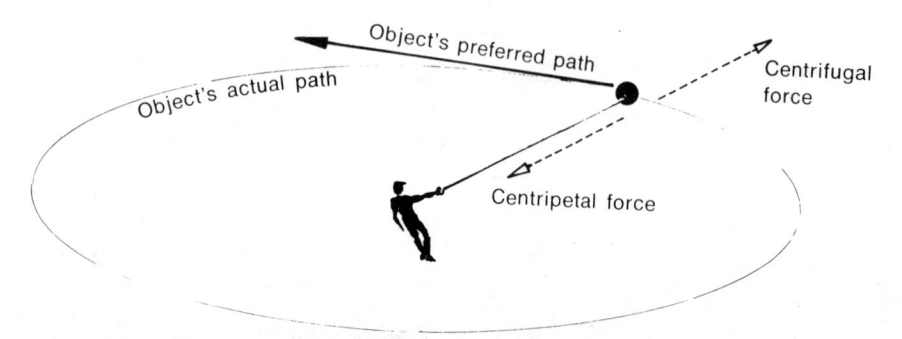

Figure 8-6 The balance of centrifugal and centripetal forces at any one point allows an object to move in a curved path. (From Thomas L. Burrus and Herbert J. Spiegel, *Earth in Crisis*, 2nd Ed., St. Louis, 1980, The C. V. Mosby Co.)

THE BALANCE OF WIND FORCES 87

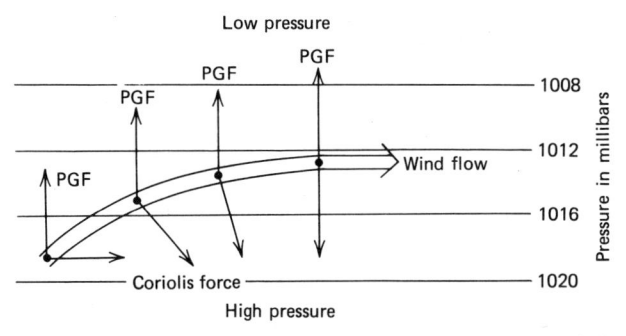

Figure 8-7 Above the friction level, straight line winds have a balance between pressure gradient force and the coriolis force. When the balance occurs the wind is said to be in *geostrophic balance*. Using this information, Buys-Ballot formulated his now famous Buys-Ballot's law relating air pressure to wind flow.

low pressure with wind speeds increasing with time. However, as air begins moving, the coriolis force tends to deflect the moving particles to the right. At each successive step, as shown in Figure 8-7, the coriolis force provides a larger deflection perpendicular to the increased wind flow, ending up with a still greater alteration. This deflection continues until the pressure gradient force is equal and opposite to the coriolis force, and a balance has been achieved. At this point the air moves in a straight line parallel to the isobars with a constant wind speed and fixed pressure gradient. The constant wind speed produces a fixed coriolis force and the balance remains with a continuous straight line path. However, when the pressure gradient changes, an imbalance occurs as before, and the atmosphere reacts to counter it once again.

Step 2: Gradient Wind (Wind Moving Parallel to Curved Isobars). In the second stage centrifugal force is added as a third force in the balance attempt, and it always acts outward from the center of a curved path. By looking at weather maps it is quite obvious that isobars seldom lie in a straight line; therefore centrifugal force must be taken into consideration when analyzing wind flow. Figure 8-8 represents curved flow around high and low pressure areas. In Figure 8-8a the pressure gradient force is inward toward the low center while coriolis is outward at that point. The third force, centrifugal, is in the same direction as coriolis since it always extends outward from the center. The centrifugal force adds to the coriolis to balance the pressure gradient or:

$$PGF = CF + \text{Centrifugal}$$

By plotting these balances of forces point by point on a diagram, and knowing that the coriolis force must be to the right of the direction of motion, the counterclockwise parallel circulation of a low above the friction level falls into place.

In the high-pressure system of Figure 8-8b, the pressure gradient and centrifugal force are both outward. Thus, the coriolis force must be inward to achieve the balance:

$$PGF + \text{Centrifugal} = CF$$

This balance determines the counterclockwise circulation parallel to the isobars. If the same balances were to occur in the southern hemisphere, the only change would be the wind direction. Since the coriolis force deflects to the left in the southern hemisphere, the direc-

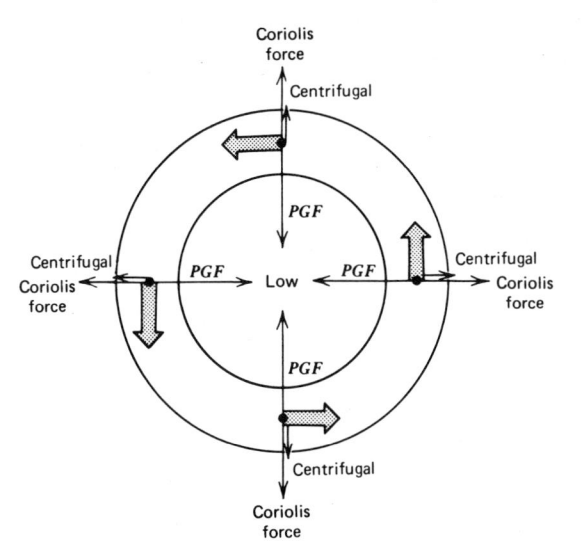

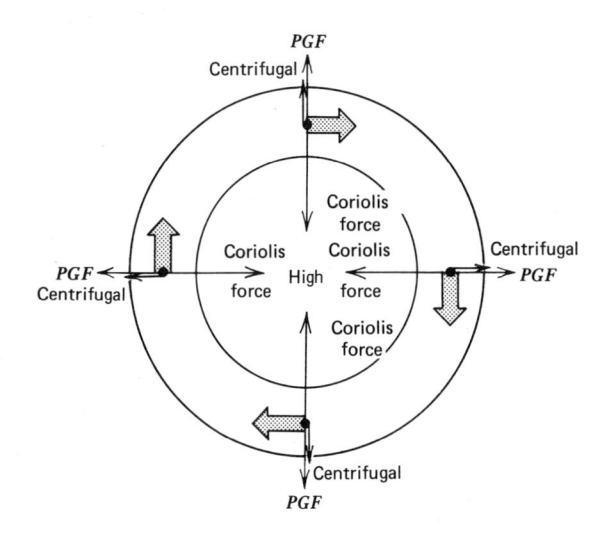

Figure 8-8 The balance of forces diagram around (*a*) a low pressure system, (*b*) around a high pressure system.

88 WINDS OF THE WORLD

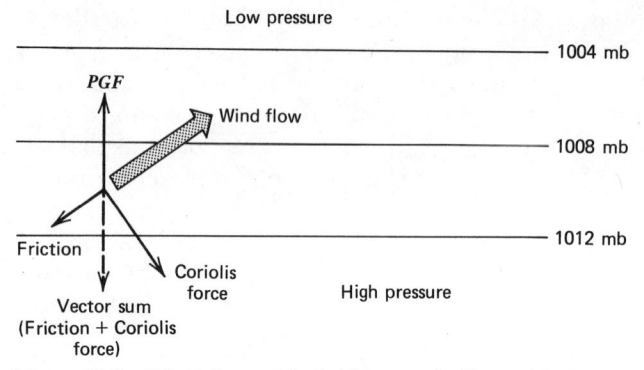

Figure 8-9 Wind flow with the frictional effect added results in cross isobar flow from high to low pressure.

tion would be clockwise for a low and counterclockwise for a high.

Step 3: Straight Line Wind Flow Below the Friction Level. Below 2000 feet, friction, the final force, must be accounted for if a balance of forces is to exist. We have seen that without friction, equalization of forces occurred when the pressure gradient force equaled the coriolis force (Step 1). The addition of friction results in a lower wind speed, which in turn reduces the coriolis force. This reduction allows the pressure gradient force, which is not affected by the lower wind speed, to pull the air at an angle across the isobars even though all three forces are in balance (Fig. 8-9). The frictional effect on wind direction is determined largely by the roughness of the earth's surface. In mountainous areas the angle at which air is deflected from high to low pressure could be as much as 45° and the wind speed could be reduced almost to half.

Step 4: Curved Flow Below the Friction Level. In Step 2, we discovered that northern hemisphere winds blow clockwise around a high pressure area and counterclockwise around a low pressure area. In the southern hemisphere just the opposite is true. When friction is added, this movement becomes inward toward the low center and outward from the high center. With all force in balance, the air circulation would be as shown in Figure 8-10.

WIND SYSTEMS OF THE WORLD

The wind that blows at any particular time is not the result of a simple set of circumstances. Meteorologists have known for years that wind results from the complex mix of the general worldwide wind pattern, winds that are associated with migrating pressure systems, and winds that are generated by local conditions. The largest factor of the wind systems in this mix is the general worldwide pattern. This type of system, as well as the more familiar wind flow around the smaller migrating pressure systems, is commonly called *macroscale* because of its comparatively larger dimensions. *Mesoscale* systems (middle size), although prevalent throughout the year, last only a few days at a time and cover only a small area of the globe. Local winds such as land and sea breezes, valley winds, and thunderstorm winds fit into this mesoscale system. *Microscale* (small scale) winds are reserved for wind flow that lasts only minutes or less, and pertain to eddies, currents, winds, gusts, and dust devils. These scales interact in such a way that microscale winds can be superimposed on the mesoscale system, which in turn can be superimposed upon the larger macroscale winds.

THE GENERAL WIND PATTERN

It is much easier to understand how the general wind circulation is created if we start with an idealized model of the earth. In 1735, George Hadley introduced a simple model of global wind circulation. In this model he assumed the following.

1. The earth does not rotate (no day or night, no coriolis effect).
2. The earth's surface is smooth and of uniform composition (no unequal heating over small areas).
3. The earth does not tilt on its axis (no seasonal change).

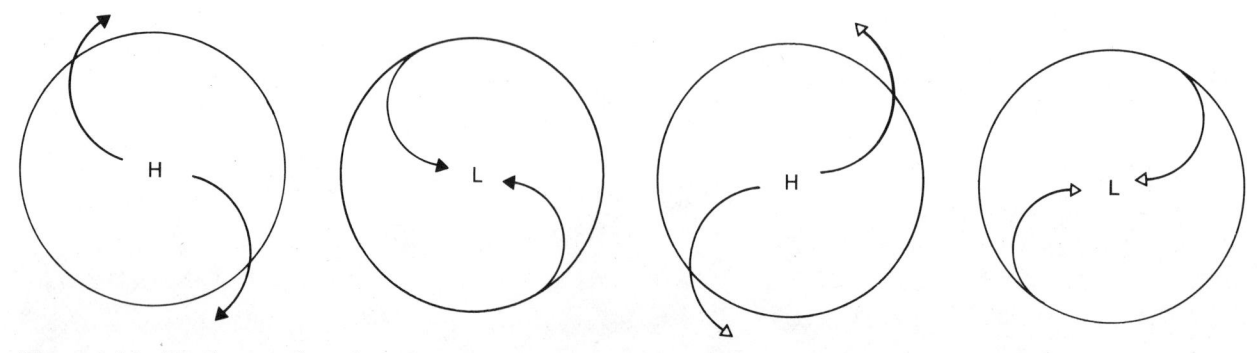

Figure 8-10 Northern and southern hemisphere winds in high and low pressure systems.

THE GENERAL WIND PATTERN

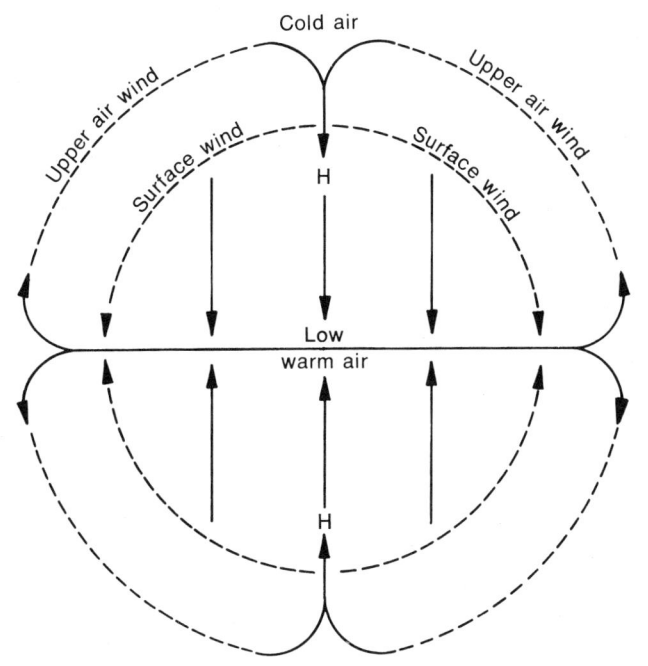

Figure 8-11 The simple one-cell Hadley circulation model, without the complicating features of rotation, axial tilt, and varying composition. Warm air from the equatorial low would rise and flow poleward and the sinking polar air would flow along the surface toward the equator. (From Thomas L. Burrus and Herbert J. Spiegel, *Earth in Crisis*, 2nd Ed., St. Louis, 1980, The C. V. Mosby Co.)

Therefore, the only difference would be unequal heating on a large scale between the poles and the equator. Hadley theorized that only a simple, large scale, one-cell circulation system would take place in each hemisphere. In such a system, air at the surface would flow toward the equator while the upper wind pattern would move from equator to poles. The idea that air flows generally from areas of higher pressure to lower pressure is not compromised, because cold air produces higher pressure than the lighter warm air at the equator. Figure 8-11 shows this circulation system, and is used as a stepping-stone to a much better idealized global wind system, which takes into account the three assumptions previously deleted.

In the early 1900s, a three-cell wind model was introduced. In this tricellular model, the equator still remains the hottest region of the earth, and the air at this point rises and flows both northward and southward to the poles. (*Note:* Since the southern hemisphere is a mirror image of the northern hemisphere, our discussion will be limited to the northern hemisphere.) The upper wind flow between the equator and 30° is deflected to the right by the coriolis effect (Fig. 8-12), which causes it to pile up and move from west to east. Here the air becomes a wind that is easily identified.

This upper air wind that moves eastward is called the *jet stream*, and can reach velocities in excess of 322 km/hr (200 mph).

As a result of this piling up of air aloft, a general *subsidence* (sinking) results. As the cooler air descends, it accumulates on the surface, resulting in both an equatorward and poleward surface flow. Once again, influenced by the coriolis effect, the equatorward surface wind moves from the northeast (southeast in the southern hemisphere). Because of its direction this belt of winds is called the *northeast trade winds* (southeast trade winds below the equator). The poleward moving air also moves to the right and is named the *prevailing westerlies*, because of its strong westerly component.

The cold polar air sinks and moves southward. This air is also deflected to the right of the direction of motion and is termed the *polar easterlies*. These polar winds travel in a region between 90° and 60°N. At 60°N, the *polar front*, that is, the warmer surface air, is forced aloft. Here again the piling up of air results in a second narrow belt of high-speed westerly winds, called the *polar front jet stream*.

The three zones that separate the prevailing wind belts are also specifically identified. Near the equator is a region called the *doldrums*. It is an area of almost calm winds except for locally generated wind flow from thunderstorms. Also called the *intertropical convergence zone*, this region is noted for its extremely low pressure, cumulus cloud formations and large amounts of thunderstorm activity. Many a sailing vessel has been trapped for days in the doldrums, with their only hope of escape being a local storm or by drifting slowly out of the area.

The second zone, located near 30°N (or 30°S), is called the *horse latitudes*. This term is derived from a story passed down through the years. Sailing ships plying their trade from Europe to the New World would take a route within the northeast trade winds region. Unfortunately, they would sometimes come too close to 30°N and find themselves becalmed in the light winds that prevailed there. In order to lighten their ships and conserve food and water, the horses they carried would be thrown overboard.

The third zone, the *polar front*, is located close to 60°N. This is only an average location, since the polar front migrates south in winter and north in summer. This zone separates the cold polar easterlies from the warmer prevailing westerlies. It is at this location that both air masses battle for dominance, and cold air clashes with its warmer counterpart. This is the region of the world where most of the world's weather fronts are born. This topic will be discussed in greater detail in Chapter 9. A summary of weather and wind conditions is listed in Table 8-1.

90 WINDS OF THE WORLD

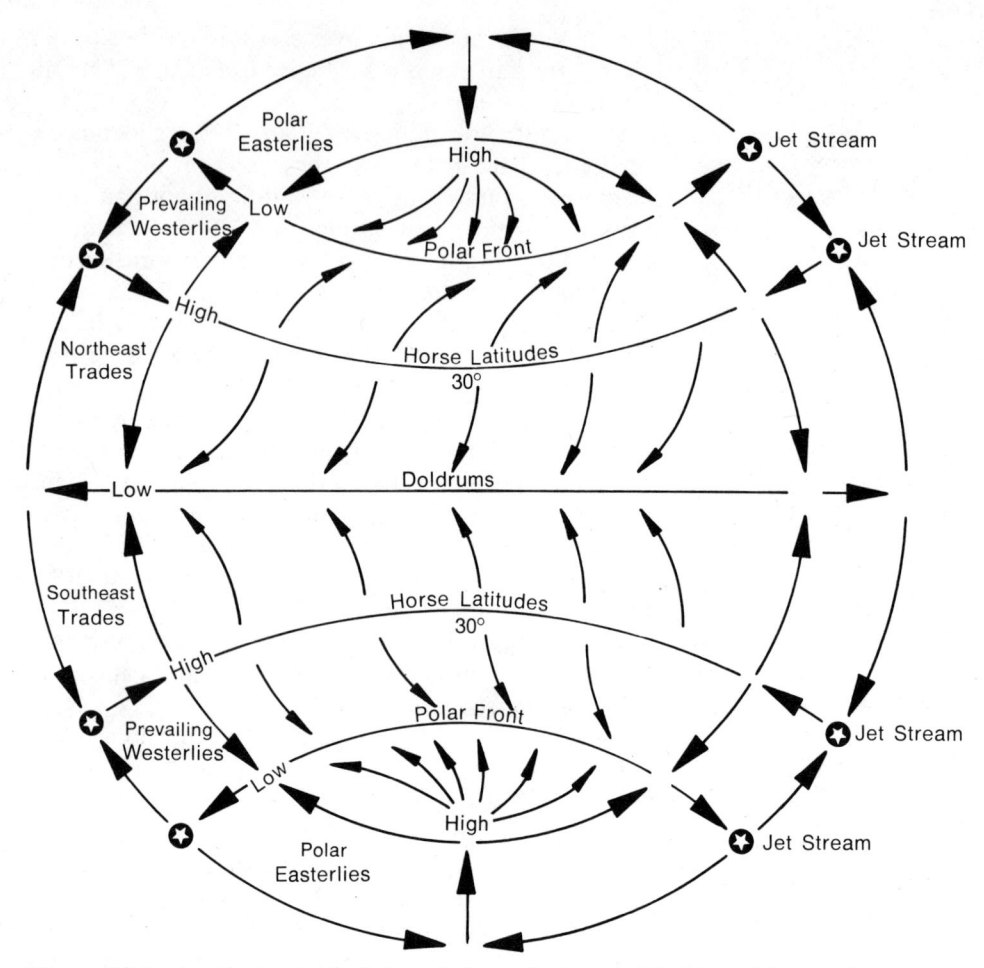

Figure 8-12 An idealized tricellular wind circulation model. (From Thomas L. Burrus and Herbert J. Spiegel, *Earth in Crisis*, 2nd Ed., St. Louis, 1980, The C. V. Mosby Co.)

TABLE 8-1 General Circulation System Weather and Winds

Region	Name	Pressure	Surface Winds	Weather
Equator (0°)	Doldrums	Low	Light	Cloudiness, precipitation, breeding ground for hurricanes.
0°–30°N and S	Trade winds	—	Northeast in northern hemisphere, southeast in southern hemisphere.	Pathway for tropical disturbances.
30°N and S	Horse latitudes	High	Light, variable winds.	Little cloudiness.
30°–60°N and S	Prevailing westerlies	—	Southwest in northern hemisphere, northwest in southern hemisphere.	Pathway for subtropical high and low pressure.
60°N and S	Polar front	Low	Variable.	Stormy, cloudy weather zone.
60°–90°N and S	Polar easterlies	—	Northeast in northern hemisphere, southeast in southern hemisphere.	Cold polar air with very low temperatures.
90°N and S	Poles	High	Southerly in northern hemisphere, northerly in southern hemisphere.	Cold, dry air.

Source. From Thomas L. Burrus, and Herbert J. Spiegel: *Earth in Crisis*, Ed. 2, St. Louis, 1980, The C. V. Mosby Co.

JET STREAM WINDS

Along the polar front where temperature differences can be quite large, a great pressure gradient can exist. From our previous discussion on pressure gradient, we know that the greater the gradient the faster the wind speed. Thus, above this region is born the *polar front jet*, a turbulent core of air with winds in excess of 240 to 480 km/hr (250 to 300 mph). The known location of this jet stream is not exact, for it meanders swiftly from one place to another in almost a wave pattern. However, its average location is estimated to be near the tropopause and its dimensions to be 40 to 160 km (25 to 100 miles) wide and 3.5 km (2 to 3 miles) deep. Although its prevailing direction is west to east, its gyrating motion can result in just about any wind direction.

The polar-jet stream can be both detrimental and beneficial to commercial or military aircraft. Flying west to east, a pilot would like to locate this jet stream, because it would greatly increase his ground speed and shorten his inflight time. However, flying east to west in the stream could prove disastrous, because inflight time could double, with a resulting unplanned fuel consumption increase.

There are several jet streams similar to but not as well studied as the polar jet. None are continuous and a few are seasonal in nature, occurring only when large temperature differences occur.

AIRFLOW AROUND PRESSURE SYSTEMS

Wind flow around pressure systems results from the interaction of the previously discussed primary and secondary forces. In the temperate zone (30° to 60°), pressure systems migrate constantly, bringing with them changes in wind direction and speed. The variability of the wind in these systems does not depend on the general wind pattern of the world, but is related to the distance and direction from the geographic center of a pressure system and its accompanying circulation.

In the northern hemisphere, the air in a low pressure system flows in a counterclockwise direction. At the earth's surface, this air spirals inward and converges at the center (Fig. 8-13). This *convergence* and rotation are called cyclonic flow, and the low is given the name *cyclone*. As the air funnels in from all directions, it piles up in the center and is forced to rise. These rising convection currents transport both heat and moisture from the surface of the earth to the higher layers of the troposphere. This upward movement allows air to cool adiabatically, reach saturation, and form a cloud cover, which in many instances produces precipitation.

High pressure areas located in the northern hemi-

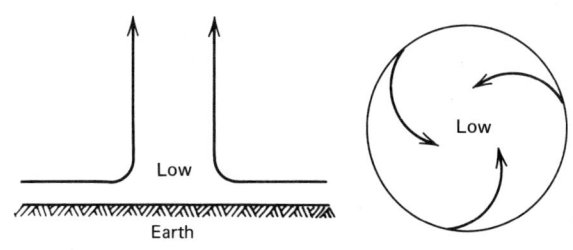

Figure 8-13 Vertical profile and surface diagram of a low pressure system in the northern hemisphere.

sphere have a circulation that spirals clockwise and outward from its center (Fig. 8-14). This surface *divergence* is directly opposite to the wind flow in a low, so the name *anticyclone* is applied to the high pressure area. The air moving down to take the place of the diverging air is said to be *subsiding*, and as it moves down it warms up slightly through the adiabatic process. This warming produces drier air, few clouds, and good weather. Anticyclones generally cover up to several thousand miles and are capable of transporting massive amounts of air into an area. In winter, huge masses of arctic air are pushed out of central Canada as far south as Florida, producing intense cold waves over the interior of the United States. These air masses are cold and dry, and produce very little cloud cover and negligible precipitation.

Although the presence of a high would lead one to believe that sunny skies and good weather will always prevail, there is a hidden flaw in that thinking. In some instances the high creates a *subsidence inversion*, which does not allow air from the surface to rise. Subsidence inversions create a high level of pollution close to the earth's surface and are particularly troublesome when a high pressure system stagnates over an area for seven or eight days (Fig. 8-15). This is quite common over the northeastern United States in fall and winter. One such stagnating anticyclone occurred over the eastern part of the United States in 1966 on Thanksgiving, bringing with it one of the most notorious episodes in air-pollution history. Similar experiences have been shared by

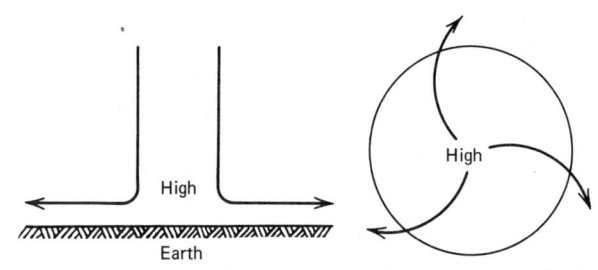

Figure 8-14 Vertical profile and surface diagram of a high pressure system in the northern hemisphere.

92 WINDS OF THE WORLD

Figure 8-15 Smoke streaming horizontally and its resulting pollution as the result of a subsidence inversion. (Courtesy Dick Jaffe)

the Meuse Valley in Belgium, in London, in Los Angeles, and in Donora, Pennsylvania.

LOCAL WIND SYSTEMS

Superimposed upon the general wind circulation and air movement associated with pressure systems are the winds that are generated by local conditions. Local winds are usually caused by temperature differences and topography, and are confined to small areas.

Land and Sea Breezes

The fundamental reasoning behind this local wind is the differential heating of land and water bodies. Land heats up and cools down faster than water, and in tropical locations the land and sea breezes are almost daily occurrences. In the midlatitudes they are only summertime phenomena, and they rarely occur in the polar region.

During the day, the incoming solar radiation heats both land and water, with the land heating up faster. This produces a temperature difference in the overlying air, wherein the cooler air lies over the water. The warmer air that lies over the land weighs less and produces a small local low pressure system. Over the water the air weighs more, and a small local high pressure system is created (Fig. 8-16). The resulting pressure gradient sets up a small circulation system with air moving from high to low pressure (the system is too small for coriolis to influence the flow). This onshore wind is called the *sea breeze* and it increases in intensity throughout the early afternoon hours and during the summer. The sea breeze can be quite strong and has been known to penetrate as much as 50 km (approximately 30 mi) inland.

At night, a reversal takes place. The land cools off faster than the water, and the accompanying air temperature drop develops a high pressure cell over the land. Now the cooler air moves from the land to take the place of the warmer rising air over the water. This offshore wind is called the *land breeze* and is the exact reverse of the daytime sea breeze. Land breezes are not as strong as sea breezes because the temperature differences between land and water are not as great during the evening hours as they are during the afternoon.

Monsoon Winds

The true monsoon winds are characterized by a change in prevailing wind direction by season. In winter a land mass cools off faster than an adjacent large water body. This difference in rate of cooling results in a high pressure system over the land and a low pressure system

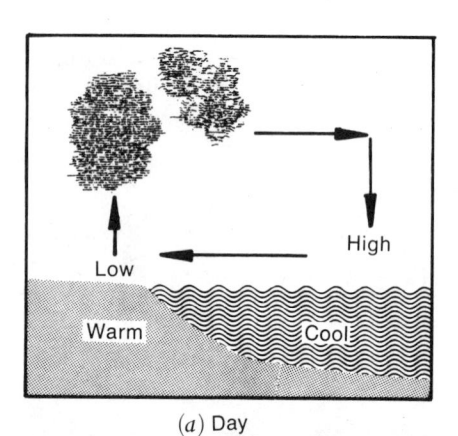

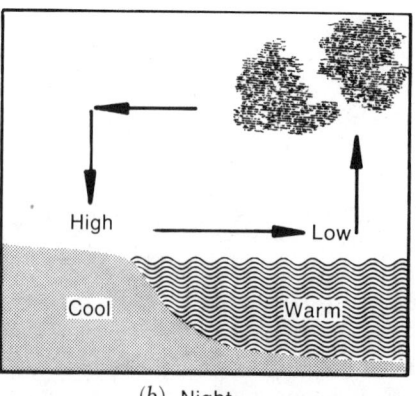

Figure 8-16 (a) During the day, sea breezes are created when air that is cooled by contact with the cooler water moves inland toward lower pressure created by air being heated in contact with the warmer land. (b) At night cooler air over the land results in a high pressure area and the air flows toward the lower pressure over the water.

over the water (remember a volume of cold air weighs more than an equal volume of warm air). The pressure gradient force that is set up as a result of the temperature differential leads to a fairly constant land-to-water air flow (high to low pressure). This wind coming off the land is usually quite dry and therefore usually cloudless, which means sparse rainfall.

During the summer the reverse takes place. The land mass heats up faster than the water and the prevailing wind blows from the higher pressure over the water to the lower pressure over the land. This water-to-land wind brings moist air, which in turn produces high temperatures and high humidities as well as widespread cloudiness and heavy rainfall over a large area.

Two regions of the world most famous for these monsoon conditions are India and Southeast Asia. Over India, equatorial air from the Indian Ocean rolls over the land mass during the summer, and cold polar air from the Himalaya Mountains to the north sweeps over the countryside during the winter. This sets up the typical monsoon weather conditions of a long rainy summer season followed by a long, generally pleasant dry fall and winter. Southeast Asia has a similar wind and weather pattern.

Mountain and Valley Winds

Caused by topographic features, the mountain and valley breezes are natural outcomes of temperature differences between adjacent valleys and mountain slopes. During the day, the slope of the mountainside heats up sooner than the shaded valley below. The air that comes in contact with this slope also heats up and begins to rise upslope. This sets up a pressure gradient between the cooler, higher pressure valley air and the warmer, lower pressure rising air. Because of the pressure gradient created, the air in the valley rises upslope and additional air pours into the valley from the open side to replace this rising air. The wind flow into the valley and the rising air are known as the *valley wind* (Fig. 8-17a).

At night the heating process stops and the air that is close to the mountainside is cooled by conduction as the slope rapidly loses heat. This cooled air sinks into the valley floor, piles up, and pushes the air out the open side. This downslope wind and outward moving air is called the *mountain wind* (Fig. 8.17b).

Chinook Wind

Whenever a strong low pressure system lies along the eastern slopes of the Rocky Mountains for any length of time, a circulation system is set up such that air is forced up and over the mountain slopes from an east-to-west direction. As the air rises on the windward side (east side), it deposits any moisture it might have in the form of clouds or precipitation, leaving the air quite dry. As the air descends on the leeward (west side) it warms adiabatically, and its final temperature is warmer than when it started (Fig. 8-18). This warm dry wind is called the *chinook wind*, which literally translated from an Indian dialect means "snow eater." If chinooks occur right after a large snowfall, they cause the snow to disappear by the sublimation process, leaving the ground bare and dry. In the Alps, the chinook wind is commonly called the *föehn* and in the river valley of Santa Ana, California it is called the Santa Ana or devil wind. It should be noted that this wind blows downhill and can be broadly termed a gravity or *katabatic wind*. Other worldwide winds can be found in Table 8-2.

WIND OBSERVATIONS

Wind Direction

Wind is classified according to the direction *from which it is blowing*; that is, a north wind is air coming from

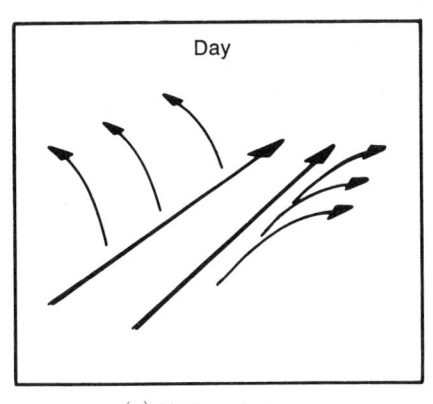

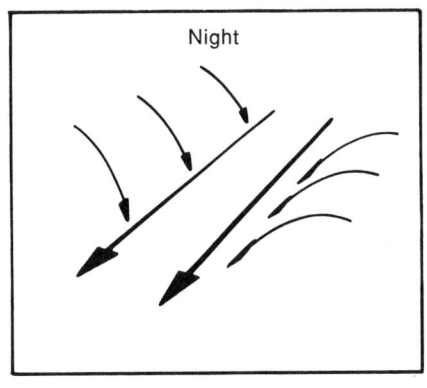

(a) Valley wind (b) Mountain wind

Figure 8-17 Mountain and valley winds. Both winds are named for their source region. (Redrawn after Thomas L. Burrus and Herbert J. Spiegel: *Earth in Crisis*, Ed. 2, St. Louis, 1980, The C. V. Mosby Co.)

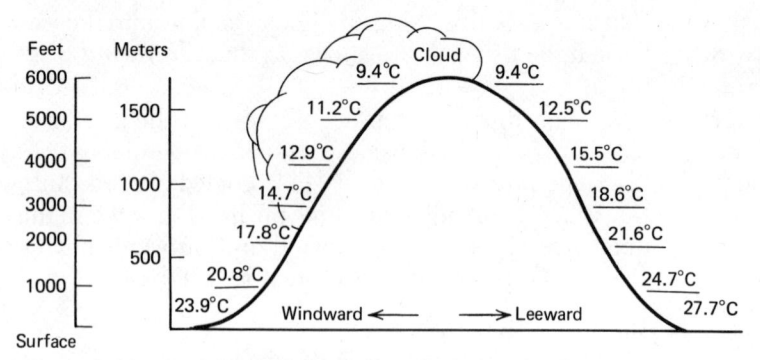

Figure 8-18 Air lifted adiabatically will cool at the dry adiabatic lapse rate (1°C/100m or 5.5°F/1000 ft) until saturation is reached. At saturation, the rising air will cool at the wet adiabatic lapse rate (0.5°C/100m or 3.2°F/1000 ft). Descending air always warms at the dry rate. In this diagram air at 75°F rose 6000 feet, but was saturated at 3000 feet. The reduced cooling resulted in a higher leeward temperature.

the north. In describing wind direction, the four cardinal points (north, south, east, and west) and four intermediate points (northeast, northwest, southeast, and southwest) are used. In many cases, an even finer division such as south-southwest is used.

Although word descriptions of direction are common, meteorologists prefer to report wind flow in degrees measured clockwise from true north. Starting at north and progressing around the circle of a compass, the degrees total up to 360. North is taken to be 360° while the number 0° is used for calm winds. Using this system, 90° would be an east wind and 270° would be air coming from the west.

To measure upper air winds a *pilot balloon* is sent aloft. This balloon is inflated with a gas that is lighter than air, such as helium. The balloon has a known ascension rate that will depend upon the type of gas used and the amount of inflation. As it ascends, the balloon is followed by a *theodolite* similar to the one pictured in Figure 8-19. By using the data from the theodolite, meteorologists can mathematically determine upper air winds.

Wind Speeds

Wind speed is the measure of the average movement of air. Generally speaking, a one-minute average is used, thus eliminating *wind gusts*, which are sudden brief increases in wind speed. For public information, all wind speeds are reported in the English system of miles per hour or in the metric measurement of meters per second. However, by international agreement, meteorologists the world over use *knots* (nautical miles per hour) as do aviation, military, and ocean-going interests. A knot is equivalent to 1.15 miles per hour.

A simple method of estimating wind speed was developed in 1805 by Admiral Francis Beaufort of the British Navy. This system relies on observations without instrumentation, and takes into account the wind effects on the natural surroundings. Separate scales are used for the effects on land and sea. Table 8-3 is the scale

TABLE 8-2 **Other Local Winds**

Local Name	Location	Type of Wind	Weather Characteristics
Bora	Northeast shore of Adriatic Sea	Katabatic	Cold and wet
Mistral	Rhone Valley	Katabatic	Cold and dry
Pampero	Central Argentina	Cold air mass	Dusty and rainy
Black Roller	Great Plains, United States	Air mass	Strong dusty wind
Santa Ana	Southern California	Katabatic	Warm and dry

Source. After Thomas L. Burrus and Herbert J. Spiegel: *Earth in Crisis*, Ed. 2, St. Louis, 1980, The C. V. Mosby Co.

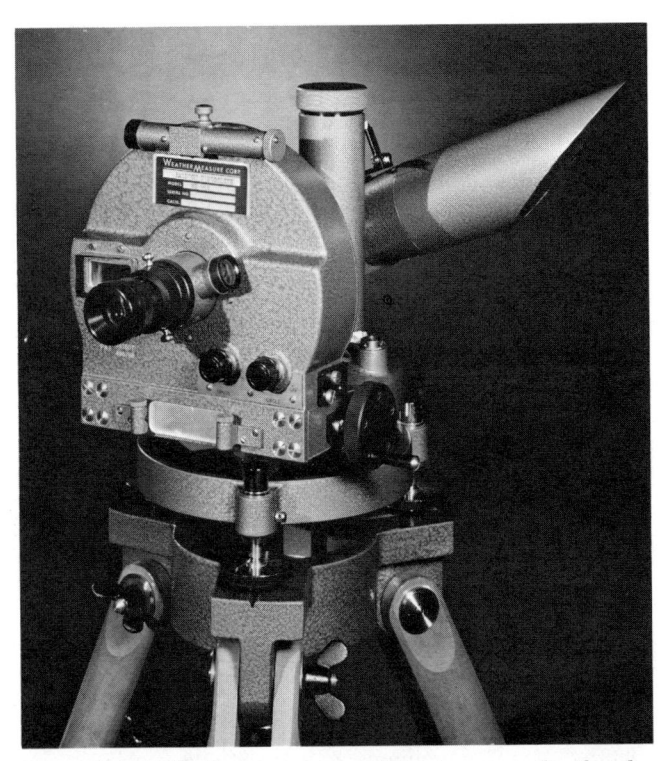

Figure 8-19 Similar to a surveyor's instrument, the theodolite furnishes the observer with angular and elevation angles that can be converted into three-dimensional locations. These locations can be used to determine wind direction and speed. (Courtesy Weather Measure Corp., Sacramento, California)

used for land observations and does not include beaufort numbers 13 to 17, which were put into the system by the National Weather Service in 1955.

Wind Measurement Systems

Surface wind can be measured by many simple or natural systems. By knowing the direction you are facing and holding up a wet finger you can approximate wind direction by feel. Other methods range from throwing grass vertically in the air and watching its drift to observing rising smoke or trees blowing in the wind.

There are of course many mechanical *wind vanes* that are more accurate. Sophisticated mechanical wind vanes operate on an electronic system and are free to move about on a vertical axis. The tail section of these vanes is heavier and has a larger surface area than does the head section. The wind's action on this larger area rotates the vane in such a way that it always points in the direction the wind is coming from. A signal is sent to a recorder such as that pictured in Figure 8-20 to record wind direction permanently, on a continuous basis.

A more elaborate system is the *anemoscope* (Fig. 8-21) which incorporates a wind vane and anemometer (used to measure wind speed) into one instrumented package. The propellers act as the anemometer, and an airplanelike structure acts as the wind vane.

TABLE 8-3 **Beaufort Wind Table**

Beaufort Number	Term	Speed Miles/hr	Knots	Observation
0	Calm	Less than 1	Less than 1	Smoke rises vertically
1	Light air	1–3	1–3	Smoke indicates wind direction, wind vane does not move
2	Light breeze	4–7	4–6	Leaves rustle, wind felt on face
3	Gentle breeze	8–12	7–10	Wind extends light flag, dry leaves and paper blow around
4	Moderate breeze	13–18	11–16	Small branches move, dust and paper rise from ground
5	Fresh breeze	19–24	17–21	Crested wavelets form on inland waters, small trees in leaf begin to sway
6	Strong breeze	25–31	22–27	Large branches in motion, telegraph wires whistle
7	Moderate gale	32–38	28–33	Whole trees in motion, walking influenced by wind
8	Gale	39–46	34–40	Twigs break, walking difficult
9	Strong gale	47–54	41–47	Ground littered with branches, slight structural damage
10	Storm	55–63	48–55	Trees uprooted, much structural damage
11	Violent storm	64–73	56–63	Widespread damage
12	Hurricane	Greater than 73	Greater than 63	Severe damage

Source. Thomas L. Burrus and Herbert J. Spiegel: *Earth in Crisis*, Ed. 2, St. Louis, 1980, The C. V. Mosby Co.

96 WINDS OF THE WORLD

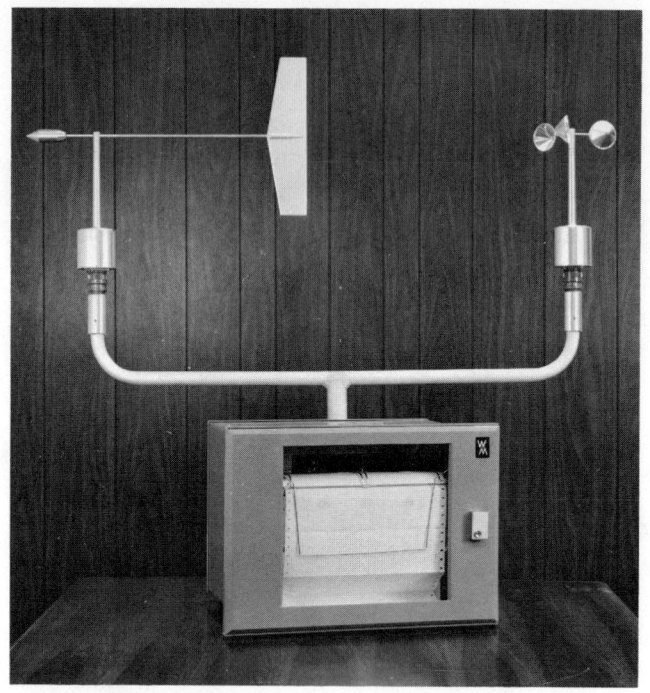

Figure 8-20 A three cup anemometer and wind vane system feeding into a wind recorder. The speed at which the cups turn varies with wind velocity; the electrical current generated is related to the rate of turning; the electrical current produced is then used to calculate wind speed. (Courtesy Weather Measure Corp., Sacramento, California)

Figure 8-21 The anemoscope, a modern day version of the anemometer and wind vane. (Courtesy Weather Measure Corporation, Sacramento, California)

TABLE 8-4 **Wind-Chill Chart**

		\multicolumn{8}{c}{Wind Velocity (mph)}
		5	10	15	20	25	30	35	40
	+35	32	22	16	11	8	5	4	2
	+30	27	16	9	4	0	−2	−4	−6
Air Temperature (°F)	+25	22	10	2	−3	−7	−10	−12	−13
	+20	16	4	−5	−10	−14	−18	−20	−21
	+15	11	−3	−11	−17	−22	−25	−28	−29
	+10	6	−9	−18	−24	−29	−33	−35	−37
	+5	1	−15	−25	−32	−37	−40	−43	−45
	0	−5	−21	−31	−39	−44	−48	−51	−53
	−5	−10	−27	−38	−46	−51	−56	−59	−61
	−10	−15	−33	−45	−53	−59	−63	−66	−69
	−15	−20	−39	−52	−60	−66	−71	−74	−77
	−20	−26	−46	−58	−67	−74	−79	−82	−85

WIND-CHILL FACTOR

Temperature can be both specific and relative. Specific temperature scales use numbers to identify degrees of hotness and coolness. Relative temperature scales are merely descriptive, such as "hotter than." A person's comfort is not necessarily reflected in how hot or cold the air is at any one time. Comfort can be related to relative humidity (as in the Temperature Humidity Index) or wind. The relative discomfort you feel as you step outdoors is a result of your original environment and a combination of outdoor temperature wind and humidity. Wind and temperature in combination are especially important at the colder end of the scale. Meteorologists use the term *wind-chill* to show how the wind affects the real air temperature and how cold a person "would" feel in such a combination. In cold calm air you feel cold because energy is being emitted from your skin's surface area at a higher rate than in warmer air (energy moves from higher to lower values). As the wind speed increases, more energy is dragged away from your skin in the same time period, and you feel colder. This feeling is real even though in both instances the temperature of the air is the same. Table 8-4 is the official wind-chill table used by the National Weather Service of the United States.

9
AIR MASSES OF THE WORLD

Admiral R. Fitzroy (1805–1865) was one of the pioneer synoptic meteorologists who discovered that low pressure areas are usually composed of two different air masses. In his publication *Weather Book, A Manual of Practical Meteorology* he describes and diagrams air flow around lows in essentially the same way it is done today.

LEARNING OBJECTIVES

After reading this chapter, you should be able to:

1. Describe the general characteristics of an air mass and explain how those characteristics were inherited.
2. Define *air mass* and *source region*.
3. Describe an air mass by its identifying alphabetized description.
4. Draw on a weather map the seven air mass source regions affecting the United States.
5. Describe the general weather changes resulting from the invasion of a particular air mass.

In the 1920s a team of researchers at a university in Bergen, Norway introduced the concept of using air masses for daily weather forecasting and map analysis. Under the supervision of the famous Norwegian meteorologist Vilhelm Bjerknes, the *air mass method* became one of the predominant techniques for short term weather forecasting and is still used today.

From that small beginning, meteorologists have expanded the air mass concept so that it now takes into account fronts and low pressure areas. It is now realized that many changes in the weather, either dramatic or slow, are the result of a new air mass pushing into a region. Thus a basic understanding of how air masses form and move is important to the further understanding of frontal formation and weather systems, both of which will be explained in greater detail in Chapter 10.

GENERAL CHARACTERISTICS OF AN AIR MASS

If the forces responsible for the movement of air become balanced, a calming of the air takes place. This stagnation results in a large mass of air that remains relatively undisturbed for a fairly long period of time. The time frame may vary from a few days to two or three weeks, depending upon how long it takes for an imbalance to set in. If the air happens to stagnate over a large land mass or water body, it will tend to take on the characteristics of that underlying surface. Two meteorological parameters, temperature and moisture content, are especially dependent on this underlying surface. If the air is warmer than the surface below, it will cool down until it reaches a state of equilibrium. Similarly, the moisture content of the air will increase or decrease until equilibrium is again reached. Using the foregoing information, the definition of an air mass can be written logically. *An air mass is an extensive body of air whose temperature and moisture characteristics are horizontally uniform.* This means that for any height chosen these characteristics would be approximately the same over a large expanse of area.

SOURCE REGIONS

In order for the air to take on the characteristic properties of the surface it overlies, the land mass or water body must be relatively uniform. This large uniform surface area is called a *source region* and has a tendency to create large stationary or slowly moving high pressure areas. These *anticyclones*, as they are called, tend to have weak pressure gradients, which in turn result in calm or very light winds. In North America, the Central Plains of Canada is an excellent example of the vast mostly level terrain needed for an air mass source region. The Sahara Desert of North Africa and the interior of Asia also represent ideal land mass source regions, as do other deserts, large ocean areas, and vast ice-covered wastelands. Each of these areas is capable of producing both uniformity temperature and moisture content.

There are six major source regions in or near North America (Fig. 9-1):

1. Interior Canada, which is covered by ice and snow as well as the frozen waters of the Arctic Ocean.
2. The large expanse of cold North Atlantic waters eastward from Greenland, Newfoundland, and Labrador.
3. The northeastern part of the Pacific Ocean with its limits extending to the northwestern coast of the United States.
4. The warm ocean waters of the Gulf of Mexico and the Caribbean Sea. This area also embraces the land portion of Florida and the Yucatan Peninsula of Mexico.
5. The warm narrow region of Northern Mexico and the states of Texas, New Mexico, and Arizona. This limited source region is absent in winter.

AIR MASSES OF THE WORLD

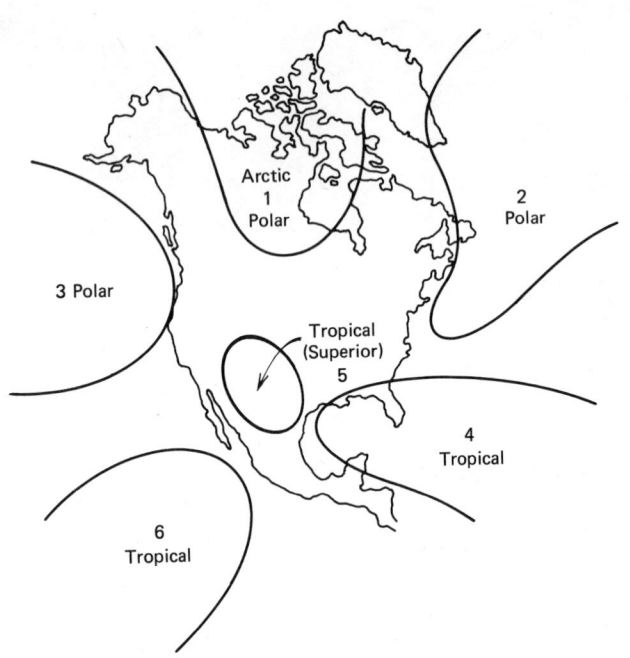

Figure 9-1 Major source regions affecting the North American continent.

6. Along the lower southwestern coast of the United States stretching westward into the Pacific lies a source region that is strong only during the winter months.

Worldwide, many of these source regions can be found close to the semipermanent high pressure systems located near the horse latitudes or on the poleward side of 60°N latitude in the semipermanent low pressure belt.

AIR MASS CLASSIFICATION

Tor Bergeron, a Swedish meteorologist, introduced a classification scheme for identifying air masses. This classification involved the geographical starting point (source region) of the air mass as well as its temperature and moisture characteristics.

Geographic Classification (Source Region)
A simple method of identifying the starting point of an air mass is by the latitude where it forms. However, it should be noted that latitudinal areas tend to vary slightly with the seasons and that the geographical boundaries may vary slightly. The four low level source regions of the Bergeron classification are as follows:

Air Mass	Location
Equatorial (E)	10°N–10°S
Tropical (T)	25°N–25°S
Polar (P)	55°–60°N or S
Arctic (Antarctic) (A)	Close to the poles

In addition to this list of surface air masses is a high level air mass called *Superior*(S). It is most frequently found in the desert southwest of the United States and in rare instances over Spain and North Africa.

Temperature Classification*
The temperature characteristics represent the degree of hotness or coolness compared to the underlying surface. If the air is colder than the land or water body over which it is traveling, the air mass is classified as cold (k). Thus cold air masses can exist with temperatures of 60° to 70°. Warm air masses are also classified in this manner.

If the air is colder than its underlying surface it will warm up from below, and unstable convection currents will result up to 10,000 feet. This type of situation sometimes produces cumuliform clouds and showery precipitation. On the other hand, if the air is warmer than the surface it is traveling over it will be cooled from below, and stable conditions will exist. Weather related to this situation will be smooth air, stratiform clouds, poor visibility, fog, and steady precipitation.

The two thermal classifications are:

Warm (w): Air warmer than the surface
Cold (k): Air colder than the surface

Moisture Classification
The moisture properties are again tied to the source region. If the air starts over land it is classified as *continental* (c) or dry, but if it originates over water it is called a *maritime* or moist air mass (m).

By using a shorthand notation method, the meteorologist can identify the major features of each air mass that moves over a specific area. This is done by using the moisture identification, then its geographical location identifier, and last the temperature characteristic. Thus a mTw air mass (maritime Tropical warm) has been formed over the tropical ocean areas and is quite moist and warm. By being able to identify and understand air mass characteristics, a weather forecaster can relate the oncoming air into predictable weather changes.

Other examples of worldwide air mass types are:

mPk: Cold wet polar air
cTw: Warm dry tropical air
E: Warm wet tropical air
A: Cold dry arctic air

Notice that the letter designation of wetness or warmness of the equatorial and arctic air is left out. Since a

---
* Although for teaching purposes, the temperature classification better illustrates the differentiation of air masses, it is now seldom used. In this text we will use it sparingly and only when a more apt description is needed.

predominance of land masses or ice fields exists in the always colder arctic, very little moisture enters the air, and a moist or warm air mass is very rare. The same holds true with the always warm equatorial air mass that forms mostly over the water region of the equator. Therefore, only one letter is needed to identify this type of air.

AIR MASSES AFFECTING THE UNITED STATES

Continental Arctic (A) and Continental Polar (cP)

These extremely cold dry air masses originate over the snow and ice-covered higher altitudes of North America, where they build up enough force to pour out across the heartland of the United States. There are three major wintertime tracks for this polar or arctic air; all of them lead to frigid conditions in the northern regions and chilly weather in the southern regions of the United States (Fig. 9-2).

The first path takes the polar air over the Great Lakes. During the winter, heavy snow showers and gusty winds whip across the lee side of the lakes and continue to batter the northeastern seaboard with wind and cold temperatures as the air mass moves offshore.

The second path predominates over the central part of the United States. This frigid blast of cold air has enough strength to reach across the Gulf of Mexico into Central America where it is called a *norther*. While traveling southward from Canada at high speeds, the *cP* or *A* air leaves in its wake an extreme temperature drop, low relative humidity, and clear skies. When this cold *polar outbreak* is in the making, the citrus growers in middle and south Florida prepare to counteract the expected freezing conditions. They know full well that the polar air has the ability of reaching into the very southern limits of their state.

The third path is a rarity. It usually occurs when a very intense high pressure area is over the Canadian midsection. This high must be strong enough to force air over the northern Rockies and when it takes place, northern and central California experience cold rainy weather and occasional snow as far south as the central California coast.

In the summer, *cP* or *A* air is much warmer because there are no snow-covered surfaces in the source region. Normally, summer heat waves of the Northeast

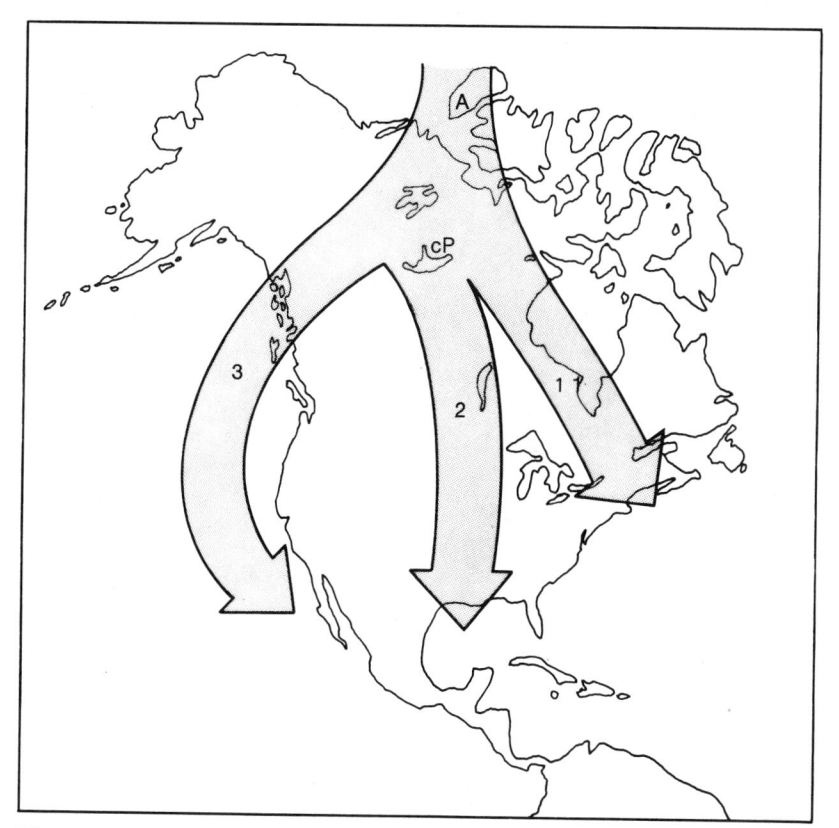

Figure 9-2 Three major tracks of continental polar and continental arctic air masses within North America. (From Thomas L. Burrus and Herbert J. Spiegel, *Earth in Crisis*, 2nd Ed., St. Louis, 1980, The C. V. Mosby Co.)

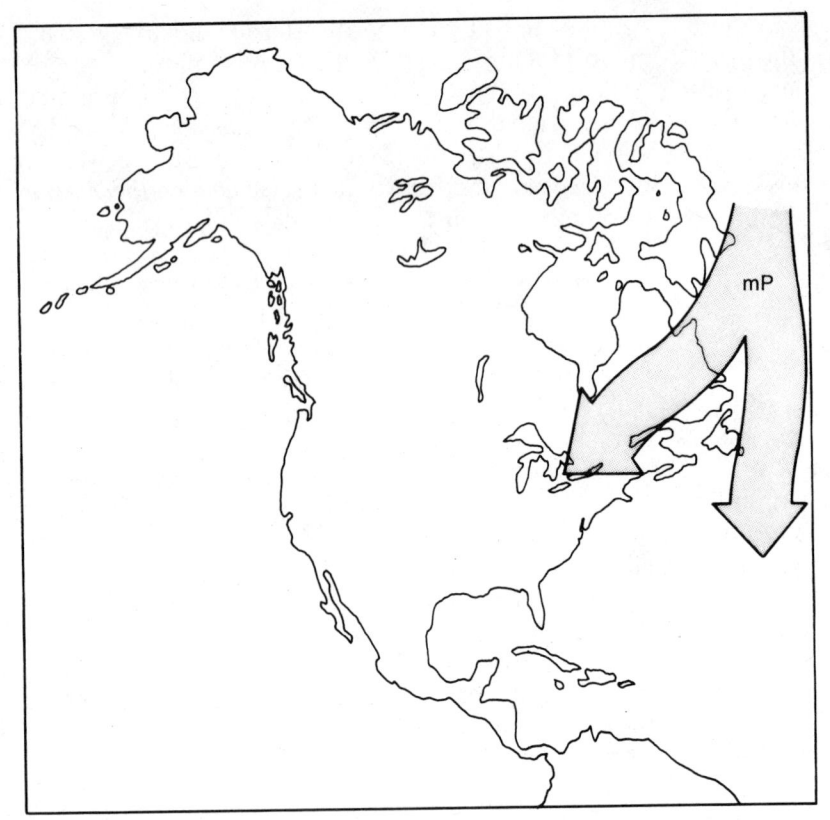

Figure 9-3 On rare occasions maritime polar air sweeps in from the north Atlantic over eastern Canada and the northeastern United States. (From Thomas L. Burrus and Herbert J. Spiegel, *Earth in Crisis*, 2nd Ed., St. Louis, 1980, The C. V. Mosby Co.)

are broken by the invasion of the *cP* air, which very seldom reaches into the southeastern part of the United States in the warmer seasons.

Maritime Polar (*mP*), Atlantic

This *mP* air mass originates in the cold Atlantic waters off the coasts of Newfoundland and Labrador. Since the prevailing upper air wind flow is normally west to east, it is very rare that an invasion of maritime polar air takes place. However, aided by a winter high pressure system over the North Atlantic, this cold wet air can penetrate into the coasts of New England, bringing with it freezing temperatures, strong gale winds, high relative humidity, and winter precipitation (Fig. 9-3). Along the New England coast these conditions are called *Northeasters*, after the northeast direction of the wind.

During the summer, the *mP*-Atlantic air mass is less conspicuous than its winter counterpart. An occasional influx of cool, moderately wet air results in partly cloudy skies, no precipitation, and pleasant temperatures.

Maritime Polar Air (*mP*), Pacific

Originating over the northern part of the North Pacific, this wet air mass greatly affects the weather pattern of the West Coast. Similar to the *cP* air discussed earlier, the *mP* air reaches the North American coast by three distinct paths (Fig. 9-4). The first path originates over Alaska and follows a relatively short overwater route. Accompanying this cold winter air flow is a rapid increase of cloudiness with numerous showers and rain squalls. This comes about because the cold, moist air is forced to rise over the coastal ranges, with the air cooling to its saturation point.

The second and third paths are long overwater trajectories with resulting warmer air temperatures and high relative humidities. Rainfall is greater along the western coast of the United States but almost nonexistent along the Continental Divide just east of the Rocky Mountains.

In the summer the influx of maritime polar air from the Pacific generally results in clear skies or only a few low scattered cumulus.

Maritime Tropical Air (*mT*), Atlantic

Although occurring during the entire year, *mT* air has its greatest influence on the eastern half of the United States in the summer (Fig. 9-5). The air mass has one of the highest temperature and moisture contents of any air mass affecting the United States and, as it moves over warmer land, it is rapidly heated and terminates in

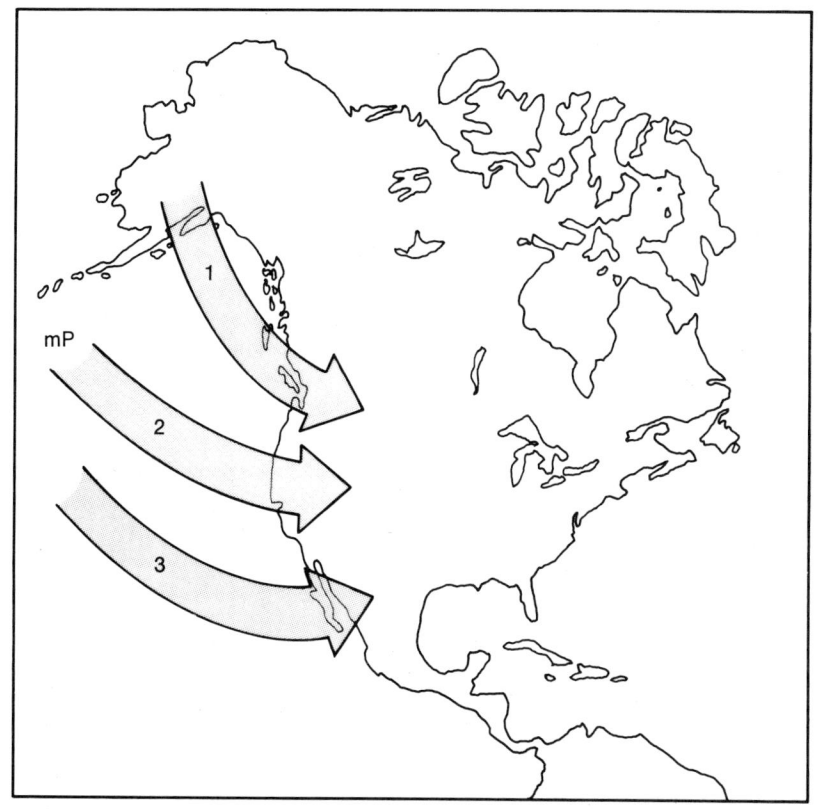

Figure 9-4 Maritime polar air invading the west coast of the United States greatly influences the weather pattern along the entire coastal section. (From Thomas L. Burrus and Herbert J. Spiegel, *Earth in Crisis*, 2nd Ed., St. Louis, 1980, The C. V. Mosby Co.)

showery cloudbursts and violent thunderstorms (Fig. 9-6).

If the air mass is particularly strong, it will push its hot sultry air northward, resulting in oppressive heat waves in the densely populated regions of the Northeast. Movement off the northeast coast finds the warm wet air overriding the cooler Labrador currents. The result is *advection fog* over areas such as the Grand Bank of Newfoundland.

Occasionally, during the summer a high pressure area located over Bermuda will extend westward over the Gulf of Mexico. When this situation arises, warm *mT* air penetrates into the southwest, and because of *orographic lifting* (air going over a mountain) huge amounts of rain will fall from thunderstorms and will flood areas which are usually parched throughout the rest of the year. A local name for this situation is *Sonora weather*.

During the winter the *mT* air mass is responsible for mild temperatures and high humidities along the south Atlantic and Gulf Coasts. Usually night and morning cloudiness give way to fair weather cumulus during the day. However, when the land has been excessively cooled by colder air, lack of sunshine, or even snowfall, the intrusion of this warm moist air results in a thick blanket of advection fog.

Superior Air (*S*)

Although mostly an upper air mass, superior air occasionally reaches the surface. Usually found over the southwestern states, this dry hot air mass is thought to be directly connected with the horse latitudes. The air is fairly stagnant and has no definite path when it does move. Therefore, only minute rainfall amounts occur over an area dominated by the superior air mass, and droughtlike conditions and sparse vegetative growth are predominant.

Maritime Tropical (*mT*), Pacific

This air mass does not have a direct influence on the weather of the West Coast in the summer. During the summer the dominant Pacific high pressure area moves so far south that only *mP* air invades the area with resulting widespread coastal fog.

In the winter *mT* air penetrates only into the lower coast of California and northwestern Mexico and then only rarely. It brings with it warm moist air which, when forced aloft by colder air, produces rainfall in record quantities (Fig. 9-7).

MODIFICATION OF AIR MASSES

As an air mass moves out of its source region, it undergoes a slow change in its characteristic properties.

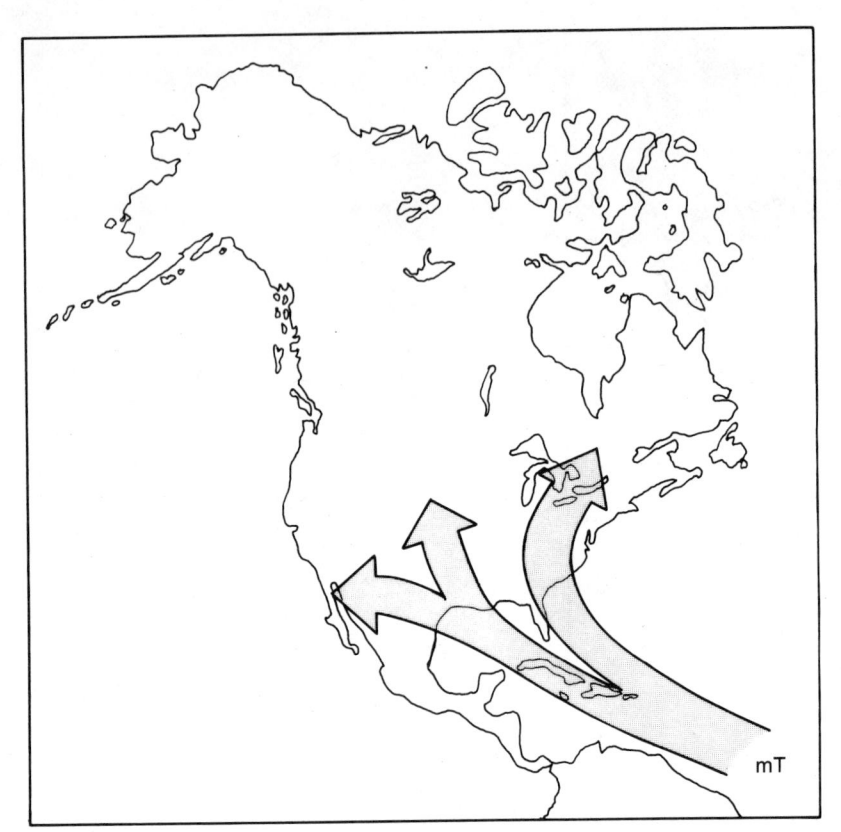

Figure 9-5 Maritime tropical air from the Caribbean is the fuel that is needed for summer thunderstorms to the entire southeastern United States. In many instances this air mass penetrates northward beyond the Canadian border. (From Thomas L. Burrus and Herbert J. Spiegel, *Earth in Crisis*, 2nd Ed., St. Louis, 1980, The C. V. Mosby Co.)

These changes are the result of the different surfaces over which the air travels. Many factors influence this modification, but the most important are the topographic, temperature, and moisture peculiarities of the earth's surface.

Figure 9-6 A mature thunderstorm with its anvil head. The low level moisture of maritime tropical air from the Caribbean and the hot land combine to produce these locally destructive storms. (Courtesy Joe Golden)

Topography

Moving air is at the mercy of topography. This is quite evident in mountainous regions where air is forced to rise on the *windward* side and then subside on the *leeward*. The resultant modification is increased precipitation on the windward side and an increase of temperature on the descending side of the mountain.

Temperature

The temperature of the air can be significantly changed by the temperature of the surface area. The temperature change of the air ultimately affects the associated weather patterns with a general modifying influence. For example, if warm air from the tropics were to move over a cold snow-covered region, the air would be immediately cooled by touching this surface. As the air progressed northward, the air would cool to higher layers, with fog becoming prevalent throughout the entire area.

In the winter Florida seldom receives frigid arctic air because of the long trajectory the air must take. This great distance aids in moderately warming the air before it reaches the southern limits of the United States.

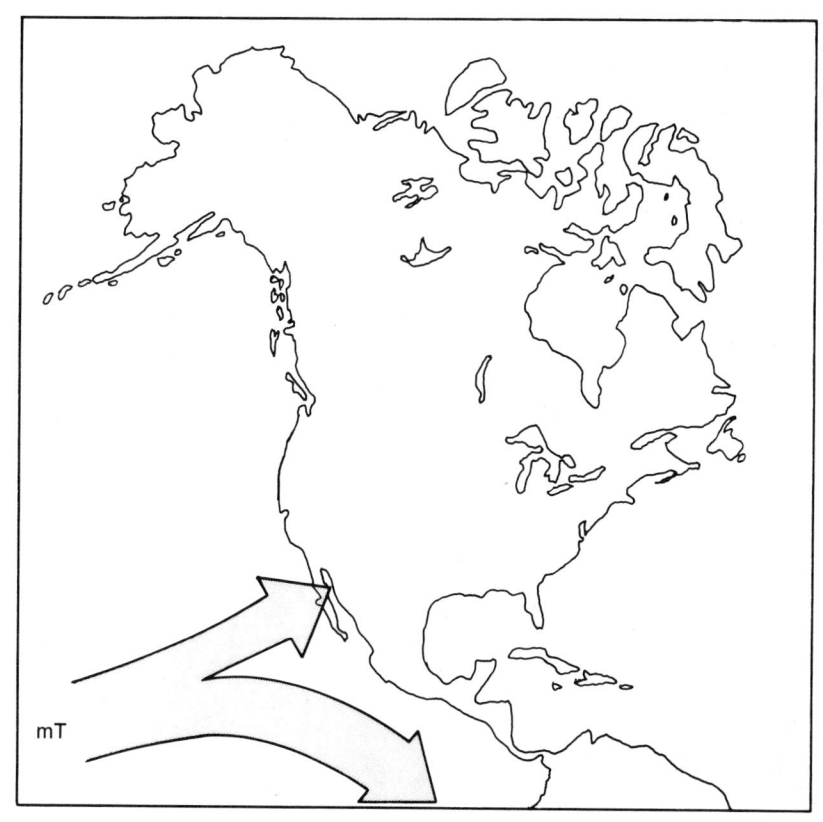

Figure 9-7 On rare occasions when maritime tropical air swings far enough north to affect the lower California coast and Mexico, record rainfalls are produced. (From Thomas L. Burrus and Herbert J. Spiegel, *Earth in Crisis*, 2nd Ed., St. Louis, 1980, The C. V. Mosby Co.)

Moisture

Moisture is added into the atmosphere by evaporation and removed by precipitation. If an air mass transits over a large unfrozen water body the air mass takes in additional moisture and becomes very wet. Any body of water as large or larger than the Great Lakes can act as a good source of moisture. As cold polar air funnels out of Canada during the winter, it picks up large quantities of moisture in its passage over the Great Lakes. On the lee side of the lakes the air is cooled and snow begins to fall. These snow flurries, as they are called, can average 3 to 4 in. per day and as much as 10 to 12 in. without the benefit of a storm being present.

10 WEATHER SYSTEMS

Vilhelm Bjerknes (1862–1951), a Norwegian meteorologist, who founded the Bergen School of Meteorology in Bergen, Norway. This school was a pioneer in synoptic weather forecasting at the turn of the twentieth century. Under Bjerknes' guidance, a staff of scientists including his son, Jacob, developed many important advances in map analysis and forecasting techniques. One such technique was the *polar front theory*, which is still the basis of our modern day weather forecasting. V. Bjerknes is often called the Father of Modern Meteorology.

LEARNING OBJECTIVES

After reading this chapter, you should be able to:

1. Explain how weather fronts are formed.
2. Define *frontogenesis, frontolysis, squall line, cyclogenesis, cyclolysis, easterly wave,* and *hurricane eye.*
3. Draw a cross section of a cold, warm, or occluded front.
4. List the general weather changes that occur with the passage of a cold or warm front.
5. Locate and describe any front in a newspaper weather map.
6. Explain the difference between warm and cold front occlusions.
7. Trace the life cycle of a cyclone.
8. Draw the six preferred cyclone tracks in the United States on a map.
9. Compare disturbance, depression, tropical storm, and hurricane.
10. Describe the growth, maturity, and decay of a hurricane.
11. Detail the weather in and around the eye wall of a hurricane.

FRONTAL SYSTEMS

One of the more important discoveries made by the Bergen School of Meteorology was the boundary zone that exists between two distinctly different air masses. The school established that because two colliding air masses usually had atmospheric characteristics that were vastly different, they would not mix together readily. In fact, the air mass with the greatest density would remain close to the surface of the earth with the lighter air rising above. The contact zone between the two air masses would be in the form of a sloping surface called a *front*. Fronts are defined as zones of transition between air masses of different densities. Frontal zones may be several miles across, and it is in this area that meteorological characteristics such as temperature, wind, humidity, and pressure gradually change to reflect the transition from one air mass to another. These changes are much more noticeable than the extremely small changes that take place in the air mass itself. In some instances, abrupt changes may take place if the frontal boundary is very narrow, but this is not the general rule.

Fronts, although seemingly only a surface phenomena because of the way in which they are depicted on surface maps, are really a three-dimensional entity starting at the surface and sloping upward. It does not matter which air mass is the strongest, for when the collision takes place, the warmer air is always displaced upward.

The Formation of Fronts

By definition, a front is composed of two different air masses. We must therefore look for areas on the earth where the wind pattern or circulation system allows such air masses to form. At lower latitudes (0°–30°) the air circulation does not enable fronts to easily form, and at higher latitudes (70°–90°) only cold air streams southward, without producing a front. However, in the midlatitudes (30°–70°) there exists a definite battleground between warm tropical air and cold polar or arctic air. This boundary zone is known as the *polar front*. This front is a worldwide discontinuous zone, meandering south or north depending on the intensity of the air masses. A second but much weaker frontal zone is the *arctic front*, which separates frigid arctic air from cold polar air; a third area, the *intertropical front* (3°N–3°S) is a convergence zone between the trade winds.

In order to have *frontogenesis*, which means the forming of a front, two conditions must exist:

1. Two air masses of different densities must lie side by side.
2. The wind must be blowing in such a direction so as to bring the air masses into constant contact with one another.

A perfect example of this phenomenon is in the formation of a *cyclone*. After a cyclone is formed, the wind circulation is in a perfect situation to form a front. Because of the counterclockwise and inward flow at the surface, cold northerly air can easily meet warm southerly air. In fact, this can and usually does take place on either side of a low pressure area (Fig. 10-1).

108   WEATHER SYSTEMS

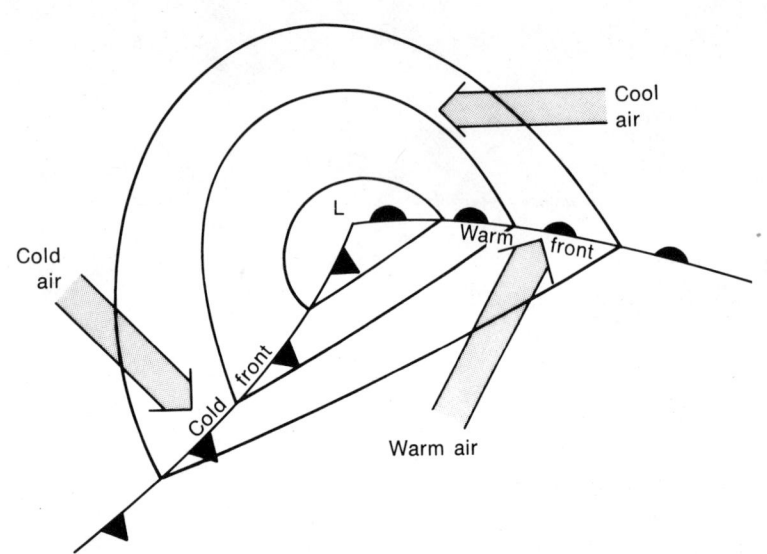

Figure 10-1 A typical wind flow arrangement around a low pressure area (L). This circulation allows colder air to clash with a warmer air mass while the warmer air is also battling with cooler air to the north. (From Thomas L. Burrus and Herbert J. Spiegel, *Earth in Crisis*, 2nd Ed., St. Louis, 1980, The C.V. Mosby Co.)

In an *anticyclone*, or high pressure area, the air circulation is outward and clockwise on the surface. Thus, air always leaves or *diverges* from a central point. Because air is always leaving, the two different air masses can never meet. Frontogenesis never takes place in a high (Fig. 10-2).

Frontal Types

The one important criteria for classifying fronts is the relative strength of the moving air masses. Although certain weather conditions prevail with particular fronts, they are the products of frontal formation and cannot be used to initially classify it. Simply put, a front

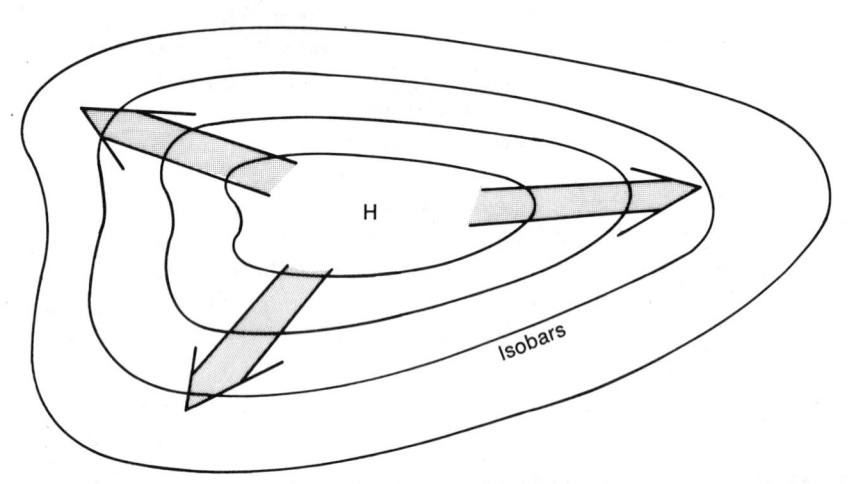

Figure 10-2 Wind flow associated with anticyclones are clockwise and *divergent* in the northern hemisphere; counterclockwise and divergent in the southern hemisphere. This circulation results in the spreading out of the air mass from the center. Therefore, contact between two distinctly different air masses can never take place within the high pressure zone (H) and fronts do not form. (From Thomas L. Burrus and Herbert J. Spiegel, *Earth in Crisis*, 2nd Ed., St. Louis, 1980, The C.V. Mosby Co.)

is classified according to which air mass is stronger; that is, which air mass pushes the other out of the way and takes its place. If a front is moving in a direction such that cold air is replacing warm air, it is called a *cold front*. If warm air invades cooler air, the front is designated a *warm front*. Two other fronts, *stationary* and *occluded*, will be discussed in detail later in this chapter.

Cold Fronts

When cold air pours southward replacing warmer air, a cold front develops. This boundary is three-dimensional in nature and can be visualized as a sloping wedge of cold air underriding a warm sector, which is being pushed up and out of an area. The slope of the front varies with the speed at which the cold air is moving. A fast-moving cold front will rise one mile in the air for every 40 to 80 miles (slope of 1:40 to 1:80), and have a forward average speed of 25 to 30 mph. A slow-moving cold front will slope one mile to 100 miles (1:100) and move as slow as 10 mph. However, in general, cold fronts move faster than warm fronts.

In a fast-moving cold front (Fig. 10-3), the warm air is pushed up sometimes quite violently, resulting in *unstable* conditions. Instability produces cumuliform-type clouds with great vertical height. Precipitation is in the form of showers and takes place in a narrow band along and just ahead of the front. In certain instances, a line of extremely violent thunderstorms develop 160 to 320 km (100 to 200 miles) in advance of the front called a *squall line* (Fig. 10-4). Pressure generally falls in advance of a cold front with a sudden rise as the front passes and the colder and more dense air arrives. The wind generally *veers* with the passage of the front, becoming strong and gusty at the *frontal surface*. Ultimately, the wind is from a northerly direction, usually northwest. The temperature starts to decrease in the warm sector because of evaporation of the falling precipitation (remember, evaporation cools things down) and takes a sharp drop behind the front. As the cold dry air moves in, rapid clearing is generated and relative humidity becomes quite low.

When fast-moving cold fronts, which lie great distances from the low pressure center, move through a region, a dry cold front may exist. It is known that the farther away a front lies from the low center the less moisture it has available. In these instances, a serious forest fire potential exists because the strong dry gusty winds dry out the earth's surface.

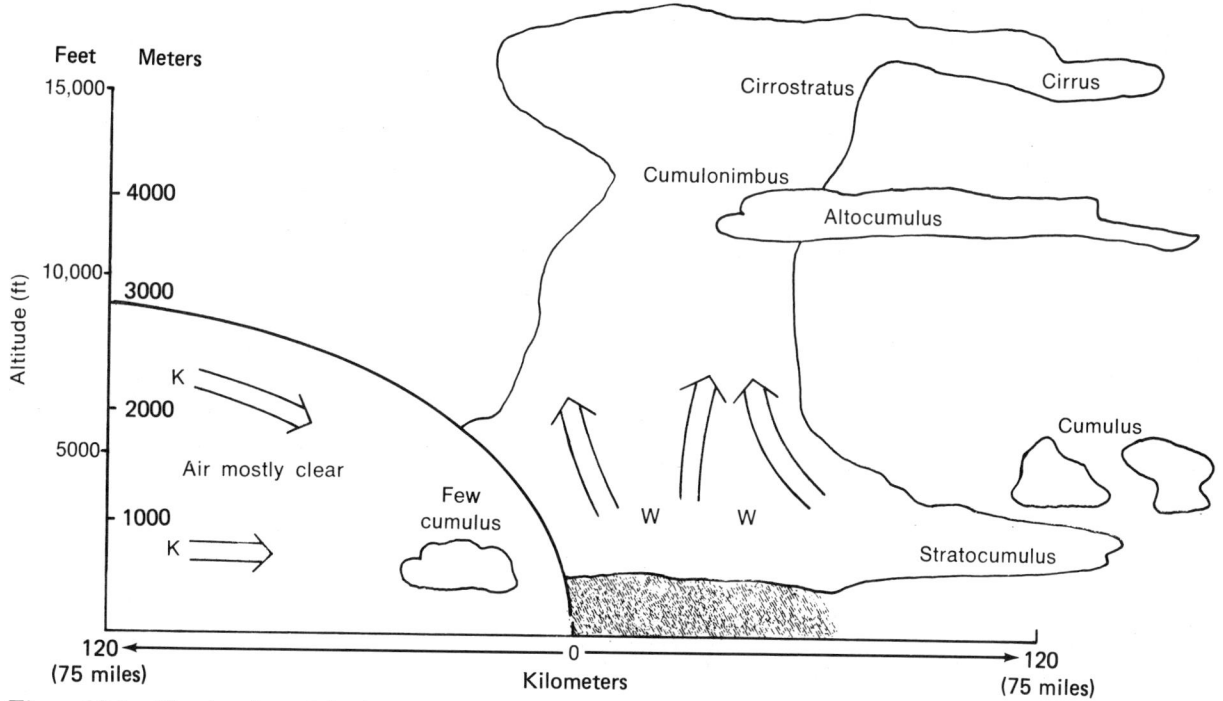

Figure 10-3 Cloud and precipitation pattern associated with a fast-moving cold front. The sloping boundary between the cold (K) and warm air (W) is the result of increased friction toward the earth's surface. (From Thomas L. Burrus, and Herbert J. Spiegel, *Earth in Crisis*, 2nd Ed., St. Louis, 1980, The C.V. Mosby Co.)

110 WEATHER SYSTEMS

Figure 10-4 A squall line produced by a fast-moving cold front. These solid lines of heavy thunderstorms and showers are often formed 50 to 300 miles in advance of a cold front by the overrunning of cold air. (Courtesy NOAA)

Slow-moving cold fronts produce *stable* air and stratus-type clouds. Rain is produced over a wide area both ahead and in back of the frontal zone. If the air, in some instances, is *conditionally unstable* embedded thunderstorms will form close to the front (Fig. 10-5). For the most part, similar changing weather conditions exist with either type of passage, but the time of change is quite different.

Warm Fronts

Warm fronts develop slowly as warm air glides slowly over the retreating colder denser air. The frontal boundary moves very slowly, and the slow advance results in *stable* conditions and the formation of stratiform-type clouds (Fig. 10-6).

With the oncoming warm front, clouds are observed from their highest to lowest category. Many miles in advance of the front the observer will sight cirrus and cirrostratus clouds. Then the lower altostratus will be seen as the warm sector approaches. Finally, the nimbostratus or rain cloud will appear when the front is 200 miles away. The precipitation will be in the form of steady drizzle, rain, or snow, which at times will be quite heavy. Still closer to the front, stratus clouds and

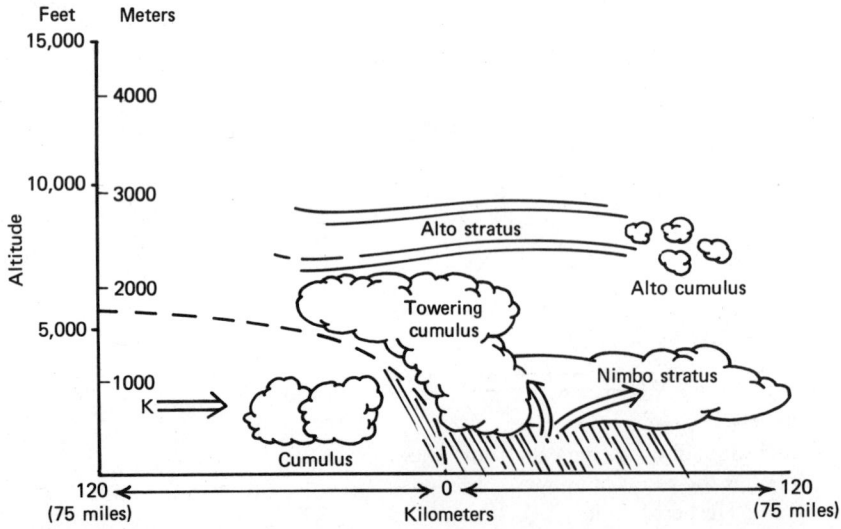

Figure 10-5 Cloud and precipitation pattern with a slow-moving cold front.

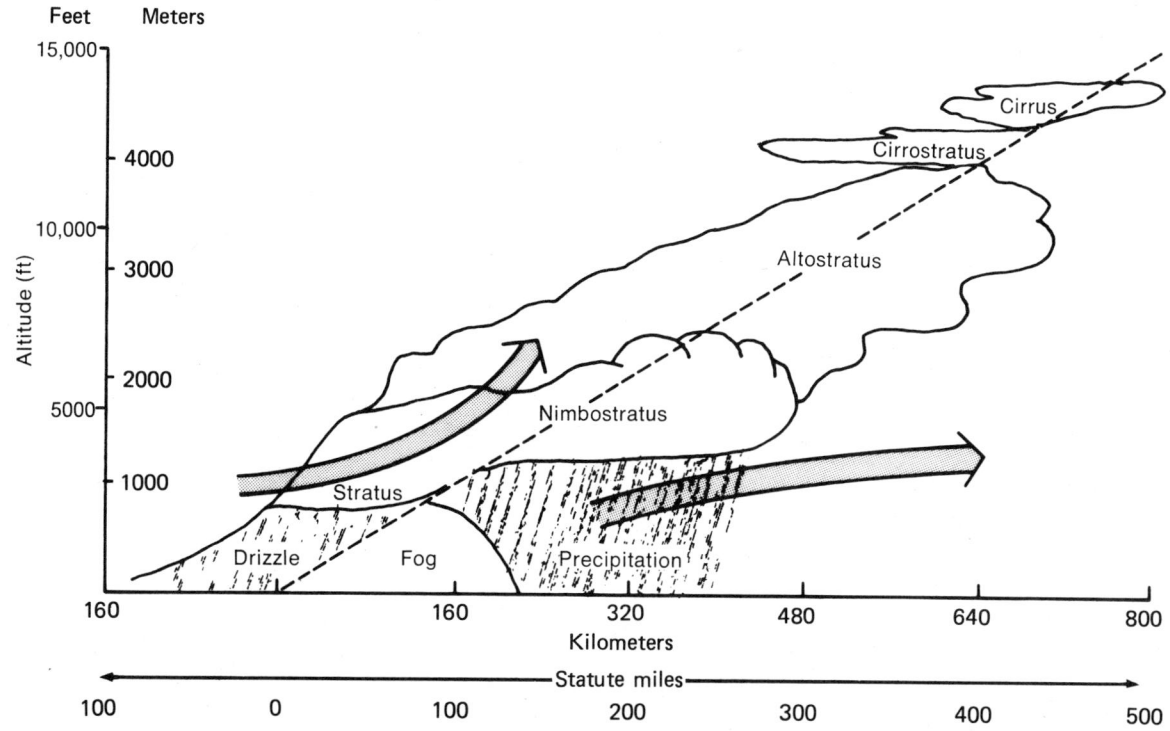

Figure 10-6 A warm front cloud and precipitation pattern of stable warm air. Note the diagonal alignment of the clouds and the more gradual transition zone between the air masses. (From Thomas L. Burrus and Herbert J. Spiegel, *Earth in Crisis*, 2nd Ed., St. Louis, 1980, The C.V. Mosby Co.)

fog will make their appearance. Clearing after the frontal passage will be very slow. The wind will shift gradually from an east or southeast direction to the west or southwest with wind speeds light. The pressure falls as the warm front approaches and levels off after its passage. If the warm air is maritime in nature, high humidities can occur and result in lingering oppressive conditions.

In rare instances a warm front will move through an area rather fast. These fast-moving warm fronts produce unstable air and thunderstorm activity embedded in the already existing stratiform clouds. These thunderstorms produce torrential downpours of precipitation that can cause serious flooding.

Stationary Fronts

Since the character and hence the name of a front is dependent upon which air mass is replacing already existing air, it is no wonder that a *stationary front* exists when there is no replacement. If the air masses on both sides of the boundary line are moving parallel to each other, the front will not move.

The weather pattern existing in such a condition depends on the sector of the front. Warm front conditions occur within the warm sector and the cold-type weather pattern in the colder part. Eventually the system either disappears or one of the two air masses become stronger and the stationary front becomes a cold or warm front again.

Occluded Fronts

Earlier in the chapter we stated that on the average cold fronts move faster than warm fronts. Therefore, eventually a cold front will overtake the slower-moving warm front, forming an occluded front. Usually when this happens the cold front wedges beneath the warm front, forcing the warm air aloft. This is called a *cold-type occlusion* (Fig. 10-7a) because the cold air in the cold front is colder and denser than the other two air masses. Although cold-type occlusions can form over the oceans, they more frequently form over land. In the United States, cold occlusions can be found forming east of the Rocky Mountains, accompanied by the widespread cloudiness and steady rain of the warm front, and a narrow band of violent weather associated with the cold front. The precipitation occurs both before and after the occluded front passes.

A *warm-type occlusion* is the result of the invading

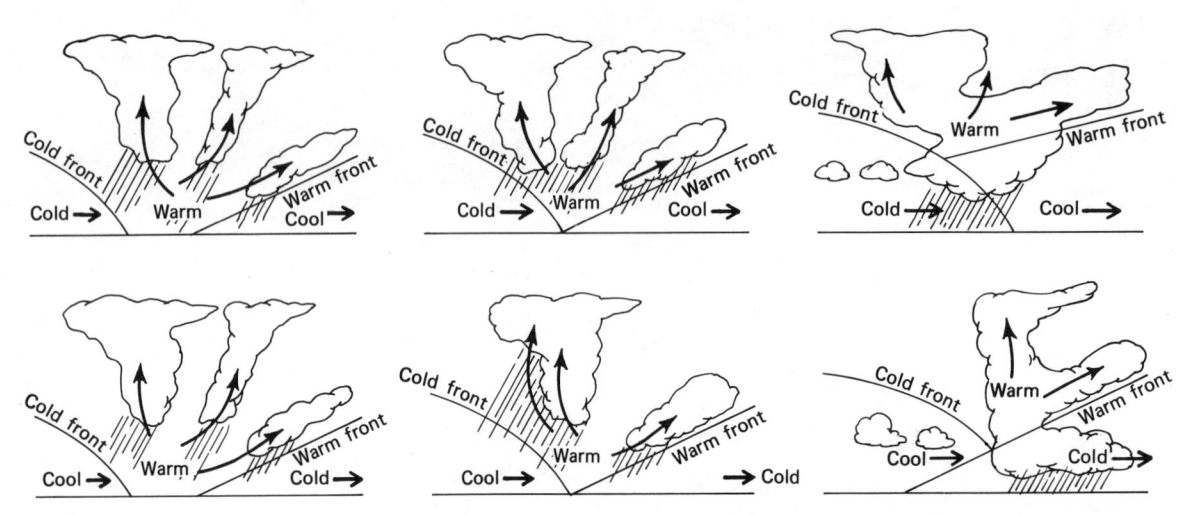

Figure 10-7 The three stages of (a) a cold-type occlusion, and (b) a warm-type occlusion. The main difference is the temperature of the invading air.

cold air of the cold front being warmer than the cold air associated with the warm front (Fig. 10-7b). The cold front slides up and over the warm front, but the warm air is still pushed up sharply. The weather associated with this type of occlusion is similar to the cold type except that the precipitation occurs prior to the passage of the frontal zone. Warm-type occlusions frequently form west of the Rockies, with the invading air being the milder maritime polar air from the Pacific Ocean. As this air is forced to rise over the mountains, it comes in contact with much colder air on the lee side and must ride aloft as an upper cold front. If this upper cold front comes in contact with moist maritime tropical air, highly unstable conditions result, producing numerous thunderstorms and a few tornadoes (Fig. 10-8).

In either case, an occluded front is the beginning of the end for the frontal system. The three air masses become well modified or mixed and eventually only one air mass emerges. Fronts cannot exist with only one air mass, and they disappear.

WEATHER MAP FRONTAL SYMBOLS

The National Weather Service uses a six-hour time interval for major map analysis. These maps display pressure centers and fronts using set designated symbols. High pressure centers are designated by a capitol H, low pressure centers by a capitol L, and fronts by a series of barbs and half moons. Table 10-1 represents the important frontal symbols found on weather maps.

CYCLONES AND ANTICYCLONES

With the onset of the fall season, cold polar air begins its domination of the northern hemisphere as the weakened warm air retreats southward. This condition is quite conducive in triggering the numerous mid-latitude winter storms called *cyclones* as more cold air

Figure 10-8 Tornadoes are sometimes triggered by the very unstable conditions produced in a warm-type occlusion, especially if the invading air is of space martime tropical air. (Courtesy National Severe Storms Laboratory, NOAA)

TABLE 10-1 Weather Map Frontal Symbols

Map Feature	Manuscript Maps	Printed Maps
Cold front	Solid blue line	▼▼▼▼
Warm front	Solid red line	●●●●
Stationary front	Alternating red and blue	▼●▼●
Occluded front	Solid purple line	▼●▼●
Cold front aloft	Dashed blue line	▽ ▽ ▽ ▽
Warm front aloft	Dashed red line	○ ○ ○
Stationary front aloft	Dashed alternating red and blue	▽ ○ ▽
Occluded front aloft	Dashed purple line	▽ ○ ▽ ○

Source: After Thomas L. Burrus and Herbert J. Spiegel, *Earth in Crisis*, 2nd Ed., St. Louis, 1980, The C. V. Mosby Co.

masses come in contact with the weakened warm moist air over the large land masses. The word *cyclone* means "the coil of a snake," and meteorologists apply the term to systems that have progressively lower pressure toward the center and a counterclockwise circulation of winds. The circulation tends to lift the air with upward movement, resulting in cooling, clouds, and precipitation. Therefore, these cyclones or low pressure areas, as they are called, are closely associated with bad weather.

Beginning in the fall, many locations in the polar region of the northern hemisphere begin their annual temperature retreat and cool the overlying air. The cooling air's density increases and an *anticyclone* predominates over the area. These are the future source regions for the cold polar outbreaks that will invade southward during the entire winter season. The anticyclone's circulation of winds is opposite to the cyclone and is clockwise and outward from the higher central pressure. This diverging air at the surface is replaced by subsiding air from aloft. As the air descends it warms adiabatically with very little cloud cover and almost no rain accompanying the system. High pressure areas are known for their generally fine weather conditions and clear, crisp air.

Cyclones

Low pressure areas can be identified by many different names, such as *extratropical cyclones*, *wave cyclones*, or *frontal low pressure areas*. Their area of frequency is between 30° and 60° north or south of the equator. Therefore, they also have the name of *mid-latitude lows*. These mid-latitude lows lie predominantly in the pre-vailing westerly wind pattern and, therefore, move in a generally west-to-east direction. Most of these storms are generated along the boundary of the *polar front* where the cold polar air meets the warmer tropical air. Here they intensify and begin their sweep over tens of thousands of square miles while rapidly deepening into an intense low pressure system with gale force winds. Although occurring in all seasons, mid-latitude cyclones are most destructive in the wintertime. Therefore, we will limit our discussion to these winter storms.

Life Cycle of Cyclones

Cyclonic development is usually associated with frontal activity, which eventually takes on a wavelike characteristic. Jacob Bjerknes from the Bergen School of Meteorology developed a model for the life cycle of such a system. Figures 10-9a to 10-9f represent the Norwegian model of a frontal wave and its accompanying cyclonic activity.

Figure 10-9a is the initial formation. It begins with an inert system made up of a stationary front that separates continental polar air from maritime tropical air. The winds, although from opposite directions, are parallel to the frontal zone. The wind speed in the colder air is greater than in the warmer sector.

Figure 10-9b represents *cyclogenesis* or the birth of a cyclone. As a result of the opposing wind directions, cyclonic shear (counterclockwise motion) develops. A simple experiment can prove this point. With a small object between your hands, move your top hand ahead of your bottom hand. The object will rotate in a counterclockwise manner. This allows the inflow of colder air on one sector of the front and warmer air on the other. Two fronts, one warm and one cold, take the place of the stationary front. The convergence of air especially the warmer air, results in a decrease of pressure and the low is formed.

Figure 10-9c represents the start of the wave formation. The cold front, which moves a little faster than the warm front, pushes closer to the warm front, and an increase of cyclonic shear takes place. This only serves to intensify the low and creates a frontal wave pattern with almost circular isobars.

Figure 10-9d is the rapid increase of the frontal wave. Pressures continue to fall around and in the central area, with the pressure gradient becoming quite steep. The low has a complete circular characteristic with air converging toward the center. Widespread cloud cover and precipitation begin.

Figure 10-9e is the mature stage of the cyclone development. At the back end of the system cold air is wedging beneath warmer air, producing an intense cold

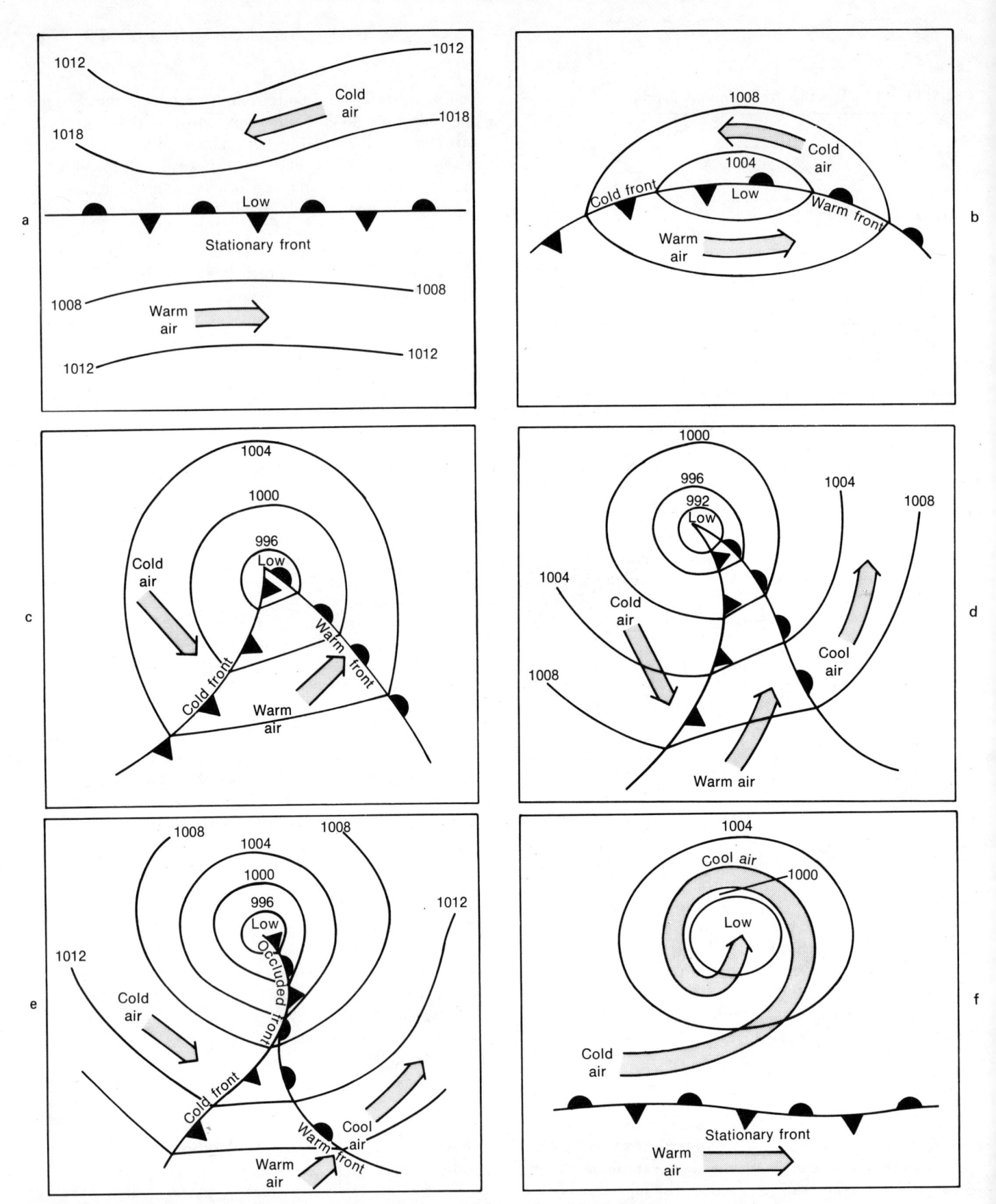

Figure 10-9 The six stages of the Norwegian cyclone model introduced by J. Bjerknes. (From Thomas L. Burrus and Herbert J. Spiegel, *Earth in Crisis*, 2nd Ed., St. Louis, 1980, The C.V. Mosby Co.)

114

front, while at the leading edge of the region warmer air is invading the cooler portion to the north and intensifying the warm front slightly. The pressure gradient has reached its maximum intensity as the cold front begins to overtake the warm front. Wind speeds increase, and the occluded front forms; this forces the warm air aloft as it is pinched from the surface.

Figure 10-9f represents *cyclolysis* or the death of the cyclone. The cold front totally consumes the warm front and the three air masses become well mixed. The occluded front loses its identity, and only one air mass spirals around the low pressure center. The pressure gradient weakens and the low eventually disappears. The entire system has lost its energy and only a stationary front or in many cases no front at all exists.

Cyclone Tracks in the United States

Winter storms pushing across the United States take preferred tracts (Fig. 10-10), which in turn depend on their point of origin or penetration on the mainland. Most of the storms striking the Pacific coastline find that the Colorado Rockies are a formidable obstacle that rarely can be overcome. Storms able to cross the Rockies degenerate only to redevelop as a *Colorado Low* (track 3). Most of the storms move northward and become the familiar *Alberta Low* (track 1), which like its counterpart to the south, seems to converge toward the Great Lakes area.

Along the eastern seaboard, winter storms, called Hatteras Lows, originate off the North Carolina coast (track 6) and whip along, skirting the coast with its gale winds and mountainous snowfalls. These are the legendary *nor'easters* (northeasters) that send temperature plummeting while reducing visibility to near 0. This most dramatic type of circumstances is called a *blizzard*, which leaves in its wake 30-ft snow drifts, below

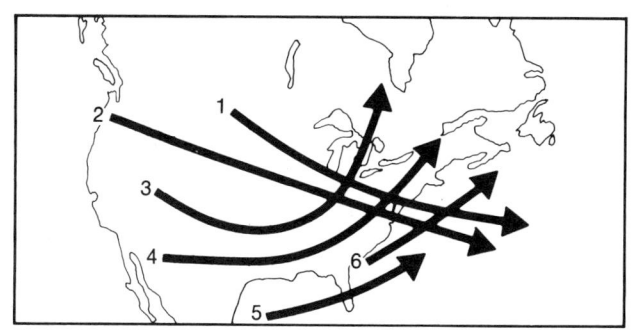

Figure 10-10 The preferred tracks of cyclonic storm systems across the North American continent. (From Thomas L. Burrus and Herbert J. Spiegel, *Earth in Crisis*, 2nd Ed., St. Louis, 1980, The C.V. Mosby Co.)

Figure 10-11 Digging out of the heavy snowfall that hit Adams, New York in February, 1977. (Courtesy NOAA)

zero temperatures, millions of dollars in damage, and many storm-related deaths (Fig. 10-11).

TROPICAL STORMS AND HURRICANES

On Sunday, August 17, 1969, winds in excess of 322 km/hr (200 mph) slammed into Gulfport, Mississippi. In its wake, the hot, sultry winds left more than 200 dead, thousands injured, and damage of nearly 1.5 billion dollars (Fig. 10-12). Extensive flooding due to the *storm surge*, which is a small but equally destructive inflow of water caused by high winds and low pressure, occurred along the northern coast of the Gulf of Mexico. In the case of *Camille*, so named by the National Weather Service, a 24-ft tide accompanied this most destructive of all storms, the *hurricane*.

Forming in the tropical regions of the world, tropical storm systems display a significantly different physical makeup than the mid-latitude systems that frequently migrate outside the tropics. Tropical systems cover large areas and are fairly circular while their extratropical counterparts embrace a region that is more elliptically shaped. Tropical storms also possess an *eye*, or relative calm region, in the center (Fig. 10-13) and have a westward movement, while *subtropical lows* have no eye and move in a west-to-east fashion. In the tropics, there are no accompanying anticyclones or frontal boundaries as there are in the mid-latitudes. The frequency of these systems varies according to season. The tropical storms are greater in the summer and autumn while extratropical storms occur at all times of the year.

It appears that the only features both systems have in

Figure 10-12 Storm damage left by hurricane Camille as it went inland on the Mississippi Gulf Coast on August 17-18, 1969. The bottom photograph is of the beautiful Richelieu Apartments located on U.S. Highway 90 before the passage of Camille. The top picture shows the remains of the same apartments. (Photographs by Chauncey T. Hinan)

common are that both systems are low pressure areas called cyclones and are associated with bad weather.

Classifying Tropical Cyclones

These infrequent storm systems of the lower latitudes have many regional names. In the northern Pacific they are called *typhoons*, while *baguio* is used in the Phillipines and China Sea. To most people of Australia a *willy-willy* is exactly like the *cyclones* of the Indian Ocean and the *hurricanes* of the West Indies and Caribbean Sea. However, the fanciest name in use is off the west coast of Mexico where residents call a tropical cyclone an *el cordonazo de San Francisco*, "the lash of St. Francis." Regardless of what name they are given, all of these storms are tropical cyclones producing high winds, torrential rains, and widespread destruction. In this text we will use the common American names of *hurricane* and *tropical storm* interchangeably, although these terms refer to storms of slightly different size and intensity.

There are four classifications within the National Weather service to denote the severity of tropical storms. These categories range from light winds and rain to monsoon-type downpours and destructive winds. They are based upon surface wind speeds and whether a closed wind circulation exists. The classification scheme is as follows:

1. **Tropical disturbance.** An incipient storm that has a slight surface circulation with no more than one closed isobar.
2. **Tropical depression.** Maximum winds are less than 50 km/hr (31 mph) with one or more closed surface isobars.
3. **Tropical storm.** Winds from 51 to 116 km/hr (32 to 72 mph) with closed surface isobars.
4. **Hurricane.** Wind speed greater than 116 km/hr (72 mph).

Hurricane Formation

Hurricanes usually start out as a *trough* of low pressure embedded in a belt of easterly trade winds. The *easterly wave*, as this newborn system is called, is believed to originate off the coast of a large continent such as the desert of North Africa. As the trough moves toward the Caribbean Sea, the air absorbs tremendous quantities of water vapor from this warm ocean source. Since the air is unstable, the water vapor is lifted aloft where it cools and condenses, releasing vast amounts of *latent heat*. This *latent heat of condensation* is the energy that intensifies a tropical system such as the easterly wave.

The *formative stage* of tropical cyclone development depends greatly on the temperature of the sea surface. Observations of previous storms indicate that to maintain a vertical circulation, a steep lapse rate (decrease of temperature) must be maintained. This can only be accomplished with water temperatures greater than 27°C (80°F). In addition, the rotary circulation so vital to the growth of the tropical system must be maintained by the *coriolis force*. Without this force there would be little or no spin of air particles around the low pressure center, and hurricane formation would not take place. This is the main reason why hurricanes do not form or move close to the doldrum belt (3°N to 3°S) where the coriolis force is practically zero.

The *tropical storm* or *intensification* stage is characterized by a well-developed chimney mechanism whereby air converges toward the center at the surface, spirals upward, and diverges at higher levels (Fig. 10-14). The pressure drops below 1000 millibars as the wind begins increasing around a tight ring, forming into a central eye. Immense towering cumuliform clouds spiral outward from the eye wall in narrow bands. It is believed that the interaction between high and low level winds determine whether the system will continue to develop. The theory suggests that if less air is pumped out at higher altitudes than is converging at the surface, a filling of the eye will take place and the pressure will rise, breaking up the eye and dissipating the storm.

The *mature stage* is attained when maximum winds and extremely low pressure are achieved. The cyclonic circulation is marked by towering bands of cumulonimbus clouds that produce heavy rain in the form of squalls and embedded tornadoes. The strong winds reach out to distances as great as 200 miles, especially in the most dangerous *northeast semicircle*. Close to the end of the mature stage, the central pressure and wind speeds level off. By this time, the system has usually recurved to the north and entered the prevailing westerly wind flow where it moves over colder water (less energy) or a large land area (no energy) in the middle latitudes. Being deprived of the energy needed to maintain its strength, the hurricane loses strength and its tropical characteristics.

As the *decaying stage* commences, winds die down, but heavy rain continues with the flooding of inland regions. The system takes on extratropical traits and eventually becomes a simple low pressure area. However, in certain instances, tropical cyclones in the decaying stage travel off the coast and regenerate into an intense storm system once again.

Hurricane Weather

Far in advance of a hurricane an abundance of high filamentary clouds, called *cirrus*, appear. In rapid succession, clouds similar to the approach of a warm front make their presence until the low nimbostratus occur. Bands of dense dark rain clouds coupled with ugly thick thunderstorm clouds start to spiral from the eye.

118 WEATHER SYSTEMS

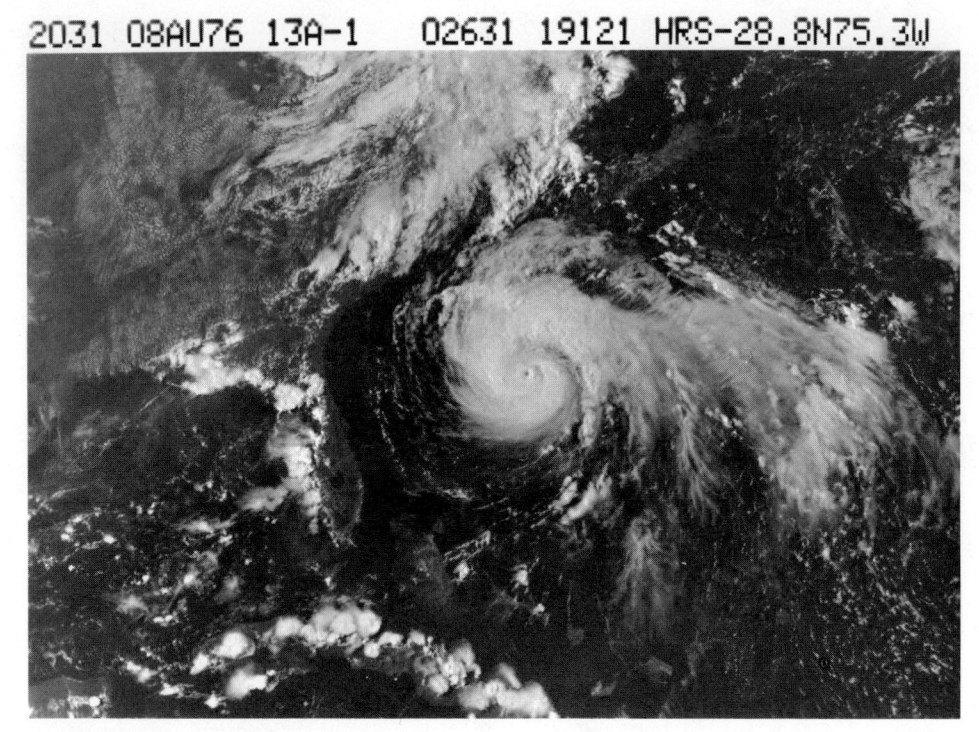

Figure 10-13 A satellite view of Hurricane Belle as it moved off the Eastern Coast of the United States in August, 1976. Notice the distinct eye and spiral bands of squall lines spinning into the central region of the storm. (Courtesy NOAA)

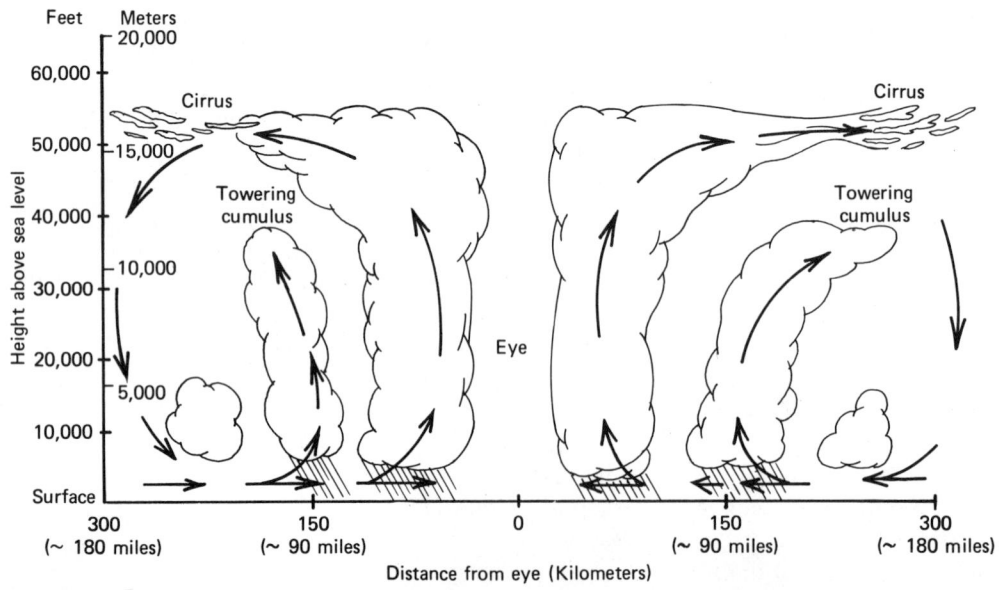

Figure 10-14 A vertical cross section through the eye of a hurricane with cloud cover and wind flow patterns.

TROPICAL STORMS AND HURRICANES 119

The *eye* is a calm center 14 to 25 miles in diameter that is associated with every mature hurricane (Fig. 10-15). The calm of the eye is certainly in sharp contrast to the wind velocities of 160 km/hr (100 mph) or more found in the *eye wall*. This wall is made up of the towering thunderstorms previously mentioned that sometimes reach as high as 24 km (80,000 ft). Updrafts spiral at extreme rates in the central cloud band while warm dry air descends gently in the eye. It is thought that this subsiding warm dry air is the mechanism responsible for clear calm conditions in the center of the hurricane.

The precipitation pattern consists of showers and thunderstorms starting 120 to 160 km (75 to 100 miles) from the center followed by heavy torrential wind-driven rains 32 to 38 km (20 to 30 miles) from the eye. The rain is supplemented by water already on the ground as it is whipped into the air and driven horizontally by the wind (Fig. 10-16). This reduces the visibility to zero in squalls and makes rainfall measurements almost impossible. Precipitation stops only within the eye, but starts again just as rapidly after the passage of the hurricane center.

The wind pattern around the tropical storm can be broken down into three regions. At 320 to 480 km (200 to 300 miles) out from the center *gale force winds* of 64 to 76 km/hr (40 to 47 mph) can be found. Usually the closer to the center the higher the wind velocity and at 80 to 96 km (50 to 60 miles) from the eye, winds can reach hurricane force of greater than 119 km/hr (74 mph). The region closest to the center, which can be from 8 to 24 km (5 to 15 miles), will have the maximum wind velocity. Although all hurricanes vary in size and intensity, the winds surrounding the eye wall can average more than 180 to 240 km/hr (120 to 150 mph) and in the case of Hurricane Camille, more than 322 km/hr (200 mph).

Strange as it may seem, wind flow in a hurricane is not the most destructive force. Rainfall in the amounts of 6 to 12 in. can cause flooding and immense damage to structures as well as loss of life. Hurricane Diane in 1955 created floods that caused $700 million in damage and took 200 lives. However, flooding from rainfall is not the most destructive element of a hurricane. The worse part is along the coast where wind-driven waves

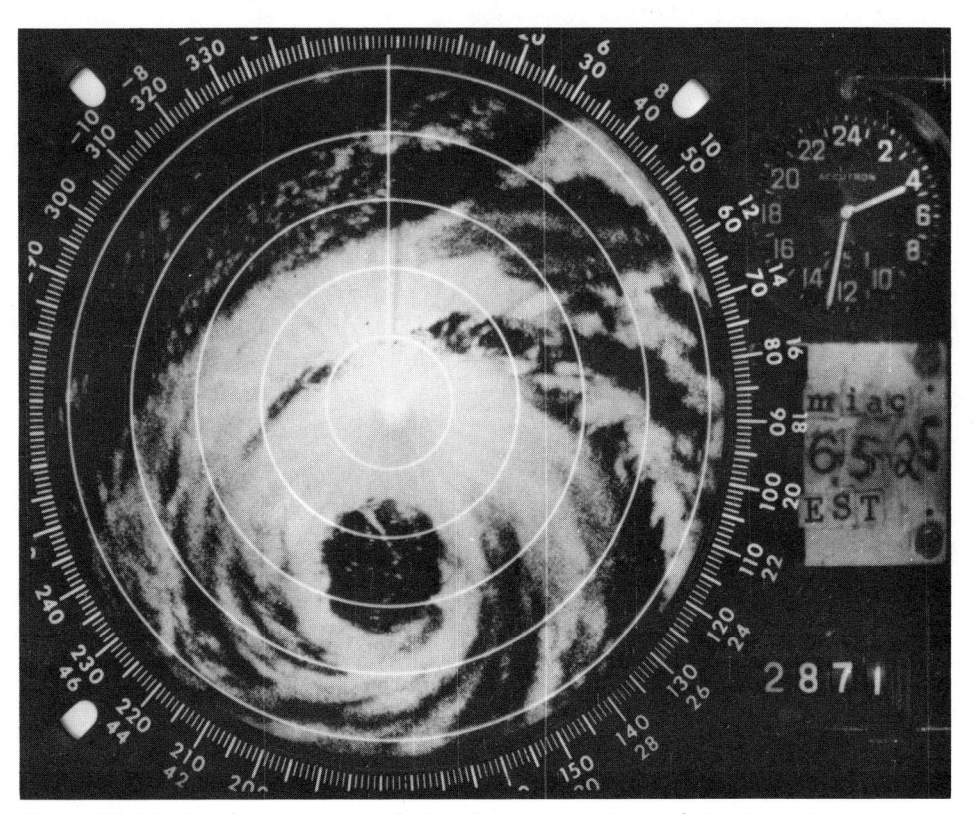

Figure 10-15 A radar scope view of a hurricane eye and the bands of thunderstorms spiraling out of the eye wall. (Courtesy NOAA)

Figure 10-16 Wind-generated waves of a hurricane. (Courtesy NOAA)

generate a *storm surge* (Fig. 10-16). Hurricane force winds are capable of creating a wall of water 15 m (50 ft) or more over the deep part of the ocean. Added to this wall is the increase in ocean surface height because of the drastic reduction in atmospheric pressure. When the storm reaches shoreward, the ocean becomes shallower and like a tidal wave the water level can increase dramatically. When the wind, waves, and buildup of water level are superimposed upon the storm surge, a destructive wall of water brings catastrophe to coastal areas. Storm surges of the past have killed 300,000 near Calcutta, India in 1737 and 200,000 in East Pakistan in 1790.

Hurricane Tracks

In the tropics of the northern hemisphere tropical storms move toward the west, especially in their development stage. Upon reaching hurricane intensity, they have a tendency to meander, although still in a somewhat westerly direction. However, often they recurve northward where they are caught in the prevailing westerly wind belt and are steered toward the northeast.

Since no two hurricanes are alike, their movements can be quite erratic. Some storms have been known to stall for days, over an area while others attained forward speeds of 80 km/hr (50 mph). In a few instances, the storm system has been observed to wander aimlessly in the Atlantic, never to touch land. In some instances hurricanes have formed loops or split into two monstrous storms. Figure 10-17 shows the average preferred tracks of the world's tropical cyclones.

Hurricane Facts

One of the greatest hurricanes in the twentieth century was the 1928 San Felipe storm. It lashed across Puerto Rico after nearly destroying the islands of Guadeloupe and St. Kitts in the Caribbean. Although it lost its strength crossing Puerto Rico, it regained its power and spread its large tentacles toward the United States mainland. The storm made landfall at Palm Beach, Florida, and raced northward up the Atlantic seaboard. The toll of this killer hurricane included 300 killed in the West Indies and nearly 2000 deaths in the United States (most occurred in the inland region of Florida). Losses due to wind and flooding exceeded 75 million dollars.

The hurricane that took the most lives hit Galveston, Texas on September 8 and 9 in 1900. It carried a storm surge of 5 feet that hit an already flooded city, drowning many who thought they were on safe higher land. More than 6000 people were killed out of a population of 38,000. Other killer storms of more recent times include Flora in 1963 and Camille in 1969.

As stated previously Camille was the most costly of all hurricanes to hit the United States; damage was estimated to be close to 1.5 billion dollars, mainly as a result of the continuous flooding along the Mississippi–Louisiana coast and the interior of Virginia.

The Naming of Hurricanes

Initially hurricanes were identified by a latitude-longitude system. This system proved to be impractical and a simpler, more efficient system was designed. Tropical storms were named for the military phonetic alphabet and the year in which they occurred. For example, if the storm was the first of the 1945 season it was called *Able-45*. Other storms for the same year would be called *Baker-45*, *Charlie-45*, etc. During World War II weathermen began naming hurricanes after females in alphabetical order. The names were short and concise and,

TROPICAL STORMS AND HURRICANES 121

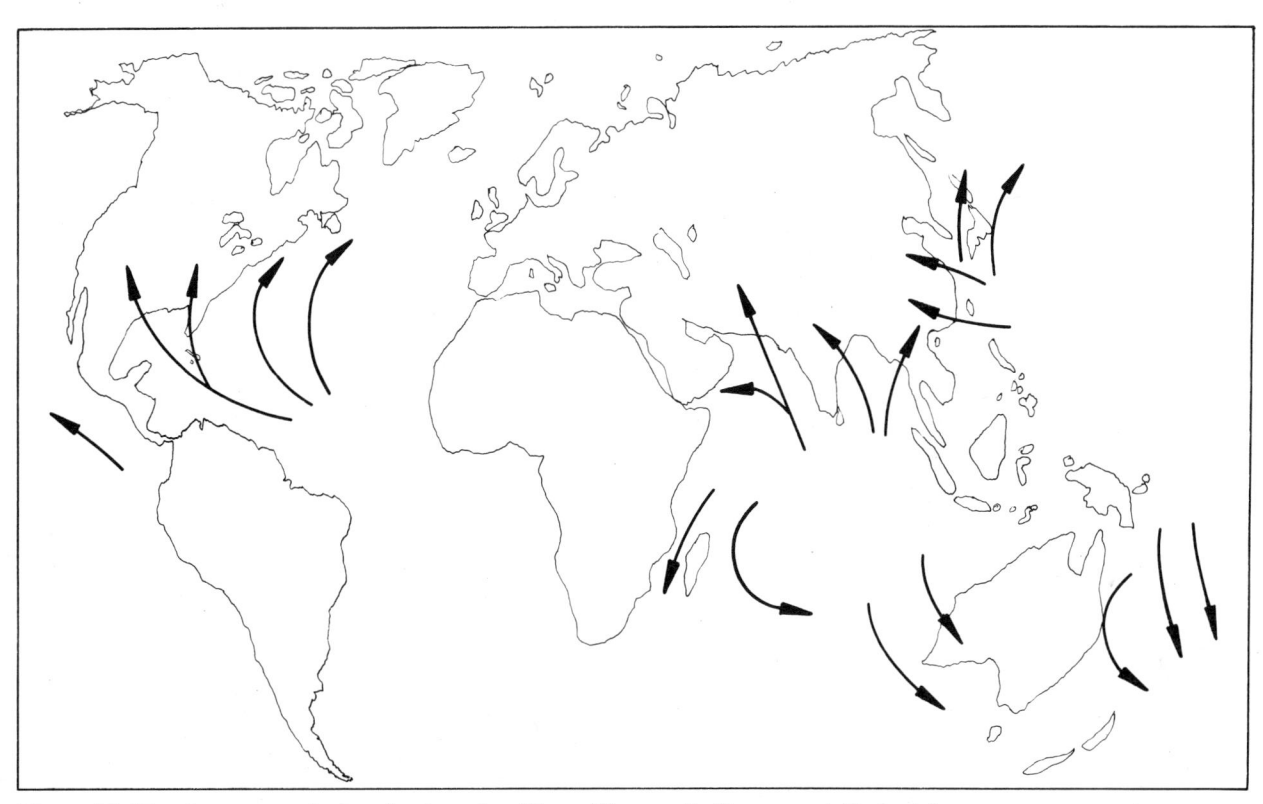

Figure 10-17 Average tropical cyclone tracks. (From Thomas L. Burrus and Herbert J. Spiegel, *Earth in Crisis*, 2nd Ed., St. Louis, 1980, The C.V. Mosby Co.)

TABLE 10-2a Eastern Pacific Typhoon Names

1983	1984	1985	1986
Adolph	Alma	Andres	Agatha
Barbara	Boris	Blanca	Blas
Cosme	Cristina	Carlos	Celia
Dalilia	Douglas	Dolores	Darby
Erick	Elida	Enrique	Estelle
Flossie	Fausto	Fefa	Frank
Gil	Genevieve	Guillermo	Georgette
Henriette	Herman	Hilda	Howard
Ismael	Iselle	Ignacio	Isis
Juliette	Julio	Jimena	Javier
Kiko	Kenna	Kevin	Kay
Lorena	Lowell	Linda	Lester
Miriam	Manuel	Marie	Marty
Narda	Norbert	Nora	Newton
Octave	Odile	Olaf	Orlene
Priscilla	Polo	Pauline	Paine
Raymond	Rachel	Rick	Roslyn
Sonia	Simon	Sandra	Seymour
Tico	Trudy	Terry	Tina
Velma	Vance	Vivian	Virgil
Winnie	Wallis	Waldo	Winifred

TABLE 10-2b Atlantic Ocean Hurricane Names

1983	1984	1985	1986
Alicia	Arthur	Ana	Allen
Barry	Bertha	Bob	Bonnie
Chantal	Cesar	Claudette	Charley
Dean	Diana	Danny	Danielle
Erin	Edouard	Elena	Earl
Felix	Fran	Fabian	Frances
Gabrielle	Gustav	Gloria	Georges
Hugo	Hortense	Henri	Hermine
Iris	Isadore	Isabel	Ivan
Jerry	Josephine	Juan	Jeanne
Karen	Klaus	Kate	Karl
Luis	Lili	Larry	Lisa
Marilyn	Marco	Mindy	Mitch
Noel	Nana	Nicholas	Nicole
Opel	Omar	Odette	Otto
Pablo	Paloma	Peter	Paula
Roxanne	Rene	Rose	Richard
Sebastien	Sally	Sam	Shary
Tanya	Teddy	Teresa	Tomas
Van	Vicky	Victor	Virginie
Wendy	Wilfred	Wanda	Walter

most important, easily remembered. In 1960 a semipermanent list of four sets of girls names was devised, and in 1971 changed to 10 sets. All names would be used over again, but if the hurricane was of particular note, its name would be retired from the list and another name added. In 1979 the lists were again changed and the names were alternated female and male. The first male named hurricane was Bob, which appeared in July, 1979 and made landfall along the Louisiana coast. Table 10-2 is a list of hurricane and typhoon names to be used in the ensuing years.

11
THUNDERSTORMS AND TORNADOES

Benjamin Franklin (1706–1790) famous statesman, writer, and scientist is known around the world for his early work with lightning. In his famous kite experiment, Franklin proved that lightning is an electric spark and that the base of thunderstorm clouds are made up mostly of negative electrons. From his experiments came the invention of the lightning rod.

124 THUNDERSTORMS AND TORNADOES

LEARNING OBJECTIVES

After reading this chapter, you will be able to:

1. Describe the three stages of a thunderstorm.
2. List the adverse weather conditions that could exist as the result of a thunderstorm and locate the approximate position of each on a thunderstorm diagram.
3. Explain lightning and thunder.
4. Describe the common types of lightning.
5. Calculate the distance you are standing from a lightning strike.
6. Describe and compare a tornado and waterspout.
7. Define *fork, streak, sheet, heat, and ball lightning*.

There are probably very few people on this earth who have not experienced the flash of lightning, the rumble of thunder, and the sight of a torrential downpour. Many a day's outing has been cancelled or severely curtailed as a local storm called the *thunderstorm* has suddenly erupted full fury. Thunderstorms are not only the most common type of local storms throughout the entire world, but also the most feared. Although lightning and thunder are always present, heavy rain, gusty winds, hail, and tornadoes can accompany the thunderstorm as it moves swiftly along at average speeds of up to 40 km/hr (25 mph).

In tropical regions thunderstorms can occur at any time of the year. In the mid-latitudes storms generally develop during spring, summer, and fall, while at the polar region, only during mid-summer do thunder storms usually form. It is estimated that throughout the world more than 45,000 thunderstorms occur every day, with most of these local storms forming during the northern or southern hemisphere summer season.

THUNDERSTORM FORMATION

Thunderstorms develop from strong cores of rising air with high relative humidities. Air is initially forced to rise by heating, orographic lifting, convergence, or the clashing of two air masses at a frontal boundary. This *mechanical lifting* is the first ingredient in a three-step process. If the thunderstorm is to develop, the air must reach *saturation* so that clouds can form. Initially, the air must contain sufficient moisture so that saturation is easily reached. In an earlier chapter we saw that there were two ways to reach saturation. First, air is cooled and, second, more moisture is added to the system. In thunderstorm formation, both processes take place, the cooling process being the most important. As the air is mechanically lifted it cools adiabatically. This cooling causes condensation to begin, which results in the formation of small vertical cumulus clouds, such as those pictured in Figure 11-1. Thus, sufficient moisture or a cooling mechanism make up the second ingredient in thunderstorm formation.

The last ingredient is called *unstable air*. The cooling of the mechanically lifted updrafts produces condensation, which releases latent heat. This release of energy partially offsets the cooling that is due to the adiabatic process and the air becomes warmer, less dense, and buoyant. This instability allows the air to rise on its own, increasing the updrafts at lower levels, which permits fresh moist air to be drawn in for additional cloud growth. When all three ingredients are present, the thunderstorm is nearly certain to go through its life cycle of *cumulus* (birth), *maturity* (violent), and *dissipating* (death) stages.

Figure 11-1 Convective cumulus clouds formed over a city due to the heat island effect. Temperatures over a city are hotter during the day than the surrounding countryside or an adjacent water body.

THUNDERSTORM FORMATION 125

✷ Ice crystals
· Water droplets

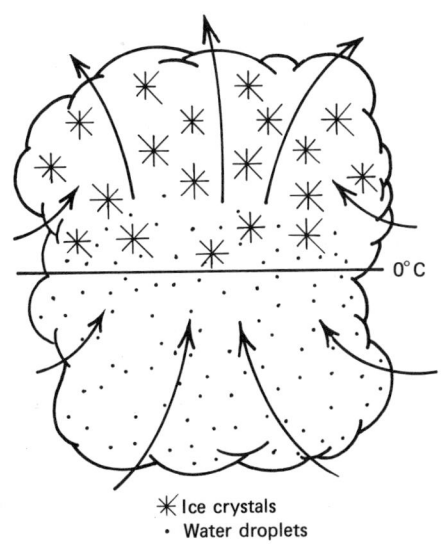

Figure 11-2 The cumulus stage of a thunderstorm. The arrow from the side of the cloud represents the *entrainment* of fresh moist air from the environment. This is added fuel for cloud growth.

Phases of Thunderstorm Development

In the one to two hours that a thunderstorm has to live, it can be identified by three distinct developmental forms with separate meteorological characteristics. It is impossible to see an abrupt change in the cloud because the transition from one stage into another is smooth. In some instances, a boiling effect will take place, but the exact instant the cloud becomes a mature thunderstorm cannot be detected.

Cumulus Stage. All thunderstorms begin as puffy, white, cumulus clouds. At most locations, these cumulus clouds are the result of the mechanical lifting process with convection currents due to heating being the major source. As solar radiation heats the ground, the ground heats the air layer just above it. The heated air rises and the first step of the three-step process has begun. It should now be quite obvious why the summer season has the most thunderstorms, or why the tropics can produce them at any time of the year.

The distinctive features of the cumulus stage are the cauliflowered-topped clouds that produce no precipitation at ground level and the *updrafts* that are found throughout the entire cloud (Fig. 11-2). Cumulus clouds can extend up to 4600 m (15,000 ft) and grow outward to a diameter of nearly 1.6 km (1 mile). The updrafts cause convergence at the surface, which adds additional moisture to the growing cloud. Initially the cloud droplets are small, but larger droplets accompany cloud growth. These droplets are carried aloft by updrafts where they can become supercooled water droplets or small pellets of ice called *sleet*. Water vapor is also carried aloft, above the freezing level, and the *sublimation process* that takes place produces ice crystals, the necessary ingredient for snow, hail, or rain.

Mature Stage. If moisture availability and instability are present, the cloud will continue to grow vertically. At the point at which the water droplets, ice crystals, and small ice pellets can no longer be supported by the updrafts of the cumulus stage, they fall. As they fall the frictional drag and cooling effect on the warmer air create the downdrafts which, although not as strong as the updrafts (up to 161 km/hr), can reach speeds of nearly 80 km/hr (50 mph). The downdrafts occur in a region where the updrafts are less potent and, therefore, both updrafts and downdrafts can exist side by side. With the existence of both up and downdrafts, the mature stage begins (Fig. 11-3).

As the downdrafts become strong enough, they pour out of the base of the leading edge of the cloud, resulting in the familiar rush of cool air that precedes a thunderstorm passage. Close behind this wind is the heavy rain as the large water droplets and ice crystal products are pulled downward from as high as 60,000 ft. If conditions are just right, hail will follow the torrential rainfall; this usually signals the arrival of the center or core of the thunderstorm. The wind, rain, and hail cool the ground below, which inhibit updrafts. The outpouring

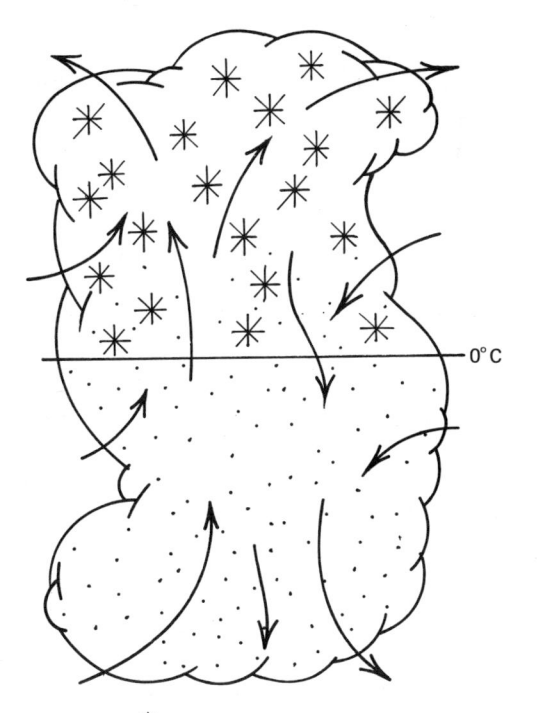

✷ Ice crystals and snow
· Water droplets

Figure 11-3 The mature stage of a thunderstorm. Notice the wind flow vector out the top of the thunderstorm. Hail can be thrown out the top as far away as 10 miles.

126 THUNDERSTORMS AND TORNADOES

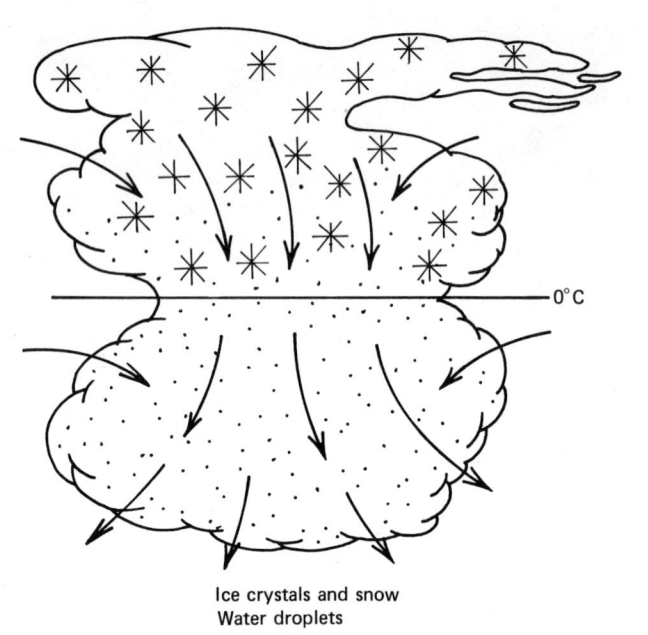

Figure 11-4 The dissipating stage of a thunderstorm. Entrainment is all downward and a resupply of moisture is negligible.

of the cool downdraft spreads out in advance of the thunderstorm and can account for as much as a 8.5°C (15°F) drop in air temperature. Usually, just as the cloud goes into the dissipating stage, an anvil top will form as the upper part spreads out because of the upper level wind flow.

Dissipating Stage. When the updrafts are less frequent and weaker than the now predominating downdrafts, the third and final stage of the thunderstorm life cycle begins. The downdrafts continue to deplete the cumulonimbus cloud of moisture and the updrafts are not strong or numerous enough to bring in a sufficient supply of moisture to continue the process. Cut off from this supply, the condensation process slows down, and the energy released by condensation is no longer available to fuel the thunderstorm. The storm rapidly disintegrates, and all downdrafts appear in the cloud (Fig. 11-4). Rainfall diminishes to a light rain or drizzle and the cloud fades away, leaving a sometimes hazy appearance or low stratiform clouds that float lazily in the sky.

LIGHTNING AND THUNDER

The Lightning Process

With over 1800 thunderstorms occurring at the very moment you are reading this line, the risk of being hit by lightning is much greater than being in a hurricane or a tornado. Each year property worth millions of dollars is damaged by lightning or lightning-related events such as forest fires, aircraft damage, and structural destruction. The latest statistics assembled by the National Center for Health Statistics indicate that lightning kills, on the average, 150 Americans per year; the number of deaths per year is much higher for the rest of the world.

Lightning is a by-product of a mature thunderstorm, although it can also occur in the dissipating stage. As the small cumulus cloud begins to grow into a cumulonimbus cloud, the up and downdrafts allow the water droplets, ice crystals, ice pellets, and hail to interact with one another. As they collide an energy exchange between the particles takes place. Although meteorologists are really not sure how this comes about, they do know that the result is particles of unlike charges. These particles distribute themselves in concentrated areas of the thunderstorm with positive charges in the upper layers and a large negative area close to the base of the cloud (Fig. 11-5). The accumulation of charges continues until a large imbalance, called an electrical potential, occurs between the charges within the cloud, two different clouds, or the cloud and ground. At this point a discharge of negative electrons takes place that allows nature to achieve a balance of charges once again. This discharge is called lightning.

Lightning is the flow of current from negative to positive potential and occurs in a step process (Fig. 11-5). It begins as a *pilot leader* (Fig. 11-5,1), which cannot be seen because of its faintness. The pilot leader sets up an initial path, which is followed by a *step leader* (Fig. 11-5,2) surging forward as much as 100 ft at a time. The step leader pauses, then repeats the sequence again along the path of least resistance until it comes close to the ground (Fig. 11-5,4). At this point a discharge streamer from the ground intercepts the leader path and completes the channel between cloud and ground (Fig. 11-5,5). Upon the stroke hitting the earth, a return stroke leaps upward at a speed approaching that of the speed of light. It illuminates the branches of the descending leader track and, because it points downward, it looks like the stroke is coming from the cloud, but it is really coming from the ground. The bright light glow of the return stroke comes from the glowing atoms and molecules energized by the stroke itself. After the return stroke ends, (Fig. 11-5,8) a secondary system of smaller strokes go toward the ground and dissipate.

Common Types of Lightning

All lightning is formed in the same manner. That is, an excess of charges produce a potential that results in a discharge. However, there are many different forms of lightning, some of which will be described below.

Forked Lightning. A very common type of lightning with a branched path and conductive bright main channel (Fig. 11-6).

Life Cycle of a Lightning Stroke

As thunderstorm induces growing positive charge in earth, potential between cloud and ground increases (1) until pilot leader starts a conductive channel toward ground (2) followed by step leaders (3) which move downward for short intervals (4) until met by streamers from ground. Return stroke from ground illuminates branches (5) and seems to come from cloud. Main stroke is followed by sequence of dart leaders and returns (6, 7) until potential is reduced or ionized path is dispersed (8). Elapsed time: about one second.

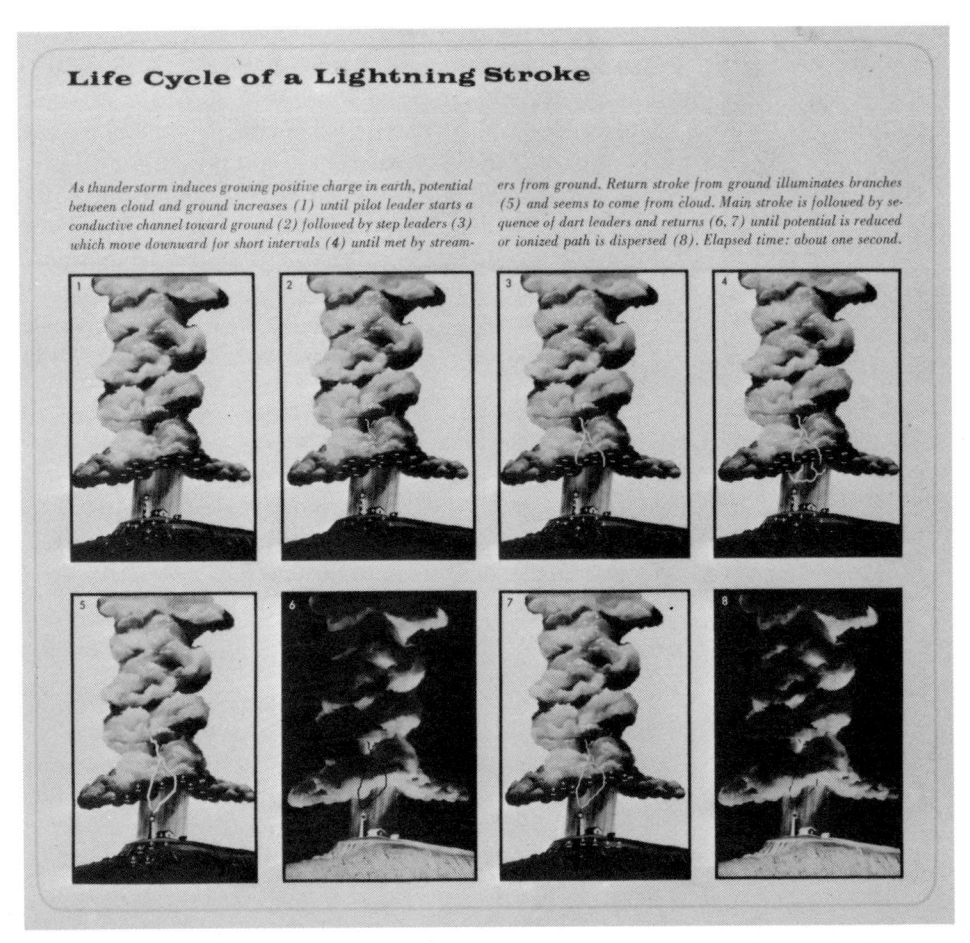

Figure 11-5 The life cycle of a lightning stroke. As thunderstorm induces growing positive charge in earth, potential between cloud and ground increases (1) until pilot leader starts a conductive channel toward ground (2) followed by step leaders (3) which move downward for short intervals (4) until met by streamers from ground. Return stroke from ground illuminates branches (5) and seems to come from cloud. Main stroke is followed by sequence of dart leaders and returns (6, 7) until potential is reduced or ionized path is dispersed (8). Elapsed time: about one second. (Courtesy NOAA)

Figure 11-6 Forked lightning with its many branches. (Courtesy NOAA)

Streak Lightning. A single path cloud-to-ground stroke without side branches. This is the most frequent type of lightning (Fig. 11-7).

Sheet Lightning. A shapeless flash of light that seems to have no form. It covers a broad area and is a result of a large lightning stroke discharge either beyond the observer's actual observation or hidden by clouds. This type is common in cloud-to-cloud discharges.

Heat Lightning. This type of lightning does not really exist. It is a term given to a light seen in the sky from a thunderstorm well beyond the horizon. Heat cannot produce lightning, and the term should not really be used.

Ball Lightning. A luminous globe or elliptical blob of glowing electrons occurring during a thunderstorm. The glowing mass seems to hang in the air or move erratically as it hisses or buzzes. It has been reported that the size can vary between that of a lemon and a basketball and that most of the time the ball just decays and dissipates. However, in one

Figure 11-7 Example of the more familiar streak lightning with small branches protruding. (Courtesy W. L. Taylor, NOAA)

case, it entered a house by the front window, floated to the back, and disappeared through an open rear window.

Thunder

Thunder is the by-product of rapidly expanding gases along the channel of lightning discharge. A lightning stroke heats up the air along its path to 10,000°C (18,000°F). The air expands outwardly in all directions with a vibrating pressure wave that results in a booming sound much like a jet airplane breaking the sound barrier. When lightning strikes close by, a sharp crack is heard. The farther away the hit, the more the sound will occur as a rumble, as the atmosphere refracts and modifies the sound. At the point of impact, lightning and thunder are simultaneous but, because the speed of light is 300,000 km/sec (186,000 miles/sec), and the speed of sound is only about 1234 km/hr (767 miles/hr) you hear the sound after you see the flash. By counting the seconds between flash and sound, you can estimate the distance between you and a lightning stroke: about every five seconds equals one mile.

THUNDERSTORM WEATHER

Only lightning and thunder are needed to classify a cloud as a thunderstorm cloud. However, thunderstorms may have torrential rains, strong gusty winds, hail, and sometimes tornadoes. All of these may appear simultaneously, and it is no wonder why the word thunderstorm strikes fear into the minds of many people.

Rain frequently accompanies thunderstorms; in rare instances, during the winter *snow* will fall. The leading edge of the *cell* will have the heaviest precipitation as the downdrafts sweep the large water droplets out of the cloud. The trailing edge and the *anvil* will produce light, steady rain sometimes for hours. Since thunderstorms are usually sharply delineated at their edges, rain can fall on one side of a street and not on the other. In fact, the localized nature of rainfall can cause one part of a city to receive seven inches of rain while another part of the city receives nothing.

Strong winds are usually found along the leading edge and in advance of a thunderstorm. The word *squall* means strong winds that last for several minutes, and is opposed to a *gust*, which indicates a short burst of increased wind. Wind squalls can have sustained winds up to 56 knots (63 mph) with gusts to 99 knots (112 mph).

Thunderstorms often align themselves in a dense, impenetrable line, which is then called a *squall line*. In many instances convective thunderstorms may line up as a squall line, or frontal thunderstorms may form in a line parallel to a cold front. In either case, thunderstorm squalls race ahead of the leading edge of the cells. In coastal areas it is common to hear someone talk about a *rainsquall*. This is a sudden burst of wind and rain that lasts for three to five minutes and does not cover large areas. In many instances navigating around the rainsquall can be done with great ease.

Hail is a precipitation product that is peculiar only to thunderstorms. These rounded lumps of soft ice are the result of supercooled water droplets growing on an ice particle as it constantly circulates through the thunderstorm. Once a drop has frozen, other droplets collide, stick, and freeze, allowing the hailstone to grow. Additional layers freeze, only to be encircled by another layer of supercooled water as the particles move at the discretion of the internal wind flow. A cutaway view of a hailstone (Fig. 11-8) reveals alternating layers of clear and opaque ice.

Eventually, the hailstones become too heavy for the updrafts to support, and they fall out of the base of the cloud. As they reach the 0°C (32°F) level they begin to melt and can reach the ground as large cold raindrops. If the hailstone is large enough or the freezing level close to the ground, the solid precipitation will hit the surface of the earth. Hailstones frequently reach the size of golf balls or baseballs in the midwestern United States (Fig. 11-9).

Not all thunderstorms produce hail. Only those cumulonimbus clouds that are high in water content and have updrafts of sufficient velocity to hold a growing ice pellet aloft are proper candidates. Hail occurs most frequently in the spring and summer when both ingredients are present, most especially in the central United States. Although Florida has more thunderstorms than any other section of the country, its hail production is quite low compared to the heartland of America.

As any farmer knows, hail can mean complete ruination of an entire planting season. The aftermath of a

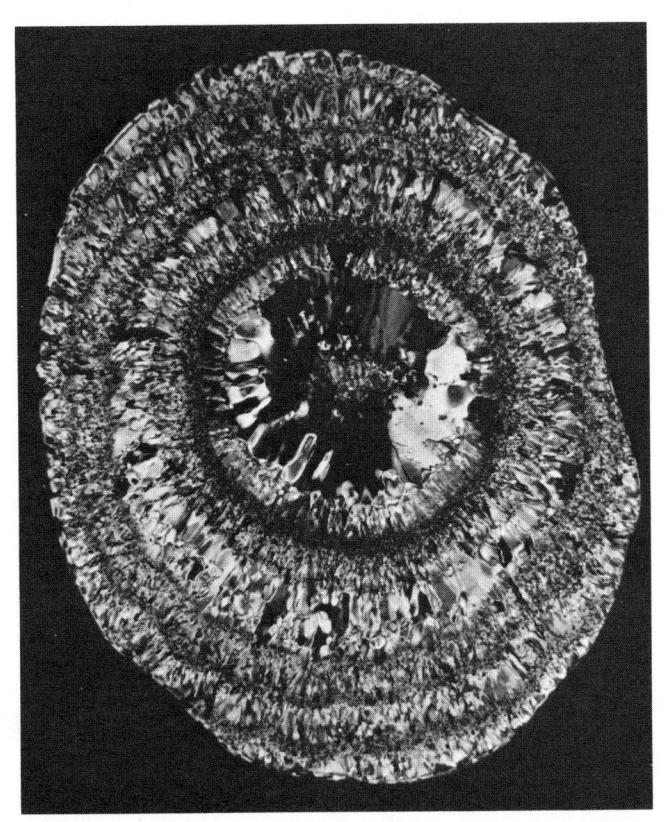

Figure 11-8 A cross section of a hailstone. (Courtesy C. and N. Knight)

hailstorm (Fig. 11-10) can be flattened fields where once lush crops grew. In certain locations, especially close to mountain ranges, hail-producing thunderstorms seem to predominate and farming is almost impossible in these areas.

Figure 11-9 Hailstones larger than golf balls are common in the midwestern part of the United States. (Courtesy C. and N. Knight)

Figure 11-10 Hail damage to a corn field. (Courtesy NOAA)

TYPES OF THUNDERSTORMS

Although all thunderstorms are the result of moisture-laden unstable air, the mechanisms that starts the lifting process may differ. Thunderstorms are classified as two distinct types—*air mass* and *frontal*. The classification is based on the triggering mechanism that produces the local storm.

The most common of all thunderstorms is the air mass thunderstorm, which is mainly caused by heating. Along the boundaries between a large land mass and water body, air mass thunderstorms are quite prevalent, especially in the summer. Warm moist air flows landward during the day because of the sea breeze or the general circulation pattern. This carries moist air to a much hotter land. The land heats up the air and convection currents are created. During a hot summer day it is not impossible to encounter a 25 to 30° rise in temperature on the land, which heats up much faster than the water. Convective thunderstorms are usually scattered and form over the land during the day and coastal area at night.

A second type of air mass thunderstorm is the mountain thunderstorm. On the windward side of mountains, the air is forced aloft, cooled to its saturation point, and unstable. Cumulus clouds quickly grow into the infamous cumulonimbus, and thunder rumbles through the countryside. Pilots are quite wary of flying within a cloud deck in mountainous areas because mountain-type thunderstorms can be embedded in other clouds and never seen until they are encountered.

Frontal thunderstorms are the result of two air masses battling with each other. They can be subdivided into warm and cold frontal types. *Warm-front* thunderstorms are less frequent and usually not as severe as the

Figure 11-11 A thunderstorm model showing the location of a tornado in relation to the wall cloud and precipitation area. (Courtesy Joe Golden, NOAA)

Figure 11-12 Mammatus cloud associated with a thunderstorm.

cold-front variety. When they occur it is usually because a fast-moving warm front is advancing toward an area of cooler air. The warm air is pushed aloft very fast, and when the thunderstorm forms it usually has high bases. *Cold-front* thunderstorms are faster moving and more violent. The cold air wedges beneath the warm air, pushing it aloft. The wedge is quite steep and the warm air is pushed to a very high altitude. The thunderstorms form parallel to the front and can extend for hundreds of miles. In certain instances the advancing cold air moves too fast for the advancing frontal zone. Pressure builds up and an outbreaking of cold air takes place aloft. This air may run ahead of the front by as much as 161 to 241 kilometers (100 to 150 miles). At this point it sinks, pushing the warm air below aloft. This triggers another line of thunderstorms called a prefrontal squall line. The midwestern United States commonly experiences these types of squall lines.

TORNADOES

Of all the destructive storm systems that meander across the face of the earth, *tornadoes* are the most violent. Derived from the Spanish word *tonar*, meaning to turn, these small intense whirlwinds spiral upward, carrying debris aloft at velocities estimated to be near 540 km/hr (335 mph). The center exhibits very low pressure readings because of the partial vacuum created by the whirling wind. The wind usually rotates in a counterclockwise direction in the northern hemisphere, although in certain instances it has been observed to spin in a clockwise manner.

Tornadoes usually form near fast-moving cold fronts, in squall lines, within the confines of a hurricane area, or in a particularly severe air mass thunderstorm. Although meteorologists are not really sure how tornadoes are formed, they believe tornadoes to be an outgrowth of energy exchange between two distinctly different air masses that collide with one another. Within these air masses are layers of air in which temperature, moisture, density, and wind flow characteristics sharply differ at each level. It is known that these components are present within fast-moving cold fronts

Figure 11-13 A funnel cloud popping out of a mammatus cloud.

Figure 11-14 The classical midwestern tornado extending from the base of a cumulonimbus cloud. (Courtesy National Severe Storms Laboratory, NOAA)

Figure 11-16 Windows blown out and the interior a shambles as the result of the extreme pressure gradient in a tornado.

attached to a middle latitude cyclone, especially in the spring. The air masses associated with this system are usually cold dry air from the Canadian Arctic region(cPk) and warm, moist air from the Gulf of Mexico or Caribbean Sea (mTw). Upon impact, unstable air triggers severe, turbulent thunderstorms. These thunderstorms produce a violent updraft with an accompanying convergence of air at the surface (Fig. 11-11).

Usually the first sign of a possible funnel formation is the appearance of *mammatus clouds* (Fig. 11-12). The surface weather feels warm, humid, uncomfortable, and still. As lightning flashes and thunder cracks the mammatus clouds grow larger, indicating *severe turbulence* and *extreme instability*, both prime ingredients for formation. Suddenly, as if from nowhere, a dark extension of the cumulonimbus cloud pouches down in a funnel-shaped appearance (Fig. 11-13). This *funnel cloud*, as it is called, bobs up and down as it extends itself earthward. The condensation of water droplets makes the funnel visible at first, then later, after touching the ground, debris turns the whirling vortex dark.

As the funnel touches the ground, it is reclassified as a tornado (Fig. 11-14). Tornadoes move at speeds ranging between 40 to 60 km/hr (25 to 35 mph) and usually in a direction from southwest to northeast as they skip along the surface. It is not unusual for these tornadoes to skip along, destroying the countryside in a hopscotch manner.

Death and Destruction

Upon impact with the surface these whirling vortices of debris-laden air destroy anything that they come in contact with. Buildings are flattened (Fig. 11-15), cars, animals and people are sucked upward and dumped many miles away, and whole forests are leveled. The destruction comes about from the very strong winds and an explosive effect created by an extreme pressure

Figure 11-15 Debris is all that was left after a tornado swept through this housing complex. (Courtesy National Severe Storms Laboratory, NOAA)

Figure 11-17 Rope tornado. (Courtesy National Severe Storm Laboratory, NOAA)

132 THUNDERSTORMS AND TORNADOES

Figure 11-18 Funnel tornado. (Courtesy National Severe Storms Laboratory, NOAA)

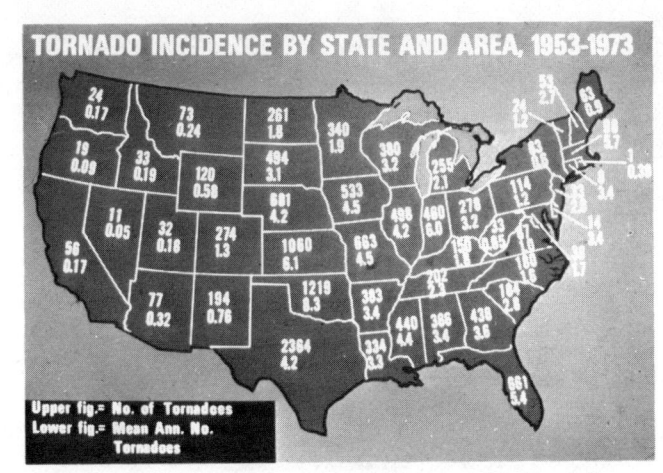

Figure 11-20 Tornado statistics. (Courtesy National Severe Storms Laboratory, NOAA)

gradient. The thin wall of wind tears at the structures, either making them collapse or weakening them. The wind is followed by the tornado cavity which, because of centrifugal force, can have a pressure decrease of more than 100 millibars. This swiftly decreasing pressure causes buildings to literally explode as the normal pressure inside the structure tries to rush to the outside to equalize the pressure imbalance (Fig. 11-16). The pressure differences and associated wind have had some very weird effects. Pieces of straw have been driven through telephone poles, chickens have been completely defeathered, but still alive, paint has been stripped from the sides of houses that were otherwise undamaged, and cars moved miles away from their original location.

Tornado Facts

There are three classified types of tornadoes; rope (Fig. 11-17), funnel (Fig. 11-18), and elephant trunk (Fig. 11-19). They can occur any time of the year in any of the 48 contiguous states and southern Canada. Tornados occur most frequently in the midwestern part of the United States; the southeast United States experiences the second greatest number (Fig. 11-20). Strangely, states west of the Rocky Mountains experience very few tornadoes because they lack the two air masses that serve as triggers and because highly unstable air occurs infrequently.

Tornadoes occur mostly in the southeast during the early spring, with the frequency increasing toward the northwest to the mid-Mississippi Valley by early summer. It has been established that tornadoes usually occur in the late afternoon or evening hours and that they can occur in families extending from the same cloud.

Figure 11-19 Elephant trunk tornado. (Courtesy National Severe Storms Laboratory, NOAA)

Figure 11-21 A waterspout off Key West, Florida. The streaks in the foreground are marine flares. (Courtesy Joe Golden, NOAA)

Many people believe that *waterspouts* are simply tornadoes over water, but this is only partially correct. There are two types of waterspouts, one that develops from a cumulonimbus cloud over the water and the second that builds upward from the water surface. Neither one is as strong as a land-type tornado, and the so-called *fair weather water spout*, which is not associated with a cloud, is the weakest of all.

In the tropics surface water is intensely heated and convection currents occur in cloudless skies. These convection currents lift water from the surface, and the waterspout builds from the bottom upward. They are very small in diameter and can turn in either direction. Like any other system their wind speeds can vary as the convective waterspout dances along the ocean surface. However, once it hits land it dissipates almost immediately, dumping all of its water and debris instantaneously.

Water spouts associated with thunderstorms start at the base of the cloud and work their way downward. They move slowly with a collar of foamy sea water extending upward 1.5 to 3 m (5 to 10 ft). The rest of the spout is fresh water condensed from the atmosphere. In a study made a few years ago, photographs of a waterspout showed a double column of water, one inside the other. The spouts always took the shape of relatively straight columns (Fig. 11-21) and have wind speeds of less than 80 km/hr (50 mph).

No matter what its wind speed, waterspouts can capsize boats, cause excessive isolated salt water rainfall, and even rain fish and frogs.

12
OBSERVATIONS, WEATHER MAPS, AND FORECASTING

James Espy (1785–1860), the first designated meteorologist of the United States. He organized the first observational network in the United States.

LEARNING OBJECTIVES

After reading this chapter, you should be able to:

1. Describe the four steps needed to make a weather forecast.
2. Decode a teletype synoptic weather observation.
3. Using the information in Appendix III, plot a weather station model from a synoptic weather code.

In order for meteorologists to forecast weather conditions for any prolonged length of time, they must go through a logical multistep process. Although this process involves many people, the goal is to predict how the atmosphere will react on a short term basis (24 to 72 hours) or produce a trend for a much longer time period (up to one month).

This process begins with the gathering of raw data by observing current weather conditions. It progresses through the steps of communications, analysis of data, retransmission of weather maps, numerical weather prediction by computer. Eventually a finished forecast is arrived at.

OBSERVING THE WEATHER

A successful forecasting program must have a well-defined system of observing stations. The data obtained from these stations serve as the basis from which all forecasts are made. Most countries of the world have an observation network, but that of the United States is by far the most extensive and well-equipped (Fig. 12-1). The observing stations of the United States include National Weather Service stations, most F.A.A. facilities, military bases around the world, military and merchant marine vessels plying the world's oceans, and even some civilian cooperative observers.

Observations are broken up into two broad categories: surface and upper air. Both types are needed to complete a three-dimensional picture of the atmosphere. Trained observers either measure or estimate all of the elements that make up what we call weather (Table 12-1). However, one major factor must always be kept in mind. In order to make a meaningful comparison of weather from station to station, all observations must be taken at the same time. It is for this reason that Greenwich Mean Time (GMT) is used by every weather service of the world.

Greenwich mean time is the time relative to the actual time at the astronomical observatory located near London, England. This time is sometimes called *Zulu* time, because the letter Z is put at the end of the numerals. All local times will either add or subtract hours in order to have the same GMT time as the time at Greenwich, England, and the local time will depend upon the time zone in which the observation is made. Time zones average 15° of longitude in width whether the location is to the east or west of Greenwich, England. Figure 12-2 shows the differences from GMT for all locations in the world, as well as the location of the International Date Line.

Surface Observations

Ten minutes before each hour, 24 hours every day, trained observers at most weather stations around the

Figure 12-1 Modern meteorological monitoring system that allows for instantaneous readings and a complete graphical analysis of data using both digital display and continuous recording charts.

136 OBSERVATIONS, WEATHER MAPS, AND FORECASTING

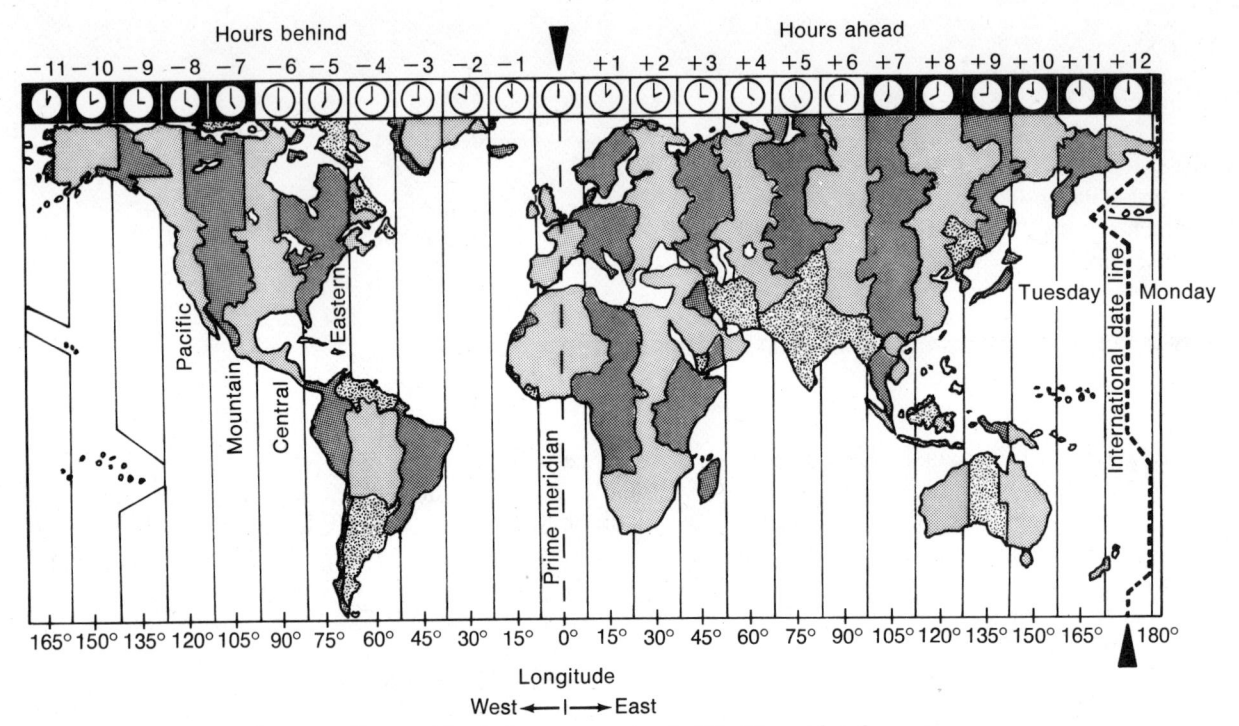

Figure 12-2 Time relative to Greenwich. A time zone extends $7\frac{1}{2}°$ either side of a main-time meridian. On the mainland of the United States there are four time zones (leaving out Alaska and Hawaii): Eastern (75°W), Central (90°W), Mountain (105°W), and Pacific (120°W). Alaska has four time zones within its spread-out area. (From Thomas L. Burrus and Herbert J. Spiegel, *Earth in Crisis*, 2nd Ed., St. Louis, 1980, The C.V. Mosby Co.)

world observe weather conditions at their particular location. At the same time ships at sea and aircraft also take weather observations. These weather conditions represent just one line in the big picture which, when put together, are used to make a forecast. These reports, which are called hourly surface reports or aviation sequence reports, are a very important tool to a pilot who must keep up with the latest factual data at all times and to the meteorologist who must be aware of changes in the weather taking place around the world (Fig. 12-3).

TABLE 12-1 Observed Hourly Weather Elements

Weather Element	How Observed
Cloud height	Estimated or measured by ceilometer or ceiling light
Amount of clouds	Estimated
Horizontal visibility	Estimated by landmarks or measured by transmissometer
Type of weather	Estimated
Barometric pressure	Measured by barometer
Temperature	Measured by thermometer
Dew point temperature	Calculated by using raw data from a psychrometer
Wind direction	Measured by wind vane
Wind speed	Measured by anemometer

Every observation that is taken at an interval of six hours (0000, 0600, 1200, 1800 GMT) contains additional information. These *synoptic observations*, as they are called, are used primarily to draw weather maps for forecasting. They contain information such as the names of the clouds, the past weather, rainfall, and snowfall. Table 12-1 lists the main elements in these observations. Ships at sea also provide additional data such as sea and wave conditions and sea water temperatures.

Upper Air Observations

Equally important to the meteorologist is an understanding of the weather of the upper atmosphere, which we briefly discussed in previous chapters. In the early days of observing, all information obtained above ground level came from manned hot air balloons or small open cockpit airplanes. Eventually, the helium free-flight balloon was developed. It could carry a small package of instruments to heights up to 30.5 km (100,000 ft.). The *rawinsonde*, a radiosonde tracked by radar, as it was called, was made up of a helium balloon, paper parachute, a miniature weather instrument system, and a radio transmitter. As was explained in Chapters 1 and 2, the radiosonde transmits back to earth signals that can be decoded into wind, tempera-

```
SA10 211700

VRB SA 1650 E30 BKN 21/2H 217/86/70/0811/017
MIA SA 1651 30 SCT 250 SCT 3H 208/87/68/0713/015
FLL SA 1651 -X E30 BKN 7 T 200/88/74/1211G16/012 TB24 T W MOVG WSW
OCNL LTGICCG RB05E31
FMY SA 1658 10 SCT 5H 203/95/70/0612/012
PIE SA 1647 30 SCT 6H 89/70/0604/016
EYW SA 1656 20 SCT 200-BKN 7 83/70/0815/012
SRQ SA 1647 10 SCT E 100 BKN 3H 89/71/0810/016
TPA SA 1652 E30 BKN 6H 220/88/73/0910/018
BOW FINA
MLB SA 1650 30 SCT E250 BKN 4H 87/72/1112/018/TCU E
TIX SA 1654 10 SCT E30 BKN 4H 84/M/E1006/022
ORL SA 1651 35 SCT 6H 0909/018/ SCT V BKN
MCO SA 1653 E35 BKN 3H 223/89/70/0911/019/HZY ALF
SFB SA 1647 30 SCT 7 88/72/0910/022
```

Figure 12-3 Teletype sequence of hourly surface observations. These reports are coded for rapid transmission.

ture, pressure, and humidity. By following the balloon by radar, wind direction and speed can be calculated for altitudes above the earth's surface.

Although rawinsondes are still the mainstay of the upper air observational network, technological advances in the atmospheric sciences has led to many other types of upper air reports. One such observation is a radar report which, like the surface report, is taken hourly. Weather radar is primarily used to track storm systems and thunderstorms, but it can also track the movement of cloud systems, measure the intensity of associated weather, and calculate the height and density of clouds. This is quite an important tool since most weather radars can go out as far as 400 km (250 miles) in all directions (Fig. 12-4).

Pilot reports are also a valuable asset for gathering realtime upper air data. Commercial jet aircraft now fly at or near 12 km (40,000 ft) while the supersonic transport (SST) and military jets fly at levels upward to 20 km (65,000 ft). By flying through existing conditions in the upper atmosphere, the pilot can provide precise weather data that cannot be obtained by any other means. Information such as the height of the freezing level, icing conditions, turbulence, wind shear, and heights of upper cloud layers can aid the meteorologist and other pilots immensely.

Special Upper Air Observations

One of the most recent developments in weather technology has been the use of satellites. Although this subject was previously discussed in Chapter 1, it is so important that it should be considered once again. With the launching of the 270 pound TIROS I satellite meteorologists had at their disposal an "eye in the sky." From over 644 km (400 miles) in space came photographs of complete weather systems and their attending cloud patterns. For the first time, meteorological data could be retrieved from areas in which there were no reporting stations, making it possible to obtain a complete global weather picture at a moment's notice.

With each new family of satellites both the quality and quantity of photographs produced had increased. At present, the GOES series of satellites (Geostationary Operational Environmental Satellite), which was first launched in 1974, is performing far above expectations. In many respects they are superior to the previous orbiting platforms because they have the following capabilities:

1. Remaining over the same earth location indefinitely, since it travels west to east at the same speed as the rotating earth.
2. Monitoring the earth's disk face nearly continuously.
3. Takes observations during the dark phase of the earth with its infrared detectors.

They are however, limited in their geographical coverage so it is not possible to observe the entire globe. Figure 12-5 is an example of a cloud photograph obtained from a Geostationary Satellite.

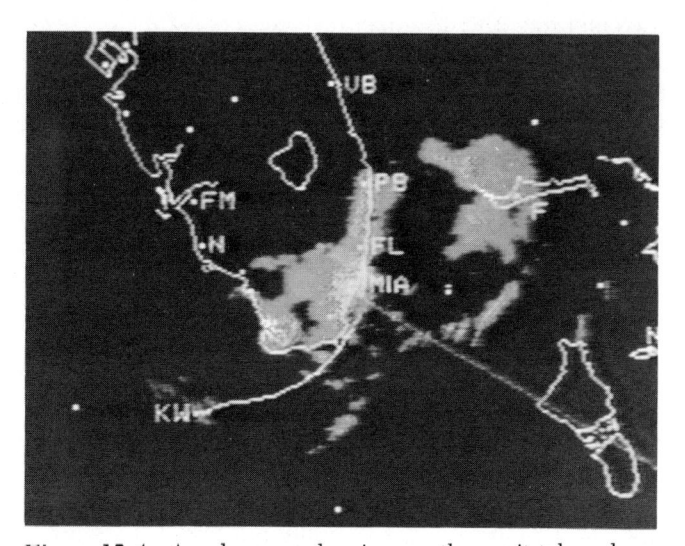

Figure 12-4 A radar scan showing weather as it takes place. Notice the line of thunderstorm activity approaching the station.

138 OBSERVATIONS, WEATHER MAPS, AND FORECASTING

Figure 12-5 Cloud coverage photograph taken by a modern weather satellite. Notice how the cloud patterns spiral around pressure systems. (Courtesy NOAA)

In addition to cloud coverage pictures, weather satellites are capable of producing vertical temperature profiles, snow and ice cover photographs, humidity measurements from the surface to the upper atmosphere, location of sea surface currents, and upper air wind patterns.

The polar-orbiting satellite such as the *ITOS* (Improved TIROS Orbiting Satellite) has automatic picture transmission (APT) capabilities and can transmit photographs immediately to a receiving station as long as it is within a 3200 km (2000 mile) range. Improvements in the processing system allows for both local closeup and global observations at the same time. Thus, the capabilities of the newer satellite systems for gathering quantitative data have enabled the scientific community to place additional emphasis on observational techniques and uses.

The GOES satellite is a case in point. Upper atmosphere wind estimates can be derived from the movement of cloud tops as they are observed from above. By taking two successive pictures and knowing the location of the cloud tops and the time frame of the two pictures, the distance traveled versus time of travel can be used to calculate movement and, therefore, wind estimates.

Meteorological rockets are another means of investigating the upper atmosphere. These rockets contain instrumented packages that explode at predesignated heights. The package floats gently down to earth by parachute, transmitting the needed data by radio. This approach, although highly sophisticated, is very expensive and therefore not widely used. The mainstay for upper air observations is still the rawinsonde balloon supplemented by satellite data.

TRANSMISSION AND COLLECTION OF WEATHER DATA

One of the most important preliminary steps to weather forecasting is the rapid dissemination of current weather to all parts of the world. This is by no means a simple task, for over 8000 land and ship stations are

TRANSMISSION AND COLLECTION OF WEATHER DATA 139

the transmission network and anyone having a teletype receiver will have access to most of these weather data (Fig. 12-6).

The coded observation, or *synoptic code*, is taken every six hours and is transmitted in a five-digit grouping with the first six groups being mandatory in the weather report. However, if additional data are available, new five-numbered groupings are added at the end. Below is the form in which the weather report is transmitted and an explanation of each significant letter or group of letters. Figure 12-7 shows the actual coded data.

$IIiii \quad Nddff \quad VVwwW \quad PPPTT \quad N_hC_LhC_MC_H$

$T_dT_dapp \quad 7RRR_ts \quad 8N_sC_hh_sh_s \quad 9S_pS_pS_pS_p$

$1d_wd_wP_wH_w \quad 2h_{85}h_{85}h_{85}h_{85} \quad 3R_{24}R_{24}R_{24}R_{24}$

$4T_xT_xT_nT_n$

SYNOPTIC WEATHER CODE

Coded Form   **Explanation**

II   Block number showing where the reporting station is located in the world. This block system has been set up by the World Meteorological Organization (WMO). North America has been assigned Blocks 72 and 74.

Figure 12-6 Teletypes are used extensively to receive current weather data. Pictured above are two receiving teletypes within a typical weather data configuration.

simultaneously taking surface observations every hour. Meteorologists employ a numerical code to reduce the time needed to transmit these weather observations. Each one or two number code describes a particular weather situation, and can be transmitted much faster than a word description. The code, which is standardized worldwide, is then transmitted by radio, microwave link, or teletype. The teletype is the mainstay of

```
SMUS1 KA 211800
ROA
72411 81806 66028 22219 6647/ 17720 69804 73620 20036 46760
LYH
72410 81607 74218 23517 854// 13610 69898 70740 20007 46660
RIC
72401 82006 66612 24122 5552/ 12614 63176 70010 47255
ORF
72308 82706 63802 24721 2567/ 08803 63237 70020 47452 TIDE MISSING
WAL
72402 71812 74021 24022 10976 12810 63236 47756
MGM
72226 82704 48052 22429 25572 20807 63152 48571
CHA
72324 82305 40618 22823 3532/ 20805 69983 70520 20005 47471
TYS
72326 82214 56052 22524 854// 19103 69877 47768
AVL
72315 81606 66032 23023 25657 18812 69474 47461
GSO
72317 81205 66299 24617 855// 17807 69928 71110 20011 49610
AHN
72311 61805 61012 23027 55608 17708 69944 48169
GSP
72312 82410 66022 23622 855// 16714 69891 47463
CHS
72208 61709 66021 24627 50878 13810 63227 48063 TIDE PLUS 005
MEI
72234 83205 61022 24128 31607 22008 63130 48266
MCN
72217 71108 59051 23028 72500 20808 63101 48570
SAV
72207 61009 61021 24828 00908 16803 63230 48365 TIDE PLUS 005
MOB
72223 61904 61021 22532 21608 18803 63148 49069 TIDE MINUS 006
TLH
72210 70000 58052 22629 72700 22810 63202 20022 48671
```

Figure 12-7 Typical synoptic observation reports from a teletype printout.

140 OBSERVATIONS, WEATHER MAPS, AND FORECASTING

iii	Station number within its assigned block. There can only by 1000 stations within each block and, since there are more than 1000 weather reporting stations in North America, two block numbers are assigned. See Figure 12-7 for station numbers.
N	The total amount of cloud cover by all clouds in tenths of sky cover. See Table 1 in Appendix III.
dd	The true direction the wind is blowing from in the nearest tens of degrees. Example: 03 = 30° 15 = 150° (*Note:* 00 = calm and 36 = 360°.)
ff	The wind speed in whole knots. One knot is approximately equal to $1\frac{1}{8}$ miles per hour. If the wind is over 99 knots, 50 is added to the wind direction and the speed is reported after 100 has been subtracted from it.
VV	Horizontal visibility plotted to the nearest fraction of a mile up to $3\frac{1}{8}$ and then in whole miles up to 10.
ww	The present weather at the reporting station during the preceding hour. There are 99 different situations and these can be found in Table 2 of Appendix III.
W	The past weather. Weather that has taken place in the past five hours preceding the one hour of the most recent observation. See Table 3 of Appendix III.
PPP	The sea level atmospheric pressure in millibars (tens units and tenths). If the number is a low number, add a 10 in front of it; if it is high, add a 9. Example: 015 = 1001.5 992 = 999.2
TT	The temperature of the air in either whole degrees Celsius or Fahrenheit. At present all United States land stations or ships report temperature in Fahrenheit. If a report is received in Celsius, it should be converted to Fahrenheit so that all observations are comparable. If the temperature is below 0, 50 is added to the absolute value. If over 100, just the tens and units are reported. Example: $-8°F = (-8 + 50) = 42$: $115°F = 15$.
N_h	The cloud cover of the lowest of either a classified low or middle cloud. See Table 4 of Appendix III.
C_1	The name of the low cloud as reported by its description in the International Cloud Atlas. See Table 5 of Appendix III.
h	The height of the lowest cloud above the ground. See Table 6 of Appendix III.
C_M	The name of the middle cloud as reported by its description in the International Cloud Atlas. See Table 7 of Appendix III.
C_H	The name of the high cloud as reported by its description in the International Cloud Atlas. See Table 8 of Appendix III.
T_dT_d	The dew point temperature in whole degrees Celsius or Fahrenheit. See TT for directions.
a	The tendency (up, down, or steady) of the barometer in the past three hours. See Table 9 of Appendix III.
pp	The net change of pressure in units and tenths of millibars in the last three hours.
$7RRR_ts$	The precipitation group. To report the amount of liquid or solid precipitation in the last six hours.
$8N_sC_hh_s$	The amount, height, and genus of clouds in the sky.
$9S_pS_pS_pS_p$	Any special phenomena needing further description. A table is available for this code, although not published in this text.
$1d_wd_wP_wH_w$	Direction, period, and height of ocean waves. Like wind, waves are reported in the direction *from* which they are coming.
$2h_{85}h_{85}h_{85}h_{85}$	Altitude of the 850 mb pressure surface in hundreds, tens, and units of geopotential meters.
$3R_{24}R_{24}R_{24}R_{24}$	The past 24-hour precipitation in tens, units, tenths, and hundredths of inches.
$4T_xT_xT_NT_N$	The maximum (T_xT_x) and minimum (T_NT_N) temperature over a predesignated time period to the nearest whole degree.

Upper air observations are taken every twelve hours, aviation weather reports every hour, radar observations every hour, river reports, special storm reports, and many other special reports; all of which need rapid dissemination. If to these reports are added all the different forecasts that are also sent worldwide, you can see why coding is necessary in weather dissemination.

DRAWING WEATHER MAPS

The third step in the weather forecasting scheme is to redistribute this information in the form of weather maps. This is the function of predesignated national and regional weather centers. In the United States, Washington, D. C. is the national center, and from this point teletype data and weather charts are transmitted on a 24-hour basis.

It is quite understandable that, with millions of pieces of weather data funneling into any one weather office, the meteorologist could not possibly come up with a mental picture of the atmosphere unless he could visualize what was happening. This, of course, is the reason for weather maps. Maps are plotted for both the surface of the earth and the upper atmosphere up to heights of

DRAWING WEATHER MAPS 141

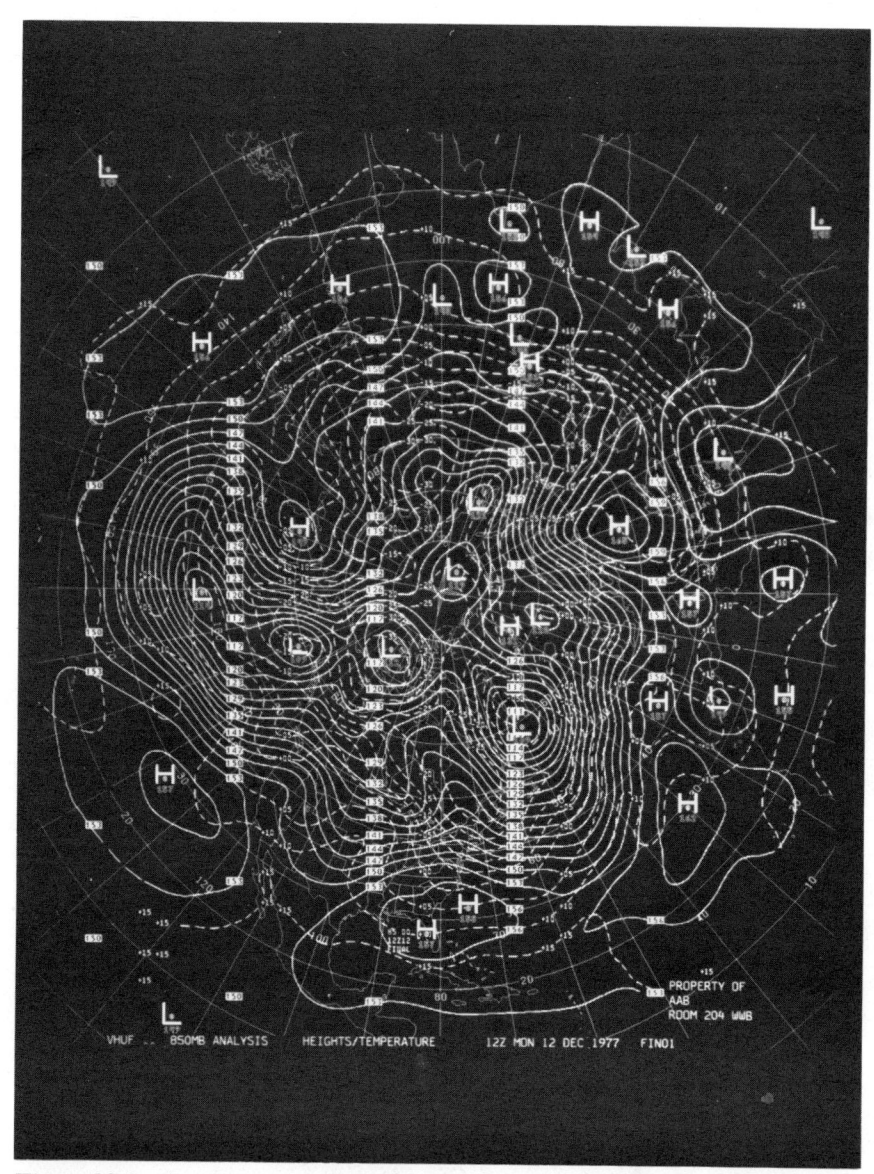

Figure 12-8 A computer-analyzed upper air chart of the heights and temperature at 850 millibars of pressure (approximately 5000 feet). (Courtesy NOAA)

100 millibars. These maps allow the meteorologist to visualize a three-dimensional view of a large area especially those located within his area of responsibility.

Upper Air Maps

For the most part, upper air charts are completed by a preprogrammed computer after it has digested the observational data from the teletype (Fig. 12-8). This upper air chart is then sent back to all weather stations by a *facsimile machine*, an instrument capable of reproducing pictures, maps, and diagrams exactly (Fig. 12-9). This method of producing upper air maps saves much valuable time and money. Before the advent of the computer and facsimiles, each weather station had to plot and analyze its own upper air maps.

Surface Maps

Similar to upper air maps, the compiled summary of surface observations must be put in a logical sequence. This was first done by a German meteorologist named Brandes, who prepared charts of different meteorological measurements.

Today, using the basic techniques of weather mapping that were developed by the early pioneers, meteorologists can produce surface weather maps that depict massive weather systems moving over the surface of the earth. Once again, a system had to be developed so that thousands of bits of raw data could be plotted on a small map. Thus, the plotted station model came into wide use.

The station model is simply a way of presenting the

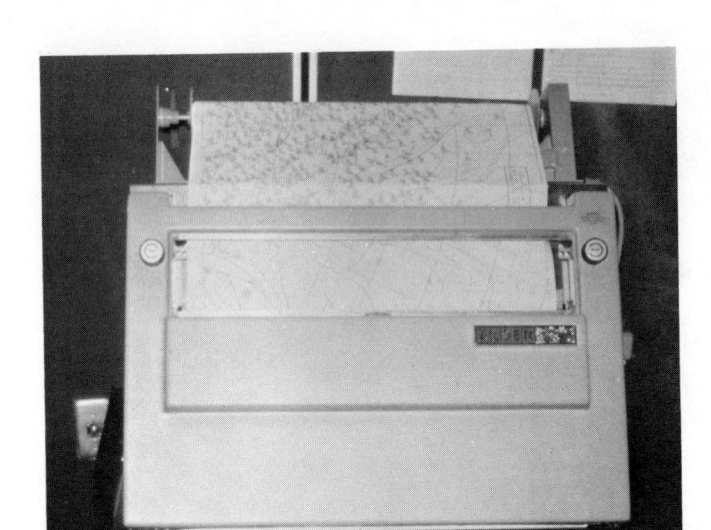

142 OBSERVATIONS, WEATHER MAPS, AND FORECASTING

Figure 12-9 A weather facsimile for the reproduction of weather maps and satellite pictures.

weather data and all nations strictly adhered to it. Therefore, a meteorologist should be able to read a surface weather map anywhere in the world. Figure 12-10 shows the abridged station model used by the National Weather Service.

Map Analysis

Surface weather maps are drawn either by the meteorologist or by computer. In either case, the familiar lines and symbols are placed on the map in such a way that the meteorologist can easily interpret the weather situation at a glance. The primary lines drawn are called *isobars*, which are lines of equal pressure. These smooth curved lines, which are drawn almost parallel to the wind flow, depict the circular patterns of high and low pressure systems. After the isobars are drawn, areas of continuity are separated by a second type of line called a *front*. This line represents a zone of transition between two air masses of distinctly different character-

istics such as cold dry air on one side and warm, moist air on the other. The symbols for frontal representation on the surface map are presented in Table 10-2, as well as the color code for each. It is important to note that the frontal symbols point in the direction in which the invading air is moving or, in the case of a stationary front, the direction in which the air would move if it could. Also represented on a surface map are areas of shading such as green for precipitation, yellow for fog, and red for thunderstorms. High and low pressure areas are colored blue and red, respectively. Using color allows the meteorologist to rapidly pick out those areas of significance. Pressure systems, isobars, and fronts are all interrelated and are shown in Figure 12-11.

WEATHER FORECASTING

By utilizing all the weather data presented on surface and upper air maps, as well as supplementary information from weather satellites, computer models, radar observations, and in some cases, reconnaisance aircraft, meteorologists make the last and most important step, the *forecast*.

Meteorologists trace past surface charts on the most current map in order to determine the past history of pressure systems and frontal boundaries. From examining these past movements, they make estimates of future movements, taking into account all available information. Generally, systems move in a westerly to easterly direction across the United States. These paths, although sometimes erratic in their movement, can be generalized with some reliability. Other facts used in forecasting are pressure tendencies at reporting stations, wind flow, cloud cover, upper air winds, and temperature disparities.

Numerical Forecasting

Weather forecasting today depends heavily on computers and mathematical models of the atmosphere. *Numerical Weather Prediction* (NWP), as it is called,

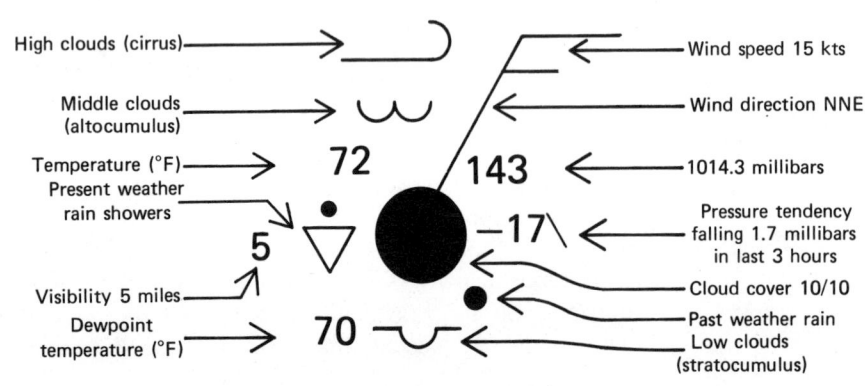

Figure 12-10 Data plotted around a station model.

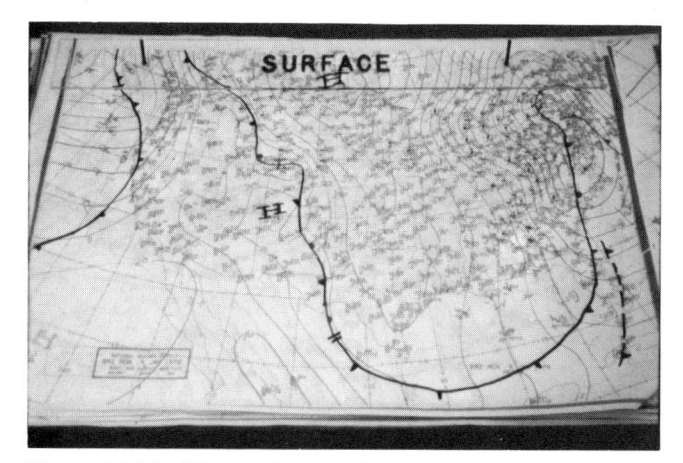

Figure 12-11 The surface weather map.

employs the use of the computer and the known physical laws governing the atmosphere. These laws are described by complex mathematical equations that can only be solved by use of the computer. After all the raw data are gathered, the computer deletes observations it decides are in error and progresses toward solving the equations governing the atmosphere, thus arriving at a forecast of frontal and pressure systems. The NWP method is quite fast when compared to the previous multistep synoptic process, and is believed to be more accurate than other forecasting methods.

Both man-made and numerical weather forecasting techniques are employed by the National Weather Service. However, neither one is completely reliable. It must be realized that the atmosphere is not a static entity. Constant changes are taking place within all weather elements. In addition, because many areas are still devoid of weather observations, and because mathematical models of the atmosphere are still undergoing improvement, errors can very well enter into the forecast.

All in all, the National Weather Service does an outstanding job in weather forecasting. On a short term basis (12 to 24 hours) its percentage of accuracy is about 80%, while intermediate range predictions (24 to 48 hours) are around 70% accurate. Long range forecasting, which are called extended forecasts, can range from a five-day forecast to a 30-day outlook. However, the longer the period the lower the accuracy. In fact, the 30-day outlook will only indicate whether a weather element will be above or below its normal in a given area.

THE FUTURE OF FORECASTING

Everyone throughout the world depends on the weather. Everyday lives are governed by wind, rain, and temperature as we plan our vacation, harvest our crops, travel by air, or even decide what food to eat. Most of the time the decisions we make are based on information supplied by a meteorologist, and he is responsible for up-to-date accurate information. Therefore, it is quite apparent that accuracy in forecasting is the goal of any Weather Service around the world. In the United States the National Weather Service is constantly striving to improve its forecasting techniques by exploring new methods and developing new programs to provide the weather data needed for decision making.

Recently the National Weather Service has developed the *AFOS* system (Automation of Field Operations and Services) (Fig. 12-12). It is designed to allow the meteorologist to retrieve, accumulate, and transmit weather data at fantastic speeds. The system replaces most manual operations as well as the common teletypes and facsimiles. It is possible to upgrade the product being produced because the AFOS system consists of a closed-linked network of stations. These stations include all Weather Service Forecast Offices in the continental United States, the National Meteorological Center, National Hurricane Center, National Severe Storms Forecast Center, and all River Forecast Centers, which in turn are linked to a subsystem of lesser weather service offices.

With many computers connected into a single system, weather data from any point can be stored at any selected site. These data, which includes observations and maps, can be called up for instant viewing on a display console, which is similar to a television screen. This technological breakthrough has allowed the meteorologist to make faster forecasts with much greater accuracy, and to disseminate timely warnings to the public.

Other programs, such as the *World Weather Watch*, were established by the World Meteorological Organization. The goal of this international undertaking was to set up a more complete worldwide observational net-

Figure 12-12 The new AFOS system.

work. It was hoped that this would relieve one of the major problems facing forecasters: large nondata areas. A second international project was the *Global Atmospheric Research Program*, better known as GARP. The objective of this project is to enlist the support by all participating nations in improving numerical weather prediction methods and the accuracy of long range forecasting.

GARP experiments, which have already completed their observational phase are *GATE* (GARP Atlantic Tropical Experiment), to determine the role of cumulus convection in the large scale wind flow; and *FGGE* (First GARP Global Experiment), to provide a detailed global data base for the development and improvement of models used in long range forecasting and climatic prediction.

Such experimental programs could conceivably make obsolete that often-used phrase by Benjamin Franklin, "Everybody talks about the weather, but nobody does anything about it."

13
CLIMATE

Wladmir Koeppen (1846–1940), a German biologist who produced a system of relating climate to vegetation. Since vegetation growth is closely related to rainfall and temperature, the system classified climate according to these two meteorological elements. The Koeppen system for climatic classification is the most widely used method for describing long term weather conditions.

LEARNING OBJECTIVES

After reading this chapter, you should be able to:

1. Distinguish between weather and climate.
2. Explain the mechanisms that control climate.
3. Briefly describe the average winter and summer temperature or precipitation distribution of the United States.
4. Explain the mechanisms believed responsible for deserts, rainforests, monsoons, and glaciers.
5. Compare climate and crop production.
6. Explain how climate affects human existence.
7. State how carbon dioxide levels have previously and presently affected the way we live.
8. Define *heating degree day*, *cooling degree day*, *upwelling*, and *El Nino*.

Weather represents the total effect of meteorological elements such as pressure, wind, temperature, humidity, precipitation, and cloud cover at any particular place at any instant in time. As common experience tells us, these elements change, often quite dramatically, over short periods of time. *Climate*, on the other hand, is the compiling of weather conditions over fairly long periods of time (usually greater than one year).

However, the climate of a region is not determined solely by its long term average properties but also by its variability from average conditions, such as seasonal and monthly variations. This is also consistent with our common everyday experiences, which tell us that the average is often composed of combinations of high and low values.

Nearly all actions of mankind are affected by the climate. It has a powerful influence on the way we conduct our lives, the homes we live in, the clothing we wear, the food we eat, the way we spend our leisure time, and the economic well-being of the community (local or national) in which we live. The importance of climate and its variations is further emphasized by the demands that an increasing world population makes on vital necessities such as food and energy. As such demands approach supply capabilities, society finds itself evermore sensitive to climatic variations.

CLIMATE CONTROLS

Climate varies from place to place with the largest variations occurring in a north-south direction. Almost everyone is familiar with the mild winter climates of the tropical regions, which make them desirable as vacation areas, while at the same time the mid-latitude regions are suffering low temperatures and perhaps snow. This latitudinal effect is due to the annual variation of incoming solar energy, as already explained in Chapter 3. Figure 3-9 is a chart of the daily solar radiation of the top of the atmosphere. In this chart we see that the tropical regions between about 20°N and 20°S receive a fairly steady large amount of insolation. On the other hand, the polar regions have a large variation of incoming energy—zero during the winter months and a maximum during the summer solstice. The mid-latitudes show an intermediate variation. We can also see that a maximum occurs during the summer and minimum during the winter. This latitudinal and seasonal variation of solar energy represents a major control on climate. In fact, if the earth was a nonrotating, perfectly smooth homogeneous surface, the climate in each latitude belt would be completely determined by the incoming solar radiation. This fact was recognized by the early Greeks who used the annual cycle of the sun's inclination to devise the first known climatic classification system. They simply divided each hemisphere into three broad latitudinal zones: the "winterless," "intermediate," and "summerless" corresponding to tropical, mid-latitude, and polar regions.

The earth, however, does not have a smooth homogeneous surface and, while the incoming solar radiation is a dominant climatic control on the surface temperature, there are other important factors that help to determine the climate of a given region. These controls are the distribution of land and water, mountain barriers, altitude, ocean currents, and the large scale circulation of the atmosphere. The latter two items are themselves influenced by some of the preceding controls (e.g., the distributions of continents and oceans influence the intensity of the large scale circulation). Nevertheless, it is convenient to consider them as cli-

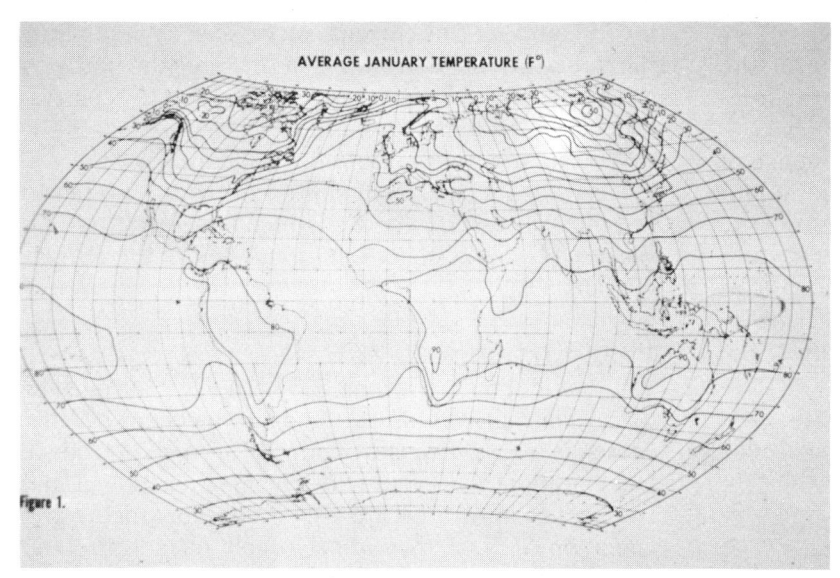

Figure 13-1 The average sea-level temperature for January (°F). (Courtesy NOAA, from *Climates of the World*)

matic controls since, as was described in previous chapters, they have a profound effect on the weather elements that make up the climate of a particular region.

Except for the addition of the large scale circulation of the atmosphere, these factors are essentially the same as those in Chapter 4 in the discussion of the controls on surface temperature. This should not surprise us since these controls are essentially permanent features and, consequently, ideally suited to influencing the climatic distribution of temperature. Because of the interrelationship of temperature with pressure, wind and, to a certain extent, moisture, these controls also affect the distribution of those parameters.

In the following sections we will discuss the global distributions of temperature and rainfall (pressure and winds were considered in earlier chapters), two of the most important of the climatic elements.

DISTRIBUTIONS OF CLIMATIC ELEMENTS

Temperature

The average sea level temperature for January and July are shown in Figures 13-1 and 13-2. The highest temperatures are located in the tropical regions; not exactly on the equator but shifted toward the summer hemi-

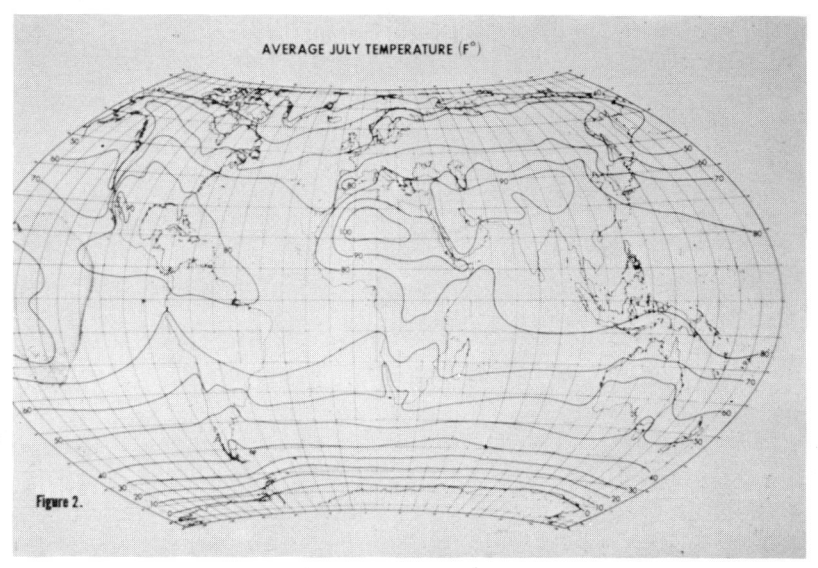

Figure 13-2 The average sea-level temperature for July (°F). (Courtesy of NOAA from *Climates of the World*).

sphere in agreement with the maximum solar energy input of Figure 3-9. The greatest variation occurs in the north-south direction. Over the oceans and in the tropics the isotherms show very little east-west variation. However, in the vicinity of land areas the influence of the land-ocean contrasts are quite evident. Thus in January, in the northern hemisphere, the isotherms in the Pacific Ocean take a large dip southward over North America. The latitudinal dip of the isotherms are not as great in western Europe mainly because the lack of extensive north-south mountain ranges permits the milder warmer maritime air to be carried further inland. The southward movement of the isotherms over North America and Asia are related to the weather disturbances that bring cold continental Arctic air toward the equator.

In July in the northern hemisphere, the water is cooler than the land masses, and the isotherms tend to bend toward the pole when crossing the boundary from water to land. This is particularly noticeable in North America.

In the southern hemisphere, during both seasons, the isotherms tend to be oriented along latitude lines. This is because of the large expanse of homogeneous ocean surface and shows the influence of solar energy on the temperature pattern.

The influence of ocean currents is also evident in Figures 13-1 and 13-2. The sharp northward bend of the isotherms off the east coast of North America in January is due to the warm Gulf Stream. As discussed in Chapter 4, the Gulf Stream carries warm water across the Atlantic Ocean toward the British Isles. The warming influence of this current can be seen by comparing the January temperatures (Fig. 13-1) along the northern coast of Great Britain (about 40°F) with the temperature along the Labrador coast ($-10°F$).

We see another example of the influence of ocean currents in the sharp dip southward of the isotherms just off the west coast of the United States in July. This is caused by the cool north-to-south flow of the California current.

Precipitation

The distribution of precipitation is largely controlled by the availability of moisture and the mechanisms that cause air to rise vertically; both are needed to produce condensation and eventually precipitation. Local and large scale circulations are important in producing precipitation because they affect supply of moisture and create vertical motions. Orographic features, such as mountains, also are important controls because they can mechanically force air to rise causing condensation and precipitation.

Figure 13-3 shows the worldwide annual precipitation in inches. As might be expected, the tropical regions exhibit large amounts of precipitation mainly because of the great amount of moisture available. This is the result mainly of the large scale convergence of air in the equatorial region that causes upward motion. This region is known as the *intertropical convergence zone**

---

* The intertropical convergence zone frequently coincides with the equatorial trough region discussed in Chapter 7, and is often used interchangeably.

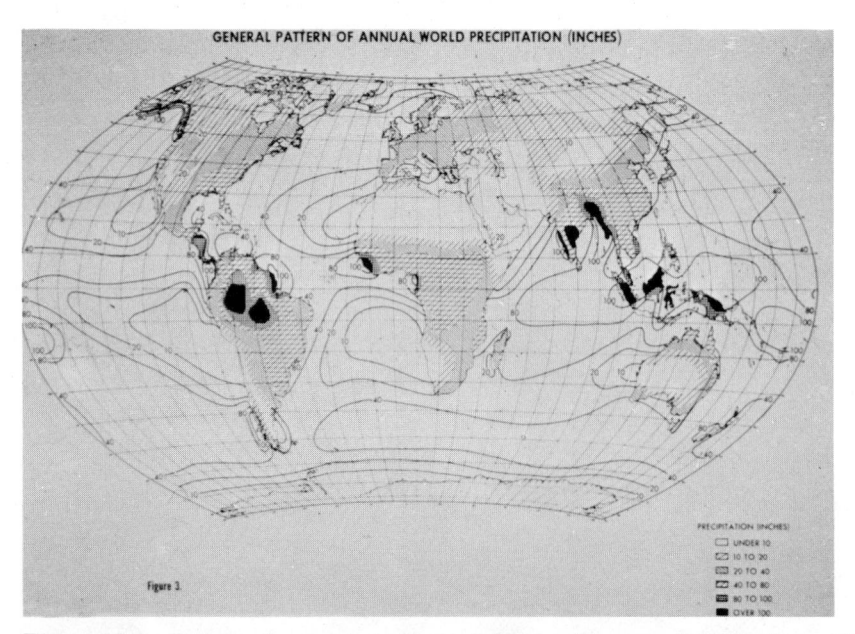

Figure 13-3 Pattern of annual world precipitation. (Courtesy NOAA, from *Climates of the World*).

DISTRIBUTION OF CLIMATIC ELEMENTS   149

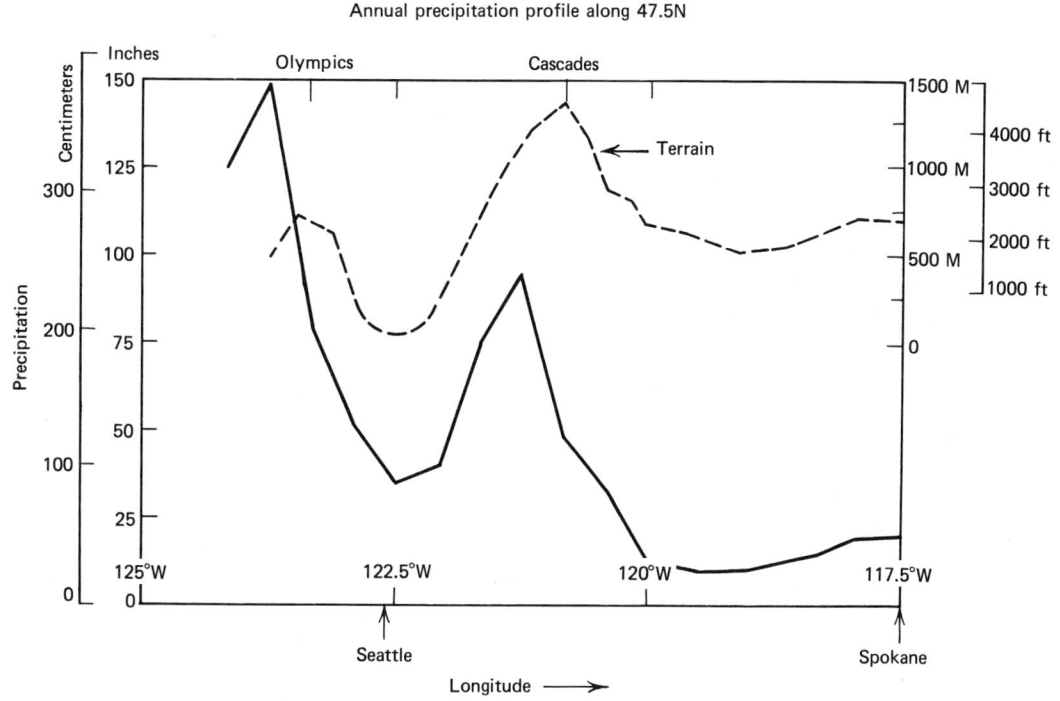

Figure 13-4 Annual precipitation profile from west to east along 47.5N. Also plotted is the elevation profile. Notice how the maximum in precipitation occurs just before the highest mountain peaks, clearly showing the effect of orography on precipitation.

and is where the trade winds of both hemispheres meet. This large scale control of precipitation is influenced by orographic effects, such as the Andes in western Brazil and by local convection caused by intense solar heating, as in Indonesia and in the Amazon River Basin. It is also influenced by seasonal shifts in the latitude of the convergence zone. This shift is most pronounced over the land. It shifts from north of the equator during the northern hemisphere summer to south of the equator during the northern hemisphere winter.

Located between latitudes 20 to 30° north and south are some of the driest areas on earth. These dry regions are caused by the domination of large scale high pressure belts that have downward motion associated with them. This results in circulation that prevents condensation and precipitation from taking place.

There are two areas noteworthy for their high precipitation. The first is the Pacific Northwest area of the United States, where every year 150 inches of rainfall is caused by orographic lifting. In this location the mountain chains are oriented in a north-south direction, approximately perpendicular to the general westerly wind flow. Thus, the coast and Cascade ranges tend to block the flow of moisture-laden air flowing from the Pacific Ocean. The sharp rise in elevation forces the air upward on the western side of the mountain barrier, causing condensation and precipitation to take place and provides the area with the needed ingredient for lush growth of vegetation. However, on the leeward side we find rain-deprived areas and deserts. The effect of this mountain barrier on precipitation is shown in Figure 13-4, which illustrates the east-west variation of rainfall in the Pacific northwest. On the coast lies the Olympic mountains. Their high annual rainfall (about 150 in.) results in a vegetation growth called the *rain forest* (dense vegetation and towering trees). See Figure 13-5.

As we move eastward toward Seattle, we find that the precipitation decreases to about 30 inches/year. Progressing further east we encounter the Cascades and an increase in rainfall to about 100 inches/year. Finally, eastward across the Cascades, the rainfall dramatically decreases to about 10 inches/year, before starting to increase again.

The second area of note is the west coast of India, whose maximum of rainfall is a result of a large scale differential heating between land and sea. Recall that the different properties of land and ocean cause the oceans to heat and cool more slowly than the land. This means that ocean surface temperatures have only a small seasonal variation whereas land surfaces undergo large temperature changes. These differences result in ocean surface air that is relatively cool in summer and warm in winter as compared to air over land surfaces. The warm air, which is less dense than its surroundings, rises and allows the cooler ocean air to replace it. Thus, there is a tendency for an inflow of air over conti-

150 CLIMATE

Figure 13-5 Rain forest in the Olympic National Park, Washington. (Courtesy of the U.S. Department of Interior, National Park Service).

nents during the summer and an outflow of air from continents in winter. This is a large scale response to the heating and is known as a monsoon effect. The best-known example of such a circulation is the Indian monsoon shown in Figure 13-6. There the outflow of air takes place during the winter months and inflow during the summer months. The inflowing air during the summer draws in considerable moisture, and rainfall during the monsoon period is quite abundant. Figure 13-7 shows this dramatic change in monthly precipitation for Bombay, India, which is characteristic of a monsoon regime. Note the high precipitation from June through September, the period when the monsoon circulation is well developed. In those four months about 94% of the annual precipitation falls. The fact that this concentration of rainfall is limited to a few months of the year has an important economic implication for the region. For example, should the monsoon rains fail to develop, the water necessary for agriculture will be unavailable—crops will fail and serious food shortages and economic disruption will follow. Unfortunately, such a situation has happened in the past and may occur in the future. It is one example of climate variability and the significance of such an influence.

CLIMATE VARIABILITY AND CHANGE

The geologic record has provided us with evidence of considerable global temperature variations with a range of about 5°C between cold and warm periods. These periods were of variable length but averaged on the order of 10,000 years. The cold periods are known as *glacial periods* and were characterized by extensive ice sheets; the warm periods, known as *inter-glacials*, were believed to be warmer than our present global temperatures. In fact, many scientists believe that the world is now in a warm interglacial period.

Within these long climatic regimes there exist variations of temperatures on much smaller time scales. For example, between 1000 and 1200 A.D. it has been estimated that global average temperatures may have been as much as 1°C higher than they are today. Between about 1450 and 1850 A.D. global temperatures were estimated to be about 1°C below average twentieth century values. Although 1°C does not appear to be a large reduction in average temperature, the expansion of glaciers in North America, Europe, and Asia took place during this period. Other effects such as more severe winters and shorter summers were felt in Europe. This period is known as the *Little Ice Age*.

During the twentieth century surface temperature from about 1900 to 1940 has shown an upward warming trend of about 0.5°C. From 1940 until the present average temperature has been decreasing.

The above discussion indicates that climate is not a static phenomena. It is variable and changing on a variety of time scales. If we include seasonal variations, then climate can be considered as having variations ranging from tenths of years to hundreds of thousands to millions of years. Consequently, when we define climate by averaging weather variables, it is important to specify the period over which the averages are taken.

As a practical matter the World Meteorological Organization has established 30 years as the period over which averages are to be made. These averages are called climate "normals" and are updated every 10 years. The current climatic normal period is 1951 to 1980 and the next will be 1961 to 1990.

Scientists study the variability in climate by examining departures of climatic variables, such as temperature and precipitation, from these normal values. Depending on the length of the time period being studied, longer or even shorter time periods are used to calculate a climatic normal. The important point is that climatic variations usually refer to a clearly defined normal period. Examples of such variability are shown in Figure 13-8, which is the January temperatures for Washington, D. C., Miami, Florida, and Anchorage, Alaska for the period 1971–1978. The temperature is shown plotted as a departure from the normal (1941–1970) temperatures.

Notice the very large departures in January, 1977. Miami and Washington experience much below normal temperatures while Anchorage is about 11.0°C above normal. In general, this month was quite abnor-

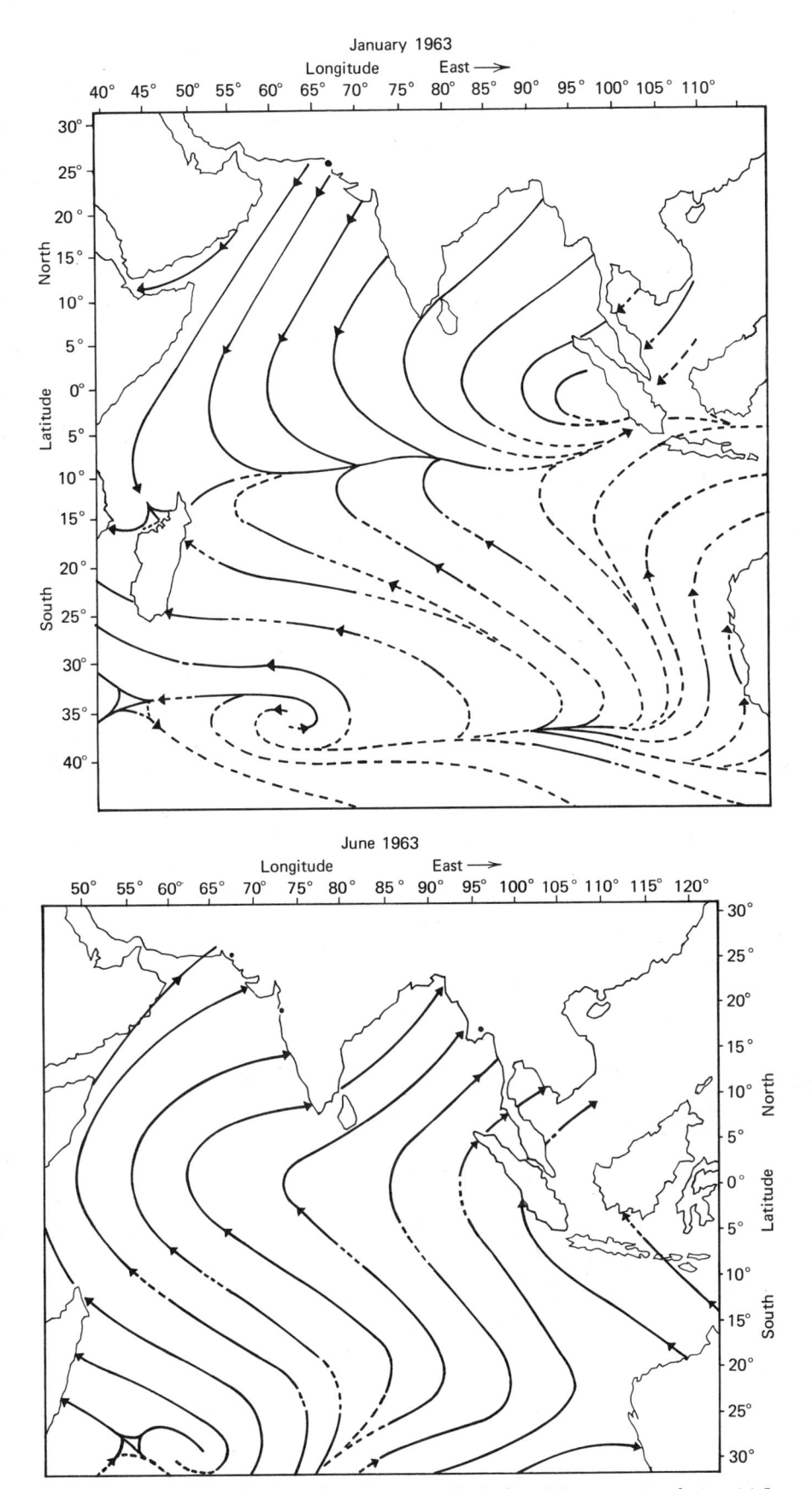

Figure 13-6 The sea-level air flow depicting the Indian Monsoon circulation (*a*) January, (*b*) June. The lines with arrows are streamlines and show the direction of wind flow. No streamlines are shown over the continent because the elevations are above sea level. Note how the air flows away from the continent in January and onto the continents in June. (Courtesy National Science Foundation)

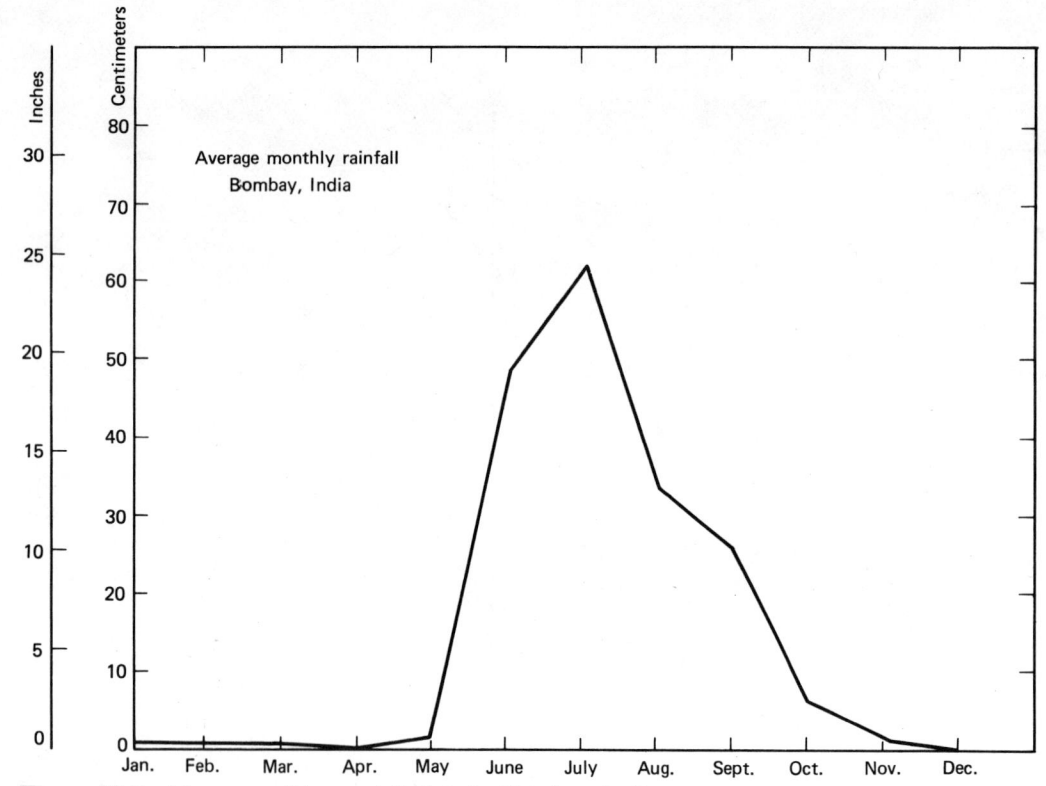

Figure 13-7 Mean monthly precipitation for Bombay, India.

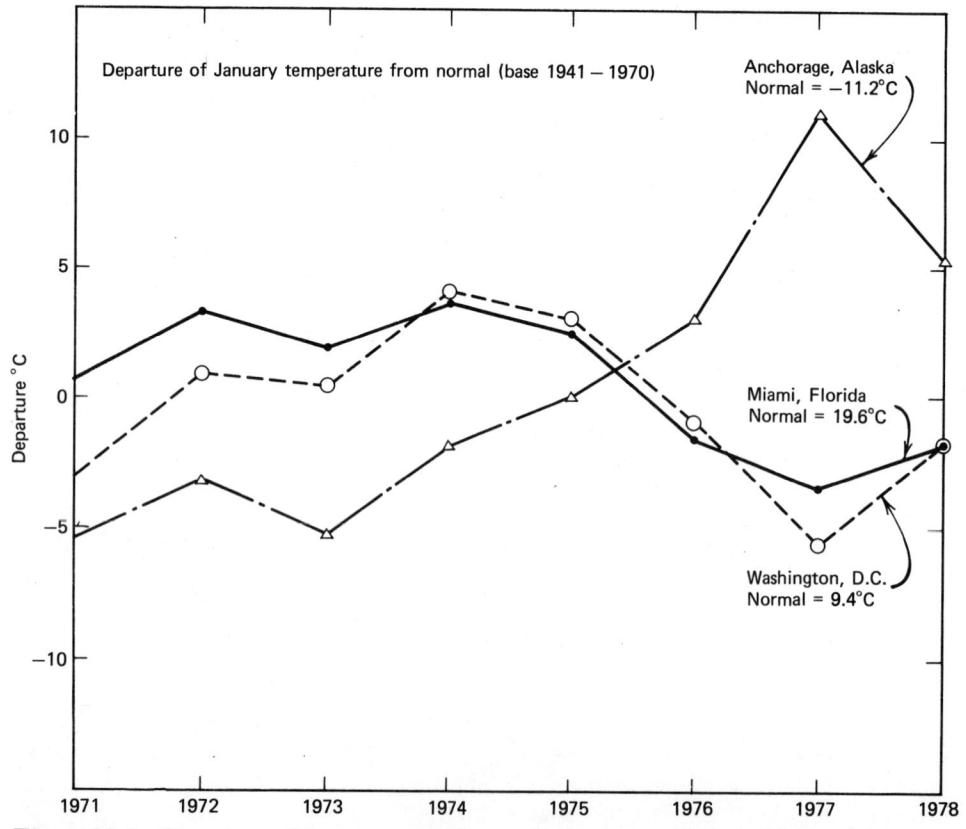

Figure 13-8 Departure of January temperatures from normal (Base 1941–1970) for Anchorage, Alaska, Washington, D.C., and Miami, Florida.

mal. It was the coldest month on record east of the Rocky Mountains, and the extremely cold weather caused many economic and social hardships. As a measure of the unusual weather that prevailed snow was recorded for the first time in Miami and Palm Beach, Florida. The Ohio River was frozen for the first time since 1948 and huge ice floes jammed the Mississippi River, bringing river traffic virtually to a halt. Similar disruptions occurred throughout the country. Natural gas was in short supply because of both the great demand and the disrupted transportation that prevented gas delivery. Because of the shortages, schools and factories had to shut down and at one point almost 200,000 people found themselves temporarily unemployed.

The cold weather was particularly difficult for Florida because it severely damaged citrus and vegetable crops. Economic losses are difficult to judge but estimates indicated losses of 100s of millions of dollars. This agricultural loss combined with the shutdown of industrial plants, and its associated wage losses throughout the rest of the country indicates how climate can have an impact on the very life blood of a nation.

An important aspect of Figure 13-8 is the tendency for compensation to take place from one year or group of years to the next. For example, while Alaska was above normal in 1976–1978, it was below normal during the period 1971–1974 and normal in 1975. Washington, D. C. and Miami, Florida also show this same compensation. Figure 13-8 shows that compensation also occurs spatially. For example, while Alaska was unusually warm, Washington, D. C. and Miami, Florida were cold. This compensation is typical and emphasizes why long and complete data records are required to detect a significant climate change. As a practical matter the compensation effect makes any permanent climate change very slow and difficult to detect. Consequently, people are more sensitive to seasonal and annual climatic changes than to longer period variations.

CLIMATE'S EFFECT ON SOCIETY

In this section we will discuss briefly some significant aspects of climate's effect on people. We will emphasize the repercussions that climatic fluctuations have on food and energy supplies. Food supply is a problem that mankind has always faced, and energy supply is of particular concern in today's highly industrialized world.

Climate Variability and Agriculture

The efficient production of grains requires a combination of favorable temperatures and moisture supply. Currently, most of the world's grain is produced between latitudes 30° and 55° in both the northern and southern hemispheres. This area is essentially determined by the high summer temperatures equatorward of 30° and the low temperatures poleward of 55° latitude. Both high temperatures equatorward and low temperatures poleward result in a limited *growing season*. The growing season is determined by the beginning of above-freezing weather in spring and the first killing frost in fall.

Variations in climate that affect the growing season can have a major impact on the production of food. For example, R. Bryson* has estimated that a drop of 1°C in the mean annual temperature in Iceland could reduce the length of the growing season by 14 days and that a drop of 2.4°C would decrease the growing season 40 days. He also showed the effect of climate by comparing hay yields in Iceland in the late 1950s with yields during 1966 and 1967 when the average temperature was about 0.8°C lower. Yields per acre in 1966 and 1967 were 26% lower than the late 1950s despite a 10% increase in the use of fertilizer. In this instance, climate variability has had a greater influence on crop production than technology, as represented by the use of fertilizers.

This is not an isolated phenomena. During the global warming period, which took place in the first 40 years of the twentieth century, the growing season in England increased by 2 to 3 weeks and during the cooling trend since 1940 a shortening of the growing season has occurred.

Since proper amounts of moisture is the other critical parameter for food growth, the occurrence of drought conditions has a major impact on agricultural production. In North America the long and severe drought conditions of the 1930s led to extensive and costly crop losses. In Figure 13-9 we can judge the magnitude of the drought from the annual average rainfall in Kansas for the period 1921–1950. Note the large decrease in rainfall during the 1930s. The average rainfall during the period 1921–1930 was 71 cm (28 in.), during 1931–1940 58 cm (22.8 in.), and during 1941–1950 78 cm (30.8 in.). There have been other years of drought in North America but none were as widespread and persistent as the drought experienced in the 1930s (Fig. 13-10).

Variations in climate also affect the production of food from the sea. A dramatic example is the change of sea surface temperature and ocean currents off the west coast of South America, known as El Niño. El Niño is characterized by higher sea surface temperatures and large rainfall in the coastal areas of Peru and Ecuador. The most significant effect of El Niño is that the higher sea surface temperatures result in a lowering of the anchovy population. For example, during El Niño, which

---

* Reid A. Bryson, "A Perspective on Climate Change," *Science*, Vol. 184, No. 4138 (1974).

154 CLIMATE

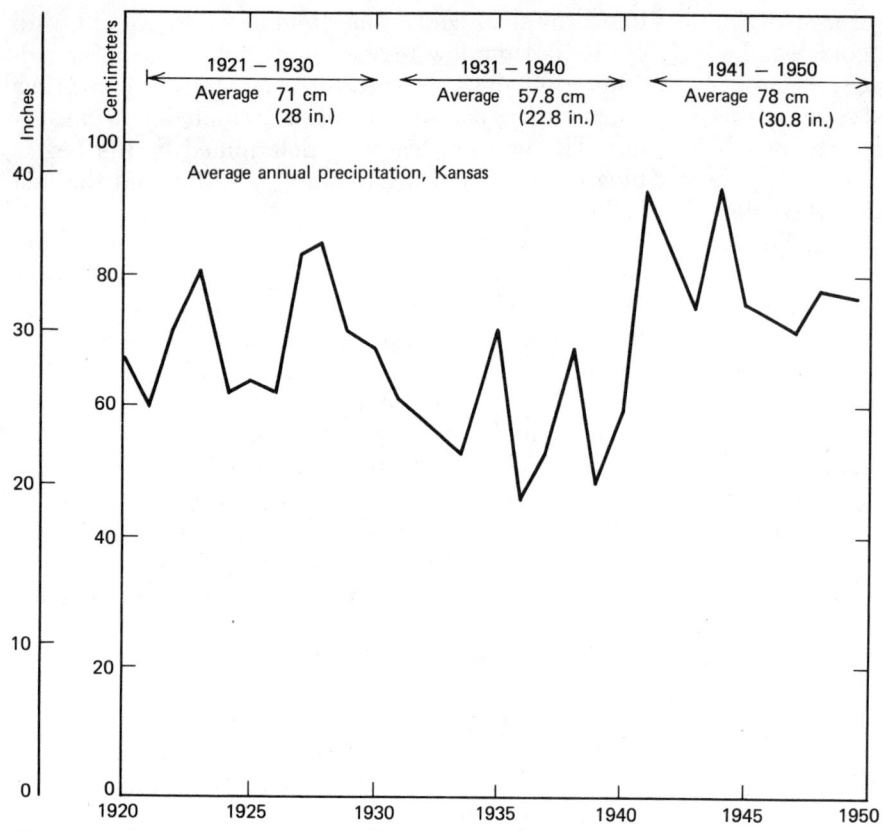

Figure 13-9 Average annual rainfall for Kansas for the period 1920 through 1950. Averages for the 1921–1930. 1931–1940, and 1941–1950 are also shown.

started in November, 1972, the anchovy catch decreased to about 2 million tons as compared to 12 million tons in 1970. Much of this catch goes into the production of fishmeal, which provides feed for poultry and other livestock. The effect of the greatly diminished catches resulted in higher prices for fishmeal, leading to higher costs of raising poultry and other livestock, and ultimately to higher costs to the consumer.

Scientists are uncertain about the exact cause of El Niño, but it appears to be related to large scale circulation changes in the atmosphere. Under normal conditions the surface winds in the eastern Pacific, which blow from the south and southeast are strong enough so that the ocean currents they produce cause an *upwelling* in a narrow region near the coasts of Ecuador and Peru. *Upwelling* is a process that brings deep cool water to the surface. During El Niño the surface winds are weaker and less upwelling occurs or ceases entirely. This allows the surface water to warm up through a combination of solar heating and the drift of warmer ocean surface water towards the coast.

In this example a climatic variation in the atmosphere affects the circulation of the oceans that, in turn, has a significant effect on an important source of food.

Figure 13-10 Drought conditions during the severe and prolonged drought of the 1930s in the United States of America. (Courtesy NOAA)

Climate and Energy

In a highly modernized industrial society, energy is consumed both for personal (e.g., heating, cooling, cooking) and industrial use. Such a society is highly dependent on both adequate fuel supplies and reliable delivery systems so that energy is available on demand. Climate variability is an important factor in determining adequate supplies of energy. It affects the consumption, distribution and, for nonfossil fuels, the production of energy.

Consumption. The consumption of energy, particularly for home heating and cooling, is quite sensitive to temperature. This sensitivity has led to the concept of a *heating degree day* and a *cooling degree day*, as measures of fuel consumption.

As used in the United States, one heating degree day is given for each one degree that the daily average temperature is below 65°F. For example, a daily average temperature of 60°F represents 5 heating degree days. A cooling degree day is given for each degree that the daily average temperature is above 65°F. Thus, a daily average temperature of 80°F represents 15 cooling degree days.

The amount of fuel consumed for heating and cooling during the appropriate season can be predicted fairly accurately based on the average number of heating and cooling degree days. Fuel suppliers and power companies use this information in determining how much fuel to purchase. Should there be unusually cold or warm conditions a power company may be unable to provide enough fuel (gas, oil, or electricity) to serve its customers. Within recent years major urban centers on the east coast of the United States have experienced summer *brown outs*. This is a condition of reduced voltage when the demand for electricity is greater than the supply. Brown outs often occur when summer conditions are unusually hot and humid. During these times electrical consumption for air conditioners, refrigeration, and other appliances increase beyond the available supply. If such demand exists for long enough periods of time a *black out* (total electrical failure) may occur.

During abnormally cold winter months the demand on gas supplies for home heating may be so large that large industrial users may be denied this fuel. This has resulted in closing of manufacturing plants and even schools, as described in the previous section.

Distribution. Gas, fuel oil, and coal are delivered by distributors and power companies using pipelines, trucks, rail, and barge. Under adverse weather conditions these modes of delivery can be severely hampered, or even halted, as during the severe North American winter of January 1977. Thus, at the very time such fuel was in great demand, power companies and distributors were unable to get the fuel they needed. In many other instances adequate supplies existed, but they were on a barge locked in ice or on a snowbound truck or railroad freight car.

Power Production. Climate can also affect the production of energy, especially the nonfossil fuels. For example, the production of hydroelectric power is very sensitive to the amount of snow fall during the winter months. A decrease of the snow fall can result in a significant decrease in the generation of hydroelectricity during the spring. In fact, power companies have attempted to augment the snow fall by various weather modification techniques.

As fossil fuels become more scarce, more and more attention is being given to alternative sources of energy such as solar- and wind-generated electricity. The availability of energy from these sources are more sensitive to weather and climate than fossil fuels, since they depend on the distributions of cloudiness and the reliability of the wind.

MAN'S EFFECT ON CLIMATE

In the preceding sections we noted how the variations in climate can have a dominating effect on society's activities and needs. We can, at times, protect ourselves from variations in climate but, for the most part, we are subject to whatever conditions natural forces have in store for us.

However, some human activities can effect the climate, both on a worldwide and regional basis. The effects mostly have to do with our society's advanced technological state and the enormous consumptions of fossil fuels that inject both gaseous and particulate matter into the atmosphere. In earlier chapters we briefly mentioned the possible effects of carbon dioxide (CO_2) and aerosols on atmospheric temperature. In this section we discuss the effects of such pollutants in greater detail and present some possible results of man's activities on climate. At this juncture it should be remembered that scientists have only recently begun to study some of these problems and that there are still great uncertainties about their ultimate effect.

Regional Climatic Effects. As we would expect, man has had a significant and dramatic effect on climate in urban areas, where the concentration of people is quite high.

Cities affect the climate through the addition of heat by combustion and other energy uses, increased particulate content of the atmosphere, and a response to solar heating that is markedly different from rural areas. This response is caused by the large amounts of paved areas in cities. The fact that cities are warmer on the average than the countryside has given rise to the concept of the "urban heat island." The urban heat island is an area with temperatures that are significantly higher than its surroundings. In large cities the temperature may even be high enough to be detected by satellites, as shown in Figure 13-11. In this photograph the cities show up as dark areas (warm) relative to its surroundings. It has been estimated[*] that on the aver-

---

[*] W. H. Matthews, W. W. Kellog, and G. D. Robinson, Ed, *Man's Impact on the Climate*, MIT Press, Cambridge, Mass., 1971.

156 CLIMATE

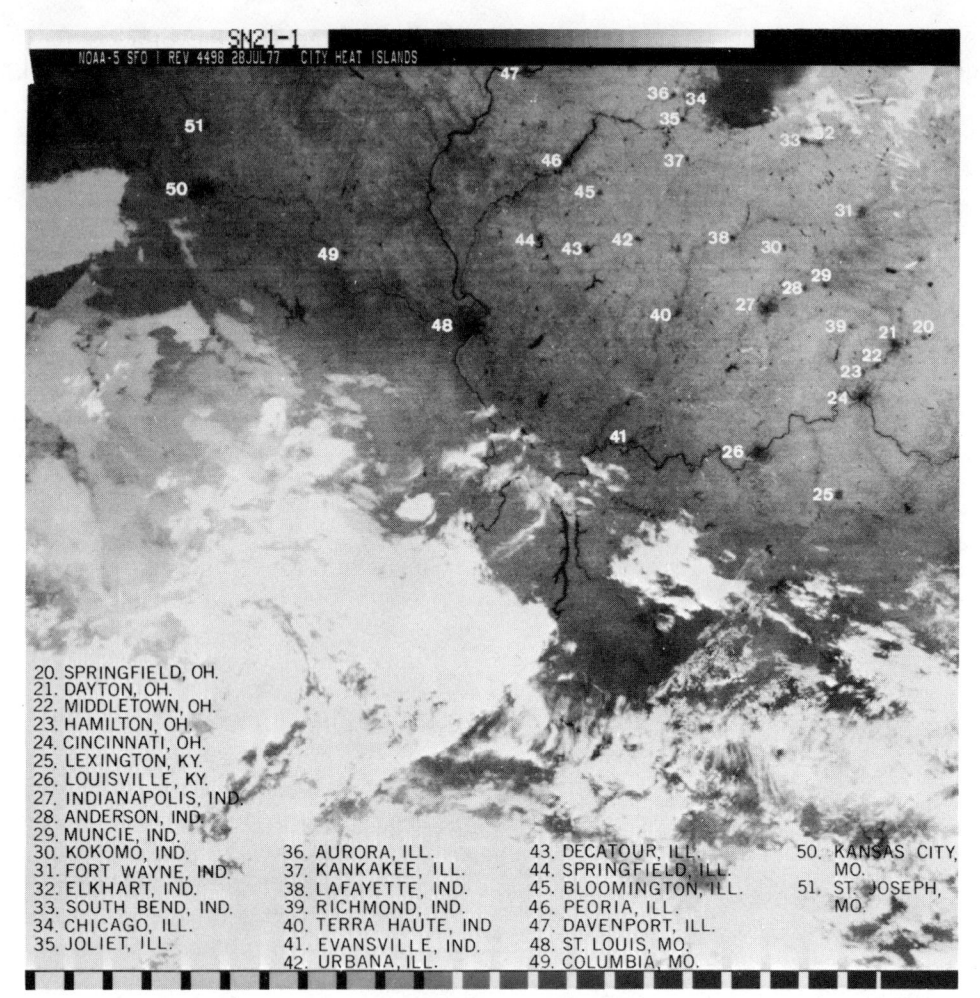

Figure 13-11 The urban heat islands as observed by satellites on July 28, 1977. The dark areas are warm relative to their surroundings. Notice how even small cities show up quite vividly. (Photograph from Matson et al., National Environmental Satellite Service, NOAA)

age, winter minimum temperatures in cities are about 1° to 1.7°C (2 to 3°F) higher than the countryside, and in some instances the differences may be as much as 10°C (18°F) between country and city.

Urban areas also discharge particulate matter that act as cloud condensation nuclei. Such nuclei have resulted in increased precipitation downwind from cities, as discussed in Chapter 2, where we presented some results of the Metromet research program being conducted near St. Louis, Missouri. Recently scientists have reported on increased precipitation downwind from Paris, France and Chicago, Illinois.*

GLOBAL OR WORLDWIDE EFFECTS

In this section we will be concerned with the dramatic effects on the earth's climate that CO_2, aerosols, and the release of heat may have. The criteria used to judge the impact of these parameters on climate are discovered by studying how they can change the heat balance of the earth-atmosphere system.

Release of Heat

In some large cities the direct release of heat into the atmosphere can exceed the amount of energy absorbed from the sun. In other words, the release of heat from human activity is in competition to the main climatic driving source, the sun. However, on a global basis, it has been estimated[†] that the total amount of heat added by human activities is about $\frac{1}{10,000}$ of the solar energy of the sun. This amount is so small that it should have a negligible effect on the overall heat balance of the earth. In other words, the earth should not experience any significant temperature change.

---

* J. Dettwiller and S. A. Chagnon, Jr., Possible Urban Effects on Maximum Daily Rainfall Rates at Paris, St. Louis, and Chicago., *J. Appl. Meteor.*, 15 (1976), 517–519.

† W. W. Kellog, Effects of Human Activities on Global Climate. *WMO Technical Note No. 156*, 1977.

Carbon Dioxide (CO_2)

Carbon dioxide is a combustion product of all fuels containing carbon, such as coal, oil, gasoline, methane, natural gas, wood, and propane. Obviously, in an industrial society with a high dependence on such fuels, large amounts of carbon dioxide are continuously being produced.

The concern with the role of carbon dioxide and its affect on climate comes about because of its ability to alter the heat balance of the earth-atmosphere system. As we described in Chapter 2, carbon dioxide is capable of absorbing and reemitting significant amounts of radiation emitted from the earth's surface and passing this heat through the atmosphere. A buildup of this gas can result in an increased absorption of energy from the earth's surface. The increased absorption will result in a greater reemission of heat toward the earth's surface and thus a higher surface temperature.

Will such reemission mean an actual increase in temperature? How much CO_2 do we have to put into the atmosphere before it is of concern? And what happens when the temperature of the surface increases? Answers to these questions are not easy to find. However, scientists using sophisticated mathematical models to simulate climate have provided us with some tentative answers.[*]

First, a doubling of the CO_2 content from the present would cause the average surface temperature to change upward about 1.5 to 3.0°C (2.7 to 5.4°F). The models also tend to show that the temperature changes in the polar regions would be greater, perhaps by a factor of 2 or 3, than the global average. We could then expect changes in the extent of glaciers, snow, and ice cover. Enormous amounts of water are tied up in glaciers and the polar caps. If they were to melt, sea level would rise and low-lying coastal areas would find themselves under water. We could also expect significant effects on rainfall and the length of the growing season. The best guess is that the total rainfall would increase and, of course, with higher surface temperatures we would expect the growing season to increase.

Note that these effects are based on models that do not simulate the real world precisely. Therefore it is uncertain that such changes will actually take place. And, even if the models are realistic, the details could be

---

[*] S. H. Schneider, On the Carbon Dioxide-Climate Confusion, *J. Atmos. Sci.*, Vol. 32 (1975), 2060–2066.

Figure 13-12 An outbreak of dust from Africa as observed by geostationary satellite on July 1, 1973. Notice how the ocean, which is normally dark, appears lighter when covered by dust. In this case the dust results in an increase in albedo. (Courtesy of the National Environmental Satellite Service, NOAA)

quite unreliable. For example, precipitation may increase, but it may increase in areas that are now desert and decrease in other areas that normally have abundant rainfall.

The next question is how likely is it that the carbon dioxide will increase and at what rate will the increase occur?

Measurements indicate that today there is about 13 percent more CO_2 in the atmosphere than 100 years ago. Estimates of the rate of addition of CO_2 to the atmosphere indicate that by about 2050 A.D. the atmospheric CO_2 content will be double what it is today.* If the models have any reality at all we can expect significant changes in the global climate by the middle of the next century.

Aerosols

Human activity also adds to the atmosphere particles known as aerosols. These particles, such as smoke, dirt, and gases are produced by industrial processes, automobiles, heating, and agricultural practices. The number of these particles exceed any of the natural particles found in the atmosphere, such as wind-blown dust or sand. Episodes of wind-blown sand and dust can cover thousands of square kilometers and are easily observed in satellite images (Fig. 13-12).

The role of aerosol particles in the heat balance is complicated and not completely understood. This is because aerosols can cause either a warming or cooling. This paradox comes about in the following way. Aerosols can both absorb solar energy, which leads to a warming, and scatter and reflect this same solar energy, resulting in a cooling. Whether or not warming or cooling takes place depends on how efficient the particles absorb or scatter energy. It is also affected by the *albedo* (reflectivity) of the underlying surface. The issue is further complicated because aerosols also absorb and retransmit long wave infrared radiation, thus having a similar effect to that of carbon dioxide. However, this effect is believed to be of secondary importance. In general, when aerosols are over a dark surface such as the ocean, they are likely to increase the albedo (see Fig. 13-12) leading to cooling, and when they are over a light surface such as snow and some land types, they will have a warming effect.

Many scientists believe that on a global basis aerosols cause a cooling, although the magnitude of this cooling is uncertain. However, most aerosols are produced in industrial areas and, therefore, the concentrations are highest over land areas where they generally have a warming effect. Since aerosols wash out with rainfall, their average residence time in the atmosphere is about 5 to 7 days, they do not get a chance to get widely distributed. Thus, some scientists believe that the net effect of aerosols will be a warming one.

Clearly scientists believe aerosols can be an important factor in affecting the earth's climate despite the large uncertainty on whether their net effect is warming or cooling. Part of this uncertainty is related to the lack of knowledge about the distribution of aerosols around the globe and an incomplete knowledge of the different types of particles found in the atmosphere and the different physical properties those aerosols possess.

In conclusion it should be emphasized that long before man became an important factor, large climatic changes occurred, some perhaps even larger than any predicted as a result of human influence. In fact, as it was already pointed out, present records indicate that the atmosphere has been cooling over the last 30 years, even though the carbon dioxide content has been increasing. It seems that natural variations may still be a dominant factor. However, if we continue to consume fossil fuels at the present rate, it may very well come to pass, in the not too distant future, that human beings will change their climate radically.

---

* W. W. Kellog, *Effects of Human Activities on Global Climate*. World Meteorological Organization TN, (1977), 156.

Appendix I
BLACKBODY RADIATION

As described in Chapter 3, a blackbody is an ideal radiating body that gives off radiation according to its temperature only. Although there is a rather broad and complicated theory of black body radiation, for our purposes we will discuss only two laws that describe the behavior of black bodies: the Stefan–Boltzman and Wien displacement laws.

The Stefan-Boltzman law states that the total amount of energy (E) emitted by a black body is proportional to its absolute temperature raised to the fourth power (T^4). In equation form

$$E = \sigma T^4 \qquad (1)$$

where σ is a constant equal to 5.735×10^{-8} watts/m² °K⁴ and T is in °K.

The Wien displacement law states that the wavelength at which the emitted radiation is a maximum is inversely proportional to the absolute temperature. That is, the lower the temperature the longer the wavelength of maximum radiation. In equation form this is expressed as

$$\lambda_{max} = \frac{C}{T} \qquad (2)$$

where C is a constant equal to 2844 μm °K.

The significance of the preceding two laws can be seen by comparing the incoming solar energy to the energy emitted by the earth-atmosphere system. The sun radiates as a black body with a temperature of about 6200°K. Therefore it emits a large amount of energy with a maximum near a wavelength of 0.48 μm. This is in the visible part of the spectrum. Alternatively, the earth-atmosphere system has a fairly low radiating temperature (about 255°K). Consequently, the amount of energy is low compared to the sun, but the wavelength is large, about 11 μm, well into the infrared part of the spectrum.

Appendix II
GAS LAWS AND THE HYDROSTATIC EQUATION

GAS LAWS

To gain some insight into the behavior of gases consider a small volume (V) that contains gas of mass (M) at a temperature (T) and pressure (P). If the volume and mass are kept constant, the pressure of the gas will respond to a temperature change, which can be expressed as

$$P/T = C \qquad (1)$$

where C is a constant. This is known as Charles' law, named after Jacques Charles who discovered the relationship. It states that a rise in temperature results in a rise in pressure so that C can remain constant. This law makes physical sense because an increase in temperature means an increase in molecular motion and in the bombardment of the molecules, with a resulting increase in pressure.

If we start over and keep the mass (M) and temperature (T) constant a relationship between pressure (P) and volume (V) can be expressed as

$$P \times V = K \qquad (2)$$

where K is a constant. This relationship, known as Boyle's law was discovered by Robert Boyle and states that when air pressure increases air volume decreases. Note that since the mass is kept constant Boyle's law can be expressed as $P/d = K'$ where K' is a new constant and d is the density of the gas. This equation states that as the density increases the pressure increases and vice-versa.

This also makes physical sense because as a volume of air decreases the same molecules are confined to a smaller space, resulting in higher density, which results in greater molecular bombardments and an increase in pressure.

By combining both laws together the equation of state, a law describing the relationship between pressure, temperature and density of air, was developed. This equation is expressed as

$$P = dRT \qquad (3)$$

where P air pressure, d air density, T air temperature, and R is a constant. It can be seen that by holding T constant we obtain Boyle's law and by holding d constant we obtain Charles' law.

HYDROSTATIC EQUATION

The hydrostatic equation is a formula that expresses how the pressure varies with height. It can best be understood by referring to Figure II-1 and starting with the definition of atmospheric pressure as the force or weight of air in a column per area. This can be expressed as

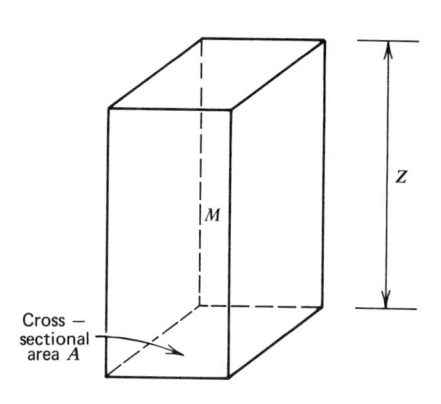

$$P = \frac{Mg}{A} \qquad (4)$$

Figure II-1 Air column of height Z, cross-sectional area A, and containing mass M. Its volume V = AZ.

162 HYDROSTATIC EQUATION

where M is the mass of air in the column, A the cross-sectional area of the column, and g the acceleration of gravity. The weight or force is, as previously stated, Mg. Multiplying both the numerator and denominator by Z, the height of the air column, we obtain

$$P = \frac{M}{AZ} \times gZ \qquad (5)$$

The volume of the column is, by definition AZ and the density (d)* of the air in the column, also by definition is M/AZ. Therefore, the above formula can be rewritten as

$$P = dgZ \qquad (6)$$

This is more convenient than the first formula because it expresses pressure in parameters (d,Z) that are easier to measure than cross-sectional area (A) and mass (M). The difference in pressure between two levels (Fig. II-2) is the difference between the weight of air above those two levels. Using the previous formula, we can write it as

$$P_1 - P_2 = dgZ_1 - dgZ_2 = dg(Z_1 - Z_2) \qquad (7)$$

where the subscripts refer to the level. This is a simple form of the *hydrostatic equation* and indicates that the vertical change in pressure is equal to the average air density (d) times the acceleration of gravity (g) times the change in height, $(Z_1 - Z_2)$, between level 1 and level 2.

* For very tall columns the density may not be the same at all heights. Under those circumstances d would be the average air density of the column.

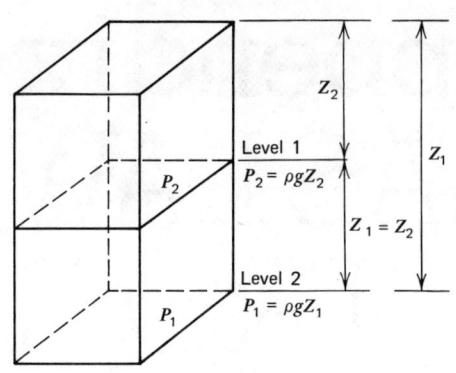

Figure II-2 A diagram showing how pressure P varies with height Z in the standard atmosphere.

An alternative but more useful form of this equation is obtained when the air density d is replaced by P and T from the equation of state (3). This results in a somewhat more complicated expression shown below.

$$\frac{P_1 - P_2}{P_1} = \frac{g}{RT}(Z_2 - Z_1) \qquad (8)$$

The term on the left side is the percent change in pressure; and g/R is a constant. Thus, this formula states that the percent change in pressure is equal to the change in height divided by the average temperature $(\overline{T})$ in the air column between levels 1 and 2. This is equivalent to the previous form of the hydrostatic equation because cold air (high density) will have a larger change in pressure than warm air (low density) for a given height change.

Appendix III
DAILY WEATHER MAP DECODING SYMBOLS AND TABLES

TABLE 1   Total Sky Cover
TABLE 2   Present Weather
TABLE 3   Past Weather
TABLE 4   Sky Coverage of Low and/or Middle Clouds
TABLE 5   Description of Low Clouds
TABLE 6   Height of Low/Middle Clouds
TABLE 7   Description of Middle Clouds
TABLE 8   Description of High Clouds
TABLE 9   Barometric Tendency

TABLE 1   **Total Sky Cover**

Code Number	Meaning	Map Symbol
0	No clouds	○
1	Less than one tenth or one tenth	◔
2	Two tenths or three tenths	◐
3	Four tenths	◑
4	Five tenths	◐
5	Six tenths	◕
6	Seven tenths or eight tenths	◕
7	Nine tenths or overcast with openings	◉
8	Completely overcast	●
9	Sky obscured	⊗

164 DAILY WEATHER MAP DECODING SYMBOLS AND TABLES

TABLE 2 Present Weather

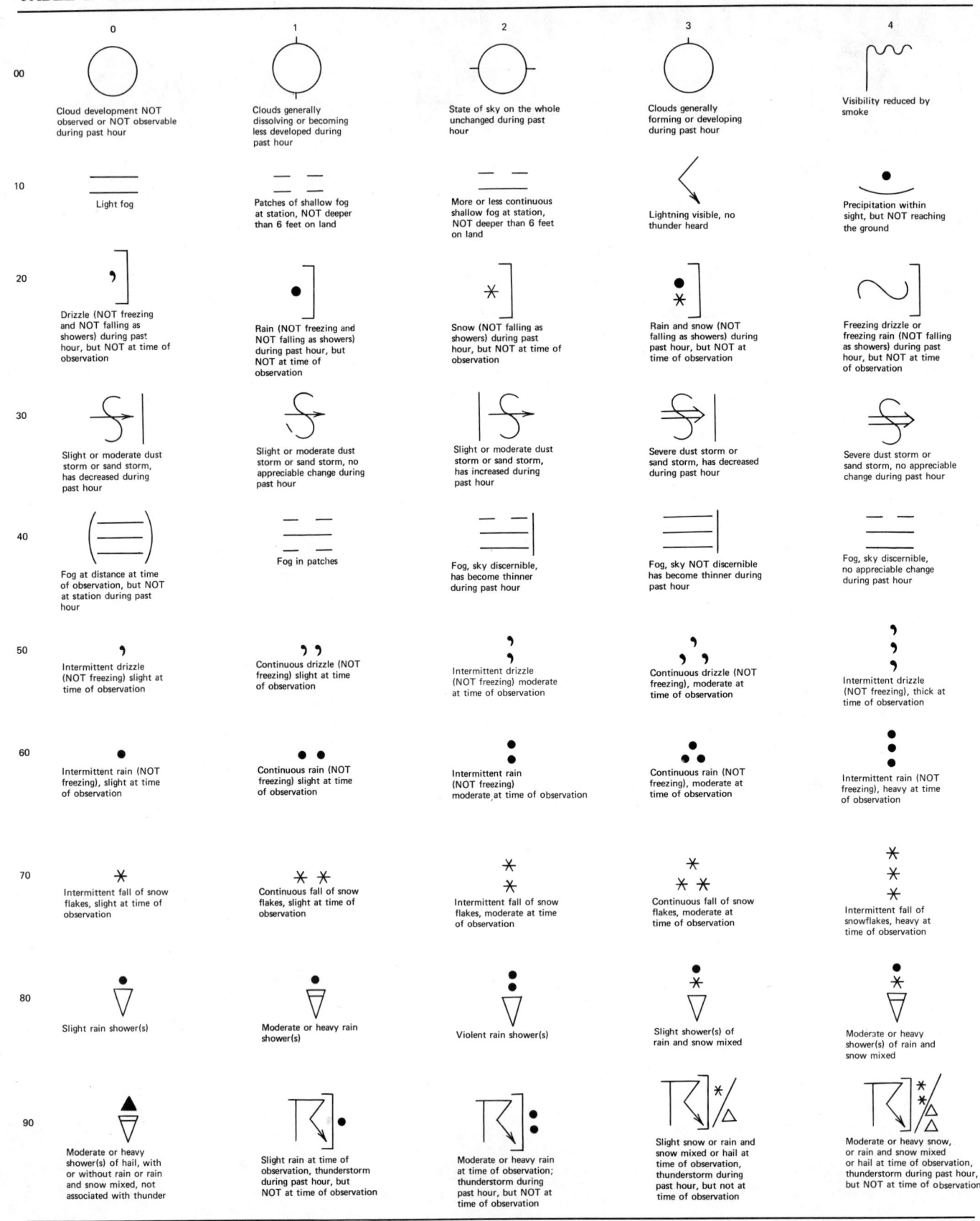

	0	1	2	3	4
00	Cloud development NOT observed or NOT observable during past hour	Clouds generally dissolving or becoming less developed during past hour	State of sky on the whole unchanged during past hour	Clouds generally forming or developing during past hour	Visibility reduced by smoke
10	Light fog	Patches of shallow fog at station, NOT deeper than 6 feet on land	More or less continuous shallow fog at station, NOT deeper than 6 feet on land	Lightning visible, no thunder heard	Precipitation within sight, but NOT reaching the ground
20	Drizzle (NOT freezing and NOT falling as showers) during past hour, but NOT at time of observation	Rain (NOT freezing and NOT falling as showers) during past hour, but NOT at time of observation	Snow (NOT falling as showers) during past hour, but NOT at time of observation	Rain and snow (NOT falling as showers) during past hour, but NOT at time of observation	Freezing drizzle or freezing rain (NOT falling as showers) during past hour, but NOT at time of observation
30	Slight or moderate dust storm or sand storm, has decreased during past hour	Slight or moderate dust storm or sand storm, no appreciable change during past hour	Slight or moderate dust storm or sand storm, has increased during past hour	Severe dust storm or sand storm, has decreased during past hour	Severe dust storm or sand storm, no appreciable change during past hour
40	Fog at distance at time of observation, but NOT at station during past hour	Fog in patches	Fog, sky discernible, has become thinner during past hour	Fog, sky NOT discernible has become thinner during past hour	Fog, sky discernible, no appreciable change during past hour
50	Intermittent drizzle (NOT freezing) slight at time of observation	Continuous drizzle (NOT freezing) slight at time of observation	Intermittent drizzle (NOT freezing) moderate at time of observation	Continuous drizzle (NOT freezing), moderate at time of observation	Intermittent drizzle (NOT freezing), thick at time of observation
60	Intermittent rain (NOT freezing), slight at time of observation	Continuous rain (NOT freezing) slight at time of observation	Intermittent rain (NOT freezing) moderate at time of observation	Continuous rain (NOT freezing), moderate at time of observation	Intermittent rain (NOT freezing), heavy at time of observation
70	Intermittent fall of snow flakes, slight at time of observation	Continuous fall of snow flakes, slight at time of observation	Intermittent fall of snow flakes, moderate at time of observation	Continuous fall of snow flakes, moderate at time of observation	Intermittent fall of snowflakes, heavy at time of observation
80	Slight rain shower(s)	Moderate or heavy rain shower(s)	Violent rain shower(s)	Slight shower(s) of rain and snow mixed	Moderate or heavy shower(s) of rain and snow mixed
90	Moderate or heavy shower(s) of hail, with or without rain or rain and snow mixed, not associated with thunder	Slight rain at time of observation, thunderstorm during past hour, but NOT at time of observation	Moderate or heavy rain at time of observation; thunderstorm during past hour, but NOT at time of observation	Slight snow or rain and snow mixed or hail at time of observation, thunderstorm during past hour, but not at time of observation	Moderate or heavy snow, or rain and snow mixed or hail at time of observation, thunderstorm during past hour, but NOT at time of observation

DAILY WEATHER MAP DECODING SYMBOLS AND TABLES 165

TABLE 2 (Continued)

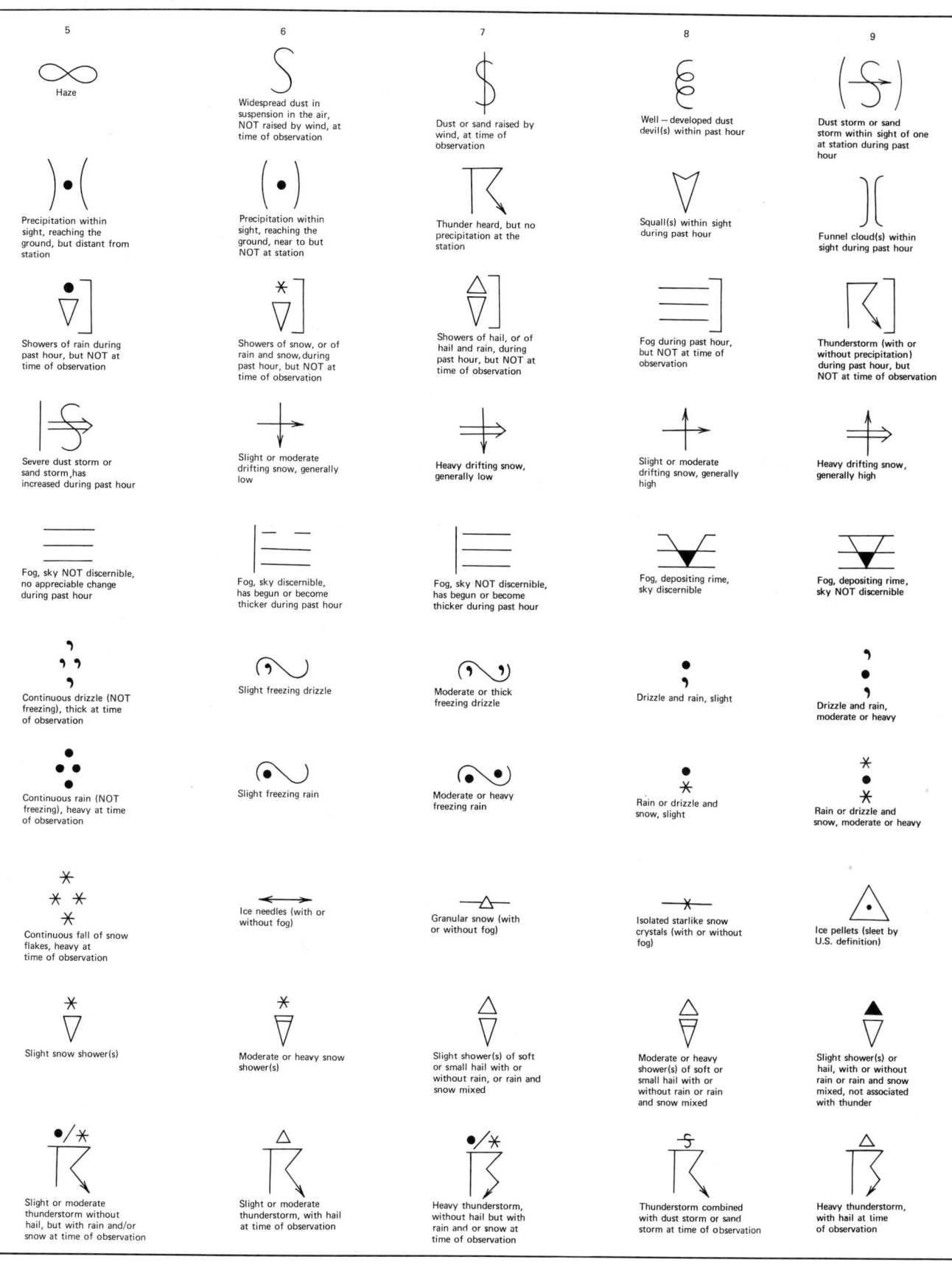

5	6	7	8	9
Haze	Widespread dust in suspension in the air, NOT raised by wind, at time of observation	Dust or sand raised by wind, at time of observation	Well-developed dust devil(s) within past hour	Dust storm or sand storm within sight of one at station during past hour
Precipitation within sight, reaching the ground, but distant from station	Precipitation within sight, reaching the ground, near to but NOT at station	Thunder heard, but no precipitation at the station	Squall(s) within sight during past hour	Funnel cloud(s) within sight during past hour
Showers of rain during past hour, but NOT at time of observation	Showers of snow, or of rain and snow, during past hour, but NOT at time of observation	Showers of hail, or of hail and rain, during past hour, but NOT at time of observation	Fog during past hour, but NOT at time of observation	Thunderstorm (with or without precipitation) during past hour, but NOT at time of observation
Severe dust storm or sand storm, has increased during past hour	Slight or moderate drifting snow, generally low	Heavy drifting snow, generally low	Slight or moderate drifting snow, generally high	Heavy drifting snow, generally high
Fog, sky NOT discernible, no appreciable change during past hour	Fog, sky discernible, has begun or become thicker during past hour	Fog, sky NOT discernible, has begun or become thicker during past hour	Fog, depositing rime, sky discernible	Fog, depositing rime, sky NOT discernible
Continuous drizzle (NOT freezing), thick at time of observation	Slight freezing drizzle	Moderate or thick freezing drizzle	Drizzle and rain, slight	Drizzle and rain, moderate or heavy
Continuous rain (NOT freezing), heavy at time of observation	Slight freezing rain	Moderate or heavy freezing rain	Rain or drizzle and snow, slight	Rain or drizzle and snow, moderate or heavy
Continuous fall of snow flakes, heavy at time of observation	Ice needles (with or without fog)	Granular snow (with or without fog)	Isolated starlike snow crystals (with or without fog)	Ice pellets (sleet by U.S. definition)
Slight snow shower(s)	Moderate or heavy snow shower(s)	Slight shower(s) of soft or small hail with or without rain, or rain and snow mixed	Moderate or heavy shower(s) of soft or small hail with or without rain or rain and snow mixed	Slight shower(s) or hail, with or without rain or rain and snow mixed, not associated with thunder
Slight or moderate thunderstorm without hail, but with rain and/or snow at time of observation	Slight or moderate thunderstorm, with hail at time of observation	Heavy thunderstorm, without rain but with rain and or snow at time of observation	Thunderstorm combined with dust storm or sand storm at time of observation	Heavy thunderstorm, with hail at time of observation

166 DAILY WEATHER MAP DECODING SYMBOLS AND TABLES

TABLE 3 Past Weather

Code Number	Meaning	Symbol
0	Clear or few clouds	Not plotted
1	Partly cloudy (scattered) or variable sky	Not plotted
2	Cloudy (broken) or overcast	Not plotted
3	Sandstorm or duststorm, or drifting or blowing snow	S/
4	Fog, or smoke, or thick dust haze	≡
5	Drizzle	,
6	Rain	•
7	Snow, or rain and snow mixed, or ice pellets (sleet)	*
8	Shower(s)	▽
9	Thunderstorm, with or without precipitation	⍋

TABLE 4 Sky Coverage of Low and/or Middle Clouds

Code Number	Meaning
0	No clouds
1	Less than one tenth or one tenth
2	Two tenths or three tenths
3	Four tenths
4	Five tenths
5	Six tenths
6	Seven tenths or eight tenths
7	Nine tenths or overcast with openings
8	Completely overcast
9	Sky obscured

TABLE 5 Description of Low Clouds

Code Number	Description	Map Symbol
1	Cu of fair weather, little vertical development, and seemingly flattened	⌒
2	Cu of considerable development, generally towering, with or without other Cu or Sc bases all at same level	⌂
3	Cb with tops lacking clear-cut outlines, but distinctly not cirriform or anvil-shaped; with or without Cu, Sc, or St	⌂ / -o-
4	Sc formed by spreading out of Cu; Cu often present also	
5	Sc not formed by spreading out of Cu	⌣
6	St or Fs or both, but no Fs of bad weather	—
7	Fs and/or Fc of bad weather (scud)	- - -
8	Cu and Sc (not formed by spreading out of Cu) with bases at different levels	⌣/⌒
9	Cb having a clearly fibrous (cirriform) top, often anvil-shaped, with or without Cu, Sc, St, or scud	⌂

TABLE 6 Height of Low/Middle Clouds

Code Number	Height in Feet (rounded off)	Height in Meters (approximate)
0	0–149	0–49
1	150–299	50–99
2	300–599	100–199
3	600–999	200–299
4	1000–1999	300–599
5	2000–3499	600–999
6	3500–4999	1000–1499
7	5000–6499	1500–1999
8	6500–7999	2000–2499
9	At or above 8000, or no clouds	At or above 2500, or no clouds

TABLE 7 Description of Middle Clouds

Code Number	Description*	Map Symbol
1	Thin As (most of cloud layer semitransparent)	
2	Thick As, greater part sufficiently dense to hide sun (or moon), or Ns	
3	Thin Ac, mostly semitransparent; cloud elements not changing much and at a single level	
4	Thin Ac in patches; cloud elements continually changing and/or occurring at more than one level	
5	Thin Ac in bands or in a layer gradually spreading over sky and usually thickening as a whole	
6	Ac formed by the spreading out of Cu	
7	Double-layered Ac, or a thick layer of Ac, not increasing; or Ac with As and/or Ns	
8	Ac in the form of Cu-shaped tufts or Ac with turrets	
9	Ac of a chaotic sky, usually at different levels; patches of dense Ci are usually present also	

* Abridged from WMO Code.

TABLE 8 Description of High Clouds

Code Number	Description*	Map Symbol
1	Filaments of Ci, or "mares tails," scattered and not increasing	
2	Dense Ci in patches or twisted sheaves, usually not increasing sometimes like remains of Cb; or towers or tufts	
3	Dense Ci, often anvil-shaped, derived from or associated with Cb	
4	Ci, often hook-shaped, gradually spreading over the sky and usually thickening as a whole	
5	Ci and Cs often in converging bands or Cs alone; generally overspreading and growing denser; the continuous layer not reaching 45° altitude	
6	Ci and Cs often in converging bands or Cs alone; generally overspreading and growing denser; the continuous layer exceeding 45° altitude	
7	Veil of Cs covering the entire sky	
8	Cs not increasing and not covering entire sky	
9	Cc alone or Cc with some Ci or Cs but the Cc being the main cirriform cloud	

* Abridged from WMO Code.

TABLE 9 Barometric Tendency

Code Number	Description	Map Symbol
0	Rising, then falling	
1	Rising, then steady; or rising, then rising more slowly	
2	Rising steadily or unsteadily	
3	Falling, or steady, then rising; or rising, then rising more quickly	
4	Steady, same as three hours ago	
5	Falling, then rising, same or lower than three hours ago	
6	Falling, then steady; or falling, then falling more slowly	
7	Falling steadily, or unsteadily	
8	Steady, or rising, then falling; or falling, then falling more quickly	

Appendix IV
COMMON METRIC CONVERSIONS

	To Convert		Multiply By	To Obtain
Length	millimeters	×	0.039	= inches
	centimeters	×	0.394	= inches
	meters	×	3.281	= feet
	kilometers	×	0.621	= miles
Volume	liters	×	1.057	= quarts
Weight	grams	×	0.035	= ounces
	kilograms	×	2.205	= pounds
Area	square meters	×	10.765	= square feet
	square kilometers	×	0.386	= square miles
Length	inches	×	25.4	= millimeters
	inches	×	2.54	= centimeters
	feet	×	0.305	= meters
	miles	×	1.609	= kilometers
Volume	quarts	×	0.946	= liters
Weight	ounces	×	28.35	= grams
	pounds	×	0.454	= kilograms
Area	square feet	×	0.0929	= square meters
	square miles	×	2.591	= square kilometers

Glossary

absolute humidity The density of water vapor.

absorption The process where energy incident on a substance is absorbed by that substance.

adiabatic lapse rate The change of temperature of an air parcel rising adiabatically. For dry air it is 10°C/km (5.5°F/1000 ft).

adiabatic process The process of temperature change as a result of a pressure change with no external sources of heat or energy.

advection fog Fogs that form when moist warm air moves (advects) over a cool surface. The air is then cooled from below, condensation is reached, and fog is formed.

aerosols Solid particles suspended in the air. They may be dust, salt particles, products of combustion, sand, etc.

Aitken nuclei Microscopic particles in the atmosphere that act as condensation nuclei for cloud drop growth.

air mass A widespread volume of air having homogeneous physical characteristics such as temperature and moisture.

albedo The ratio of energy reflected by a surface to the energy incident on that surface.

anemoscope An instrument that measures wind direction and speed at the same time.

aneroid barometer A device that measures atmospheric pressure but does not use mercury.

anticyclone A high pressure pattern characterized by clockwise wind flow, in the northern hemisphere. In the southern hemisphere anticyclones have counterclockwise wind flow.

anvil The horizontal spreading of cumulonimbus clouds. When the cumulonimbus clouds reach the tropopause they spread because vertical growth is inhibited.

aurora borealis and aurora australis (northern and southern lights). Glowing brilliant yellow, green, and red colors seen in the skies of the polar regions at night. They are caused by charged particles that are ejected from the sun into the upper atmosphere.

backing A counterclockwise change in wind direction; the opposite of veering.

baguio A Northern Pacific term for tropical cyclones. It's name is derived from the city of Baguio, in the Phillipine Islands.

biosphere The zone of earth and atmosphere where most forms of terrestial life live.

black body An ideal radiating body that absorbs all the energy incident on it and emits radiation in a way that depends only on its temperature.

black out Total electric failure occurring when demand exceeds supply for long periods of time.

blizzard A snowstorm with associated high winds that reduce visibility to zero.

brown out The condition of reduced voltage when the demand for electricity is greater than the supply.

centrifugal force A force that is equal and opposite to the centripetal force. The balance of these two forces allows for uniform circular motion.

centripetal force A force exerted on a mass in circular flow and directed inward to the center.

chromosphere A shell of gases extending outward from the sun. Temperatures can reach as high as 100,000°C.

clear air turbulence Violent and irregular random motion in clear air.

climate Average of weather periods over long intervals of time, usually greater than one year. Included in the definition of climate is its variability from average conditions.

cloud A visible collection of small water drops that do not fall to the ground.

coalescence The process of merging cloud water drops.

cold fog Fog made up of supercooled water drops.

cold front See **front**.

condensation The transformation of water vapor to liquid water.

conditional instability The situation in which an air parcel is stable until it reaches the level of free convection at which point it then becomes unstable.

conduction Heat transfer accomplished by rapidly moving molecules within the substance being heated.

convergence The process where the air flow contracts because of both speed and direction changes. A good analogy is water flowing through the neck of a funnel. The water converges from the reservoir into the funnel neck. It is the opposite of divergence.

convection The transfer of heat by the actual movement of the heated substance, such as air or water. In meteorology convection also means vertical transport through density imbalance, carrying mass, water vapor, and aerosols as well as heat.

cooling degree day A cooling degree is achieved for each degree that the average daily temperature is above 65°F.

cooling fog Fog formed when air is cooled to its dew point temperature.

coriolis force The effect of the earth's rotation on air flow. In the northern hemisphere the coriolis force deflects air to the right and to the left in the southern hemisphere. There is no coriolis force at the equator.

corona The outermost layer of the sun, only visible during a solar eclipse or with an instrument known as a coronagraph.

coronagraph A special instrument for viewing the sun's corona.

counter-radiation The radiation returned to the earth's surface by the atmosphere. See **greenhouse effect**.

cyclogenesis The process of cyclone formation.

cyclolysis The process of cyclone decay and dissipation.

cyclones Weather systems characterized by low pressure and counterclockwise wind circulation.

denitrification The process of returning nitrogen to the atmosphere.

dew Water condensed on objects near the ground (including grass). Dew forms when the temperature of the objects are low enough to cause condensation.

dew point temperature The temperature to which air must be cooled at fixed water vapor content and pressure to accomplish condensation. When the temperature is below freezing it is known as frost point, since frost forms.

divergence The process where the air flow expands because of changes in speed and direction. A good analogy is water flowing out of the neck of a funnel. As the water leaves it expands, that is, diverges. It is the opposite of convergence.

doldrums The light and variable winds found in the equatorial trough region.

drizzle A form of precipitation formed by very small water drops that fall gently and appear to drift with the wind. Drop size is less than 0.5 mm in diameter.

easterly wave A synoptic wavelike disturbance in the easterly flow of the tropics. The wave disturbance moves from east to west.

electromagnetic energy Energy transferred in the form of disturbances in electric or magnetic fields. Two theories exist about the transfer. One theory says that energy is transferred as wave disturbances; the other theory states that energy is transferred as particles. According to both theories, the energy can be transferred through a vacuum as well as through solid objects.

entrainment The process where a growing thunderstorm incorporates environmental air, generally through its side.

equation of state The mathematical relationship between temperature, pressure, and density.

equatorial trough A zone of low pressure located near the equator.

evaporation The transformation of liquid water to vapor.

evaporation fog Fog formed by evaporating water vapor into the air until saturation is achieved.

eye wall The wall of convective clouds surrounding the eye of a hurricane, which is generally cloud-free.

facsimile An instrument for transmitting analyzed weather charts electronically. This allows weather charts to be drawn at one location and sent anywhere in the United States.

flash flood A rapid flooding of rivers or creeks. Flash floods occur from very extensive and intense rainfall coupled with the inability of the stream or river to carry the runoff.

forecast A weather prediction.

freezing rain/drizzle Rain that freezes upon contact with the ground.

friction A force that acts to retard motion. In the case of air flow, friction is important near the surface.

front A boundary separating warm moist air mass from cold dry air. If the cold air pushes into the region of warm air we have a *cold front*, and if the warm advances we have a *warm front*. If the front does not move, it is called a *stationary front*.

frontal surface The boundary surface separating two different air masses.

frontogenesis The process of creating a weather front.

frontolysis The process of the decay and dissipation of weather fronts.

frost The sublimation of water vapor on the surface or objects close to the surface when the dewpoint is below freezing.

fusion The transformation from solid to liquid.

gale Winds from 17 to 20 m/sec (40 to 47 mph).

gamma radiation Radiation of extremely short wave length. They are shorter than x-rays.

geostrophic wind Horizontal wind flow as a result of the balance of the pressure gradient force and the coriolis force.

glacial periods A geologic time period characterized by low temperatures and extensive glaciers. These periods have average lengths of tens of thousands of years.

glaze A coating of ice formed by the freezing of water deposited by fog, rain, dew, etc.

gradient wind Curved, horizontal wind flow, as a result of the balance of the pressure gradient force, the coriolis force, and the centrifugal force.

greenhouse effect The absorption of radiation and its reemission by the atmosphere downward, to the earth's surface, of radiation that was originally emitted from the earth's surface. It is the principle upon which a garden greenhouse is supposed to work.

Greenwich mean time Time referenced to the zero meridian, which passes through Greenwich, England.

ground fog Fog that is formed when the ground cools radiatively, also called radiation fog. As the temperature is lowered to the dew point the air in contact with the ground becomes saturated, vapor condenses, and fog forms.

growing season The portion of the year during which vegetation can grow.

hail Precipitation of balls of ice with a diameter of 5 mm or more.

heating degree day A heating degree day is achieved for each degree that the daily average temperature is below 65°F.

hurricane A violent tropical cyclone with winds in excess of 33 m/sec (74 mph).

hurricane eye The calm, clear center of a hurricane.

hygrograph An instrument that continuously records relative humidity.

hydrologic cycle The cycle of evaporation, condensation, and precipitation. It involves both the atmosphere, within which water vapor resides, clouds form, and precipitation occurs, and the earth's surface, which is the reservoir of liquid water.

hydrosphere All bodies of water that reside on the earth's surface.

hydrostatic equation An equation that describes the variation of atmospheric pressure with height.

hygrometer An instrument used to measure atmospheric water vapor.

hygroscopic Particles that have an affinity for water.

ice fog Fog composed of ice crystals instead of water drops.

ice pellets A category of frozen precipitation. Ice pellets form when water droplets freeze upon falling through an air layer whose temperature is below freezing.

ice storm A storm composed of freezing rain. That is, rain that freezes upon contact with the ground.

interglacials Warm periods between glacials.

intertropical convergence zone The region where the northeast trade winds of the northern hemisphere and the southeast trade winds of the southern hemisphere meet.

inversion Temperature increase with increasing height; a negative lapse rate.

ions Charged atoms caused by either loss or gain of electrons.

ionosphere A layer of charged particles approximately 400 km above the earth's surface.

isobar A line connecting points of equal pressure.

jet stream A band of high winds usually found in the upper troposphere. Wind speeds can exceed 90 m/sec (200 mph).

latent heat The amount of energy needed to accomplish a phase change. *Latent heat of fusion* is the amount of energy needed to melt ice and at 0°C is 79.7 cal/g. The *latent heat of vaporization* is the amount of energy needed to evaporate liquid water or condense water vapor. It is 597.3 cal/g, at 0°C. Finally, the *latent heat of sublimation* is the energy needed to accomplish the change from solid to gas or vice-versa. It is the sum of the latent heats of vaporization and sublimation, 677.0 cal/g, at 0°C.

leeward The region protected from the direct effects of the wind, such as the protected side of a mountain range.

level of free convection The level where a rising parcel of air becomes unstable with respect to its environment.

lifting condensation level The height at which a rising parcel of air first becomes saturated.

lightning Electrical discharge from thunderstorms.

little ice age A period of low temperature between 1450 and 1850.

macroscale A term frequently used to denote large scale weather systems. Dimensions of such systems are thousands of kilometers and larger and last from days to weeks.

maximum thermometer A thermometer that shows the maximum temperature during a specific time period, such as 24 hours.

mechanical lifting The forced ascent of air through heating, mass convergence, orographic lifting, and frontal surfaces.

mesopause The transition zone between the mesosphere and thermosphere.

mesoscale A term frequently used to denote medium scale weather systems. Spatial dimensions are hundreds to a thousand kilometers and the time scale is hours to days.

mesosphere The layer located above the stratosphere; it extends from 45 to 92 km. Temperature decreases with height in this layer.

meteorology The study of the total effect of the atmospheric pressure, temperature, humidity, precipitation, wind and cloud cover at any particular place at any instant of time.

microscale A term used to denote small scale weather systems. They have spatial dimensions of meters or smaller and time scales of minutes.

millibar A unit of pressure.

mist Microscopic water droplets suspended in the atmosphere. Mist can be thought of as a very light fog. In popular usage in the United States it is the same as drizzle.

moist adiabatic lapse rate The change of temperature of a rising air parcel when condensation takes place. In contrast to the adiabatic lapse rate it is variable, depending on the amount of moisture in the air.

neutral air Air is neutral when a parcel of air neither rises or sinks when the force that originally moved it is not operating.

net radiation The difference between the absorbed and emitted radiation. It is the net radiation that determines if a body will heat or cool radiatively.

nitrogen cycle The cycle of transforming atmospheric nitrogen to plant life, decomposition, and return to the atmosphere.

noctilucent clouds Clouds that are believed to be composed of cosmic dust particles. They are located between 75 and 95 km above the earth's surface.

normal atmospheric pressure The average atmospheric sea level pressure.

normal lapse rate The average variation of temperature with height. In the troposphere it is 6.5°C/km (3.5°F/1000 ft). A positive lapse rate means that temperature decreases with an increase in height.

norther A north wind. In the United States a norther is used for a frigid blast of cold air coming from the north.

northeaster A winter storm off the east coast of North America. This storm is so named because of the strong northeast winds attacking the coastal areas.

numerical weather prediction The forecasting of weather by the numerical solution of the complex equations governing the motions of the atmosphere. These solutions are accomplished with the aid of computers.

occluded front The merging of two fronts, as when a cold front overtakes a warm front.

orographic lifting The lifting of air caused by its passage over a mountain barrier.

photolytic dissociation The process of splitting molecules through absorption of light.

photosphere A shell of relatively cool gas, around the sun about 500 km thick.

photosynthesis A plant process that combines carbon dioxide, water, and sunlight to produce sugar and oxygen.

pilot balloon A small balloon tracked by optical devices for the determination of atmospheric wind flow.

pilot leader The initial step in a lightning discharge. It is generally not seen because of its faintness. However, it establishes the path for future lightning.

polar front A boundary that separates polar air masses from tropical air masses.

polar outbreak The movement of cold polar air equatorward.

prefrontal fog Fog formed from the precipitation falling from clouds in advance of a slow moving front.

pressure gradient The change of pressure with distance. When measured from high to low values perpendicular to the isobars, it is the greatest change in the shortest distance.

pressure ridge An elongated area of high pressure.

pressure trough An elongated area of low pressure.

prevailing wind The most frequent wind direction in a given time period; day, month, year, etc.

rawinsonde A balloon-borne instrument used for measuring temperature, pressure, relative humidity, and wind direction and speed. The rawinsonde is tracked using radio or radar.

radiation fog Fog formed when radiative cooling reduces air to its dewpoint temperature. It is frequently called ground fog since this type of fog usually forms near the earth's surface.

radiosondes An instrument that measures temperature pressure and humidity of the upper air. It is carried aloft on a balloon and transmits its measurements to a ground-based receiver via radio signals.

rain Precipitation of water drops greater than 0.5 mm in diameter.

rain forest A lush dense forest growth in regions of heavy rainfall.

rainsquall A sudden short burst of wind and rain from thunderstorms.

reflection The process of turning back a part of the energy incident upon it. A mirror is a good example of the reflection process.

refraction The bending of light rays between mediums of different densities, such as the interface between air and water. It results in a change of direction of the radiation.

relative humidity The ratio of the amount of moisture in the air to the maximum amount of moisture that the air can hold. Its precise value is determined by the ratio of actual vapor pressure of the air to the saturated vapor pressure.

saturation The condition air reaches when it contains the most water vapor it is capable of holding.

saturation vapor pressure The pressure exerted by water vapor when the air is saturated.

scattering The changing of direction of incident energy by small particles. In this process the energy is scattered in all directions. It is the process that produces the blue sky; blue light is preferentially scattered downward.

sensible heat transfer The transfer of heat by conduction and convection.

silver iodide Particles added to clouds in order to aid in their development.

sleet Precipitation composed of transparent solid grains of ice formed from the freezing of raindrops.

sling psychrometer An instrument designed to measure atmospheric moisture content. It consists of two mercury thermometers, one covered with a wick that is saturated with water.

smog A mixture of smoke or other pollutants and fog.

snow Precipitation of ice crystal usually formed into snowflakes.

snow pellets Precipitation of white opaque round ice particles, about 2 to 5 mm in diameter.

solar constant The amount of solar energy per unit time on an area at the top of the atmosphere perpendicular to the sun's rays. This amount stays constant except for variations in the earth-sun distance.

solar flares Short-lived intense bursts of energy that occur in the chromosphere.

solar radiation Energy from the sun.

solidification The transformation of a liquid to a solid.

solute effect The effect that particles have on the growth of cloud drops. Particles in a cloud drop create a situation where the relative humidity can be less than 100% for further condensation. The particle is called the solute.

source region Extensive areas where air remains in the same place long enough to acquire the characteristics of an air mass.

specific heat The amount of energy required to raise the temperature of one gram of a substance one degree Celsius.

specific humidity The mass of water vapor in a unit mass of moist air.

squall line A line of thunderstorms.

stable air Air is stable when an air parcel sinks to its original location when the force that originally moved it is no longer operating. See **unstable air** and **neutral air**.

stationary front See **front**.

steam fog Fog formed when warm water evaporates into cooler air above. This fog is commonly found over water, such as lakes, rivers, and bays, when cool air moves over warm water.

step leader The initial portion of a visible lightning discharge. It has a characteristic stepwise growth at intervals of about $\frac{1}{10,000}$ of a second.

storm surge An abnormally high tide due to storm winds.

stratopause The transition zone between the stratosphere and mesosphere.

stratosphere The atmospheric layer immediately above the troposphere. It extends from 11 to 45 km above the earth's surface and temperature increases upward in this layer.

sublimation The process of a solid being transformed directly to a gas or vice-versa.

subsidence Sinking of air. Subsidence is an important process since inversions usually form in subsiding air.

sunspot Dark, cool spots on the photosphere.

supercooled water Liquid water at temperatures below 0°C (32°F).

synoptic code A code in which weather information is transmitted.

synoptic meteorology The study and analysis of weather data such as wind, pressure, rain clouds, and temperature.

theodilite An optical tracking device. In meteorology it is used to track pilot balloons.

thermometer An instrument for measuring temperature. The most common types are mercury in glass.

thermoscope The name given to the first thermometer, invented by Galileo.

thermosphere The last layer of the atmosphere. It begins at 92 km with an upper limit that is not well-defined.

thunderstorm A storm composed of cumulonimbus clouds and always accompanied by lightning and thunder; intense rain and gusty winds are also common features of thunderstorms.

thunderstorm cell The active rising portion of a cumulonimbus cloud.

tornado A violent, destructive rotating column of air usually associated with thunderstorms.

transducer A device that converts energy from one form to another. For example, conversion of sound to electrical energy.

transient pressure patterns Pressure patterns that move across the earth's surface on a daily basis.

transpiration The moisture given off by vegetation.

tropical cyclone A general expression for a cyclone that originates over the oceans. This definition includes terms indicating different intensities. In ascending order of density these are:
(a) Tropical disturbance—having only a slight surface disturbance.
(b) Tropical depression—wind speeds less than or equal to 14 m/sec (31 mph).
(c) Tropical storm—winds between 14 m/sec and 33 m/sec (31 mph and 74 mph).
(d) Hurricane—wind speeds greater than 33 m/sec (74 mph).

tropical depression See **tropical cyclone**.

tropical disturbance See **tropical cyclone**.

tropical storm See **tropical cyclone**.

troposphere The atmospheric layer closest to the earth's surface. It has an average thickness of 11 km (7 miles), and temperature decreases upward there.

trough As used in meteorology, it is an elongated area of low pressure.

turbulence Atmospheric motion that exhibits irregular and random motion.

typhoon A severe tropical cyclone in the western Pacific. It is identical to the hurricanes of the tropical Atlantic.

ultraviolet rays Radiation in the wavelength interval of 0.001 to 0.40 micron.

unstable air Air parcel that moves away from its location when the initial force exerted on it is removed. See **stable air** and **neutral air**.

updrafts Strong vertical currents usually associated with cumulus-type clouds and thunderstorms.

upslope fog Fog formed when air moves upward over sloping terrain and is cooled adiabatically to its dew point temperature.

upwelling A process that brings deep cool ocean water to the surface.

vapor pressure The pressure exerted by water vapor in the atmosphere.

veering A change in wind direction in the clockwise sense; the opposite of backing.

virga Precipitation that falls from clouds but evaporates before reaching the surface.

warm fog Fog in which temperature of the water droplets is above freezing.

warm front See **front**.

water vapor The gaseous form of water.

wave cyclone A cyclone that forms and moves along a weather front. Its initial appearance along the front takes on a wavelike character.

weather The total effect of meteorological elements such as pressure, wind, temperature, humidity, precipitation, and cloud cover at any particular place at any instant.

weather data The elements that make up the weather such as wind, temperature, precipitation, pressure, humidity, and clouds.

weather radar Radar that is specially designed to detect precipitation and large cloud drops. They are used to detect areas of thunderstorms and intense precipitation.

willy willy A severe tropical cyclone in Australia. It is called a hurricane in the Atlantic Ocean.

wind chill factor The cooling effect, on people, of any combination of temperature and wind. For example, at a given temperature the cooling effect of wind tends to make the effective temperature lower.

wind gust A sudden burst of high wind speed.

wind shear The variation of the wind, either direction or speed, in a given direction. The wind can shear horizontally or vertically.

windward The side of an obstacle that faces the direction from which the wind is blowing.

x-rays Radiation of very short wavelength: 0.00001 to 0.00015 micron.

NAME: _____

SECTION: _____

LABORATORY EXERCISE 1

MEASUREMENTS AND DIMENSIONS
REFERENCE: Appendix IV

QUESTIONS TO THINK ABOUT

1. In our system of measurement where is the decimal point assumed to be with any whole number?
2. In order to use scientific notation what two parts must a number have?
3. In scientific notation why must the powers of two numbers be the same in order to add or subtract?
4. Why is it easier to use the metric system to express dimensions in mathematical form?
5. An object's mass is the same on the earth and the moon but its weight is different. Why?

DISCUSSION

In meteorology, as well as many other scientific disciplines, the proper use of mathematics is essential. If it were not for this most important tool, many of our current theories and applications would not be readily available for practical use. Although the student who is pursuing a formal degree in meteorology must have a working knowledge of subjects such as calculus, vector anaylsis, and computer programming, the basic concepts of meteorology can be understood by using simple mathematics. Since it is not the intent of this book to prepare students for formal meteorological training, only a simple knowldege of algebra, trigonometry, dimensions, and scientific notation is needed.

Scientific Notation

One of the methods scientists use to express numbers involving many digits is called *scientific notation*. This notation saves much time and space and also aids in handling large numbers with a minimum of error. In this method, every number is expressed by a digit and a power of 10. Usually the digit lies between the value of 1 and 10 but this is not a hard-and-fast rule. Most users break the digit down to the easiest usable number.

The power denotes how many places the decimal is to be moved either to the left or the right. The movement to the right is known as a positive power while the left a negative one.

$$\begin{array}{cc} \text{Positive} & \text{Negative} \\ 10^2 & 10^{-2} \end{array}$$

Since the power is a *factor of ten*, 10^2 means 10×10 while 10^3 is $10 \times 10 \times 10$ or 1000. Therefore, the quantity 48×10^3 is equal to:

$$48 \times (10 \times 10 \times 10) \quad \text{or} \quad 48,000$$

Another short way of solving the same problems would be to determine how many decimal places to the right (positive) the decimal should be moved. In the above case 10^3 means move the decimal 3 places to the right (positive power). In a whole number or set of numbers the decimal is to the far right. Therefore,

$$48 = 48.0 \quad \text{and} \quad 48 \times 10^3 \quad \text{would be } 48.000. \text{ or } 48,000$$

Examples

1. $4.5 \times 10^2 = 450$
2. $6.732 \times 10^2 = 673.2$
3. $973.4 \times 10^{-3} = 0.9734$

In order to do the reverse process correctly, you must move the decimal the number of digits opposite to what the power you choose means. To prove whether you have done the problem right, move the decimal back according to the power chosen.

Examples

1. $574,000 = 5.74 \times 10^5$
   Proof $5.74 \times 10^5 = 5.74000. = 574,000$

2. $0.0573 = 5.73 \times 10^{-2}$
   Proof $5.73 \times 10^{-2} = .05.73 = 0.0573$

3. $1,500,000 = 1.5 \times 10^6$
   Proof $1.5 \times 10^6 = 1.500000. = 1,500,000$

Exponents

In order to properly use the scientific notation method, let us review exponents.

Addition To add numbers that incorporate exponents, the exponent *must* be the same.

Example:

$$(1 \times 10^5) + (217 \times 10^5) = 218 \times 10^5$$

If the exponents are not the same, then use the rule of scientific notation to make them the same.

Examples

1. $(3 \times 10^2) + (54 \times 10^3) = (3 \times 10^2) + (540 \times 10^2) = 543 \times 10^2$
2. $(75 \times 10^{-1}) + (22 \times 10^{-3}) = (75 \times 10^{-1}) + (0.22 \times 10^{-1}) = 75.22 \times 10^{-1}$

Subtraction Subtraction is the exact opposite of the addition procedure. Digits can only be subtracted if their exponents are the same. After we subtract, the exponent remains the same.

Examples

1. $(5.9 \times 10^{-5}) - (2.8 \times 10^{-5}) = 3.1 \times 10^{-5}$
2. $(2.7 \times 10^3) - (1.1 \times 10^2) = (27 \times 10^2) - (1.1 \times 10^2) = 25.9 \times 10^2$

Multiplication In the multiplication process the exponents do not have to be the same. The digits are multiplied and the exponents are added algebraically.

Examples

1. $(3 \times 10^3) \times (7 \times 10^4) = 21 \times 10^7$
2. $(3.2 \times 10^{-5}) \times (4 \times 10^3) = 12.8 \times 10^{-2}$
3. $(15 \times 10^{-5}) \times (4 \times 10^{-3}) = 60 \times 10^{-8}$

Division Division is the opposite of multiplication. In this mathematical process the digits are divided and the exponents are subtracted.

Examples

1. $(6 \times 10^3) \div (3 \times 10^2) = \dfrac{6 \times 10^3}{3 \times 10^2} = 2 \times 10^1$

2. $(8.8 \times 10^2) \div (2 \times 10^6) = \dfrac{8.8 \times 10^2}{2 \times 10^6} = 4.4 \times 10^{-4}$

3. $(100 \times 10^{-6}) \div (20 \times 10^{-2}) = \dfrac{100 \times 10^{-6}}{20 \times 10^{-2}} = 5 \times 10^{-4}$

4. $(7 \times 10^4) \div (7 \times 10^{-2}) = \dfrac{7 \times 10^4}{7 \times 10^{-2}} = 1 \times 10^6$

Powers and Roots To raise to a higher power a number already in scientific notation, first perform the multiplication on the number itself, then multiply the exponent by the power to which the number is being raised.

Examples

1. $(3 \times 10^3)^2 = (3 \times 3) \times 10^{3 \times 2} = 9 \times 10^6$
2. $(6.3 \times 10^4)^2 = (6.3 \times 6.3) \times 10^{4 \times 2} = 39.69 \times 10^8$
3. $(1.5 \times 10^{-3})^2 = (1.5 \times 1.5) \times 10^{-3 \times 2} = 2.25 \times 10^{-6}$

Roots of numbers, such as square root ($\sqrt{}$) or cube root ($\sqrt[3]{}$), are performed in the opposite manner. The exponent of the number is divided by the exponent of the root; the number is treated separately, just as it was in the power examples.

Examples

1. $\sqrt{25 \times 10^8} = 5 \times 10^{8/2} = 5 \times 10^4$
2. $\sqrt[3]{8 \times 10^9} = 2 \times 10^{9/3} = 2 \times 10^3$

System of Measurements

There are many ways in which a person could measure an object. It might be possible to use stones as a base for weight and say that an object weighs 5.4 stones (if we knew what 1 stone weighed). On the other hand, we might call an object 10 hands high and 18 hands wide or even 20 seconds long if we knew the definition of a second. Therefore, you can see that measurements can be whatever units we so choose, as long as we define the dimensions beforehand.

In the worldwide scientific community three systems for measurement have been adopted. All three of these systems measure the basic unit of *Length*, *Mass* and *Time*.

System	Length	Mass	Time
1. English (fps)	foot (ft)	pound (lb)	second (s)
2. Metric (cgs)	centimeter (cm)	gram (g)	second (s)
3. Metric (mks)	meter (m)	kilogram (kg)	second (s)

In the above table, mass replaces weight. *Mass* may be regarded as the amount of matter contained within an object while *weight* takes into account the attraction of the earth on the body with a certain mass.

Since most scientific study is done in the metric system, it is necessary to be thoroughly familiar with such a system. The metric system was developed in France around the late 1700s. It is a decimal system that enables us to move easily from one decimal value to the next, (as in our monetary system). The three basic measurements have been given as the *unit of length*—meter, *unit of mass*—gram, and *unit of time*—second. However, subdivisions utilizing the decimal system are also put to use. The most widely used prefixes are:

180 MEASUREMENTS AND DIMENSIONS

Prefix	Abbreviation	Meaning	Scientific Notation
milli	mm (millimeter)	one thousandth	1×10^{-3}
centi	cm (centimeter)	one hundredth	1×10^{-2}
deci	dm (decimeter)	one tenth	1×10^{-1}
kilo	km (kilometer)	one thousand	1×10^{3}

Examples

1. 1 cm = 1×10^{-2} meters = 0.01 meters = $\frac{1}{100}$ = 10 mm
2. 1 km = 1×10^{3} meters = 1000 meters
3. 1 m = 1×10^{2} centimeters = 100 centimeters
4. 1 km = 1×10^{6} millimeters = 1,000,000 millimeters

Conversion of Systems

To convert from English to metric or metric to English, it is necessary to memorize a few basic conversion factors.

1 m = 39.37 inches
1 inch = 2.54 cm
1 km = 0.621 miles
1 kg = 2.2 lbs

By using the known conversion factors and using the ratio (fractions) method, you can convert any quantity.

Example

How many meters are there in 97 inches? We know that 1 meter = 39.37 inches. Therefore,

$$\frac{1 \text{ m}}{39.37 \text{ in.}} = \frac{? \text{ m}}{97 \text{ in.}}$$

Cross multiply and solve for the unknown;

$$(?)(39.37 \text{ in.}) = (1 \text{ m})(97 \text{ in.})$$

$$? = \frac{(1 \text{ m})(97 \text{ in.})}{39.37 \text{ in.}} = 2.209 \text{ m}$$

EXERCISES

1. Express the following numbers in scientific notation:
   - (a) 0.94
   - (b) 570
   - (c) 0.000380
   - (d) 374,000,000
   - (e) 0.0047
   - (f) 932.4
   - (g) 7,870
   - (h) 0.076
   - (i) 17,000
   - (j) 0.02
2. Write the following without exponents:
   - (a) 32×10^{5}
   - (b) 3.758×10^{2}
   - (c) 872×10^{-4}
   - (d) 1.83×10^{7}
   - (e) $9.72 = 10^{1}$
   - (f) 5.8×10^{-5}
   - (g) 18.73×10^{-1}
   - (h) 0.835×10^{3}
   - (i) 1×10^{-1}
   - (j) 3.65×10^{0}
3. Solve the following:
   - (a) $(9.7 \times 10^{-4}) + (2 \times 10^{-4})$
   - (b) $(6.2 \times 10^{2}) + (3.7 \times 10^{4})$
   - (c) $(183 \times 10^{-2}) + (7.4 \times 10^{2})$
   - (d) $(92.7 \times 10^{1}) - (43.5 \times 10^{1})$
   - (e) $(7.3 \times 10^{3}) - (0.05 \times 10^{5})$

(f) $(89.7 \times 10^{-2}) - (0.004 \times 10^2)$
(g) $(17 \times 10^4) \cdot (2 \times 10^2)$
(h) $(35 \times 10^2) \cdot (2 \times 10^{-8})$
(i) $(180 \times 10^{-4}) \div (10 \times 10^2)$
(j) $(3.75 \times 10^8)^3$
(k) $(12.5 \times 10^3)^{-2}$
(l) $\sqrt{25 \times 10^8}$

4. Solve the following:
   (a) 10 m = _____ mm
   (b) 7.8 mm = _____ m
   (c) 3.2 kg = _____ g
   (d) 100 g = _____ kg
   (e) 0.072 cg = _____ mg
   (f) 1 cm = _____ m
   (g) 9.8 mg = _____ cg
   (h) 22 mm = _____ km
   (i) 0.54 mg = _____ g
   (j) 233 g = _____ cg
   (k) If a car is traveling at 60 mph, how fast is it going in ft/sec? In m/hr?
   (l) What are the dimensions in meters of your textbook? (Use the metric part of your ruler.) In inches?

5. Convert the following:
   (a) 183 in. = _____ mm
   (b) 23 kg = _____ lb
   (c) 193 ounces = _____ mg
   (d) 1000 g = _____ lb
   (e) 0.073 km = _____ in.

NAME: _____

SECTION: _____

LABORATORY EXERCISE 2

TEMPERATURE MEASUREMENTS
REFERENCES: Chapters 3 and 4

QUESTIONS TO THINK ABOUT

1. How long do you wait before reading a thermometer?
2. Why do meteorologists always measure the temperature in the shade?
3. Why does it always feel hotter over white sandy soil than a forested area with the same air temperature?
4. Why do people tend to wear dark clothes in the winter and light-colored clothes in the summer?
5. Do you think that weather, especially temperature, has a bearing on the physical makeup of human beings such as weight, height, and skin color?

DISCUSSION

The most common method of measuring the amount of energy in the air is by means of a liquid-in-glass thermometer. This type of thermometer is preferred to other devices because it is accurate, easy to handle, and inexpensive. The liquid used in most scientific standard air thermometers is mercury. Although it is a more expensive medium than other liquids such as alcohol or water, it is preferred because of its constant expansion rate, visibility, and nonevaporative feature.

A liquid-in-glass thermometer works on the premise that "nature dislikes an imbalance." When the thermometer is placed in hot water, the molecules in the water are moving at a faster rate than those in the mercury and an imbalance is setup. Immediately a flow of energy begins from hotter to colder objects in order to create a balance. This flow of energy from water to mercury results in the mercury molecules moving faster and a resultant expansion of the mercury up the thermometer path. Just the reverse takes place when the thermometer is set into a colder medium.

**EXPERIMENT 1
Shade versus Sun Temperature**

Using an ordinary thermometer, measure the air temperature in the shade, in the sun, and in the sun with a protective covering above the thermometer.
Shade _____ °F/°C   Sun _____ °F/°C   Shaded _____ °F/°C

Questions

1. Which was the highest temperature? Why?
2. Was there more than one source of energy for the thermometer to record? What were they and in what measurement did they occur?
3. What conclusions can you draw as to the temperature in the sun as compared to the temperature in the shade?

**EXPERIMENT 2
Temperature of Colored Objects**

Fill two alike containers, one with dark soil and the other with light soil. Put the containers side by side in the sun for five minutes and then stick a thermometer in each container (keep the thermometers from hitting the sides of the containers). Record the temperature of each thermometer every minute for five minutes.

Minute	Dark Soil	Light Soil
0	_____	_____
1	_____	_____
2	_____	_____
3	_____	_____
4	_____	_____
5	_____	_____

Questions

1. Which soil was hotter initially and why?
2. Which soil finished up the hottest?
3. What can you conclude about the efficiency of different-colored material?
4. Give some examples of temperature differences that are caused by colors on the surface of the earth?
5. What other factors could enter into differences of temperature on the surface of the earth?

EXPERIMENT 3
Reflectivity (Albedo) of Objects

Place a dark object, such as construction paper or black cloth, on the ground in direct sunlight. Nearby, place a piece of aluminum foil. Record the temperature six inches above both surface areas. (Be sure to shade the thermometer from the direct sunlight.)

Foil surface temperature _____

Darkened surface temperature _____

Questions

1. Why is it usually colder at night in the country than in the city?
2. Why is the desert so cold at night?
3. What type of soil would a farmer choose to till for farming and why?
4. Which areas of the world have the highest albedo? Which have the lowest?

Temperature Conversions

In addition to the two methods of temperature conversion found in Chapter 4, a third and much more convenient method may be used to convert Celsius temperature to Fahrenheit. This method uses a three-step memorization technique and in most instances the conversion can be done in your head.

Steps	Technique
1	Take the Centigrade temperature and double it: $(2 \times °C)$.
2	Take one-tenth of the answer in Step 1 away from the answer in Step 1 $\left(2C - \frac{2C}{10}\right)$.
3	Add 32: $\left(2C - \frac{2C}{10} + 32\right)$.

To memorize the steps remember *Double it—Take $\frac{1}{10}$ away—and add 32*.

Examples

1. 10°C = _____ °F?
   Step 1 10° + 10° = 20°
   Step 2 20° − 2° = 18°
   Step 3 18° + 32° = 50°F = answer

2. 50°C = _____ °F?
   Step 1 50° + 50° = 100°
   Step 2 100° − 10° = 90°
   Step 3 90° + 32° = 122°F = answer

EXERCISES

Convert the following temperatures using information in this section and Chapter 4.

	°C	°F	°K
1.	5°		
2.		68°	
3.			283°
4.		59°	
5.	−40°		
6.			0°
7.	17°		
8.			5000°

NAME: _____

SECTION: _____

LABORATORY EXERCISE 3

THERMAL CONVECTION CELLS
REFERENCES: Chapters 2 and 4

QUESTIONS TO THINK ABOUT

1. How does water vapor and energy get to the upper layers of the troposphere?
2. Why are there no clouds in the mesosphere?
3. Why does weather only occur in the troposphere?
4. What parts of the oceans behave just like the thermal properties in the atmosphere?
5. Why doesn't a solid object behave like the ocean and the air?

DISCUSSION

This experiment will demonstrate:

1. How circulation of air due to thermal properties behaves under different situations.
2. How vertical mass transport affects energy, moisture, and continuity of flow.
3. How convergence and divergence take place.

Materials Needed

1. *One* quart jar with lid or *one* 1000 ml cylinder and cover.
2. One pen flashlight, microscope light, or other source of thin pencil ray illumination.
3. Two ringstands with rods.
4. One lab clamp.
5. Two double clamps.
6. Smoke.

EXPERIMENT

A.
1. Fill the jar or container with smoke and cap the top.
2. Place the light so that it shines on one side of the bottom of the container with the smallest beam possible.
3. Observe the results after five or six minutes.

Questions

1. What results occurred after five or six minutes? Why?
2. Did you notice anything different after two minutes elapsed?
3. What resulted on the opposite side of the jar?
4. What happened at the top of the jar?
5. Can you recall two other situations in which the same effect might take place?

B.
1. Refill the jar with smoke and cap the top.
2. Replace the light at the top side corner of the container with the smallest beam possible.
3. Let the smokey turbulence settle for a few minutes.
4. Observe the results after a few minutes.

188 THERMAL CONVECTION CELLS

Questions

1. Compare your results with the first experiment. (For example, which took the longest, which was more vigorous, etc.)
2. What type of natural phenomena does this represent?

C.
    1. Fill the jar with smoke and cover the top.
    2. Place the light beam in the middle of the bottom of the jar. Make sure it is as close as possible to the glass.
    3. Observe the results after several minutes.

Questions

1. What different results were observed from the two previous experiments?
2. How many cells were observed?
3. Where does divergence and convergence take place?

GENERAL QUESTIONS TO ANSWER

1. Why is thermal convection so important to weather? Name three specific cases.
2. Using the knowledge you gained in the experiment, explain why wind blows from ocean to land in the daytime (seabreeze) and from land to ocean at night (landbreeze)?
3. Does convection play a more important role in the atmosphere or oceans? Why?

NAME: _____

SECTION: _____

LABORATORY EXERCISE 4

MOISTURE IN THE AIR
REFERENCE: Chapter 5

QUESTIONS TO THINK ABOUT

1. Is there always the same amount of water in the air?
2. How does water get into the air and what are the sources for these processes?
3. On what physical principles are measurements of moisture in the atmosphere based?
4. What is the meaning of dew point temperature, relative humidity, saturation, and condensation?
5. Why is relative humidity used in weather broadcasts when scientists prefer to use absolute humidity in their experiments?

DISCUSSION

Of all the gases found in the atmosphere, water vapor—that one wet gas—is considered to be the most important. It is the raw material that provides the main ingredient for the formation of clouds, dew, frost, and fog. Clouds provide precipitation that falls to the ground and renews our fresh water supply.

Water vapor content within the air can be described in many different ways. In this lab exercise we use the terms dew point, relative humidity, and vapor pressure to describe the true moisture content in the air and explore the relationship between them.

EXPERIMENT 1
Dew Point Determination

One of the most direct methods of measuring the dew point of the air is to visually observe when condensation begins and to record the temperature. This point is also known as the saturation point and has a relative humidity of 100 percent.

Materials Needed

1. Shiny beaker
2. Thermometer
3. Water at room temperature
4. Ice

Fill the beaker with water until the water level reaches 3 to 4 inches. Put the thermometer into the beaker, making sure it does not touch the sides. Gradually add the ice to the water stirring the mixture constantly with the thermometer. At the first sign of water droplets on the outside of the beaker read the thermometer. This will be the dew point temperature.

Record your answer here. _____

Questions

1. By what process did the beaker itself and the air touching the beaker cool?
2. Why is this method of calculating dew point, although very close, not accurate?
3. Would the dew point reading have changed if the beaker were made out of glass? Why?

190 MOISTURE IN THE AIR

4. What effect would take place if you were too close to the beaker and were breathing on it?
5. Does stirring the water faster or slower alter the reading? Why?

EXPERIMENT 2
Use of the Psychrometer

The most widely used instrument for measuring water vapor in the air is called sling psychrometer. This instrument was described in Chapter 5.

Materials Needed

1. Sling psychrometer
2. Water source

Saturate the muslin wick with water, being careful not to wet the uncovered thermometer. Sling the instrument until 2 readings of the wet bulb are the same. Record both dry and wet bulb to the nearest tenth of a degree. Use the table provided to find the dew point.

Location	Dry Bulb	Wet Bulb	Dew Point
Classroom	_____	_____	_____
Hall	_____	_____	_____
Outside in sun	_____	_____	_____
Outside in shade	_____	_____	_____

Questions

1. What was happening to the water on the wick as the instrument was spinning in the air? Where did the energy come from for this process to occur?
2. Why did the temperature of the wet bulb finally stop decreasing?
3. What does a large wet bulb depression (difference) tell you about the air?
4. What does the spinning do? Do you have to spin the instrument to calculate dew point?

ADDITIONAL CALCULATIONS
Vapor Pressure, Saturation Vapor

Pressure and Relative Humidity

These terms have already been defined in Chapter 5. Vapor pressure and saturation vapor pressure are used to calculate relative humidity. The vapor pressure is found in Table 1 by going down the air temperature column until the dew point temperature is reached (interpolate for readings not exactly on this table) and then sliding over to the saturation vapor pressure column. The saturation vapor pressure is found by using the air temperature reading and following the same process.

Relative humidity is calculated using this formula:

$$R.H. = \frac{e}{e_s} \times 100$$

where

e = vapor pressure
e_s = saturation vapor pressure

Location	Vapor Pressure	Saturation Vapor Pressure	Relative Humidity
Classroom	_____	_____	_____
Hall	_____	_____	_____
Outside in sun	_____	_____	_____
Outside in shade	_____	_____	_____

Questions

1. Why was the relative humidity different in the two outside locations?
2. Why do we use the air temperature reading to calculate the saturation vapor pressure?
3. What is another way of calculating relative humidity? (Hint: Refer to Chapter 5.)

EXERCISES

	Dry Bulb	Wet Bulb	Difference	Dew Point	Relative Humidity	Vapor Pressure
1.	50°	45°	_____	_____	_____	_____
2.	_____	_____	8°	51°	_____	_____
3.	80°	_____	_____	_____	_____	0.432
4.	65°	_____	10°	_____	_____	_____
5.	_____	_____	_____	_____	80%	.360
6.	90°	_____	_____	82°	_____	_____
7.	70°	_____	0°	_____	_____	_____
8.	_____	50°	8°	_____	_____	_____
9.	80°	60°	_____	_____	_____	_____
10.	53.4°	_____	2.2°	_____	_____	_____

TABLE 1 Saturation Vapor Pressure in Inches of Mercury and Temperature of Dew Point in Degrees Fahrenheit (Barometric Pressure, 30.00 Inches)

Air Temperature °F.	Saturation Vapor Pressure In.	\multicolumn{14}{c}{Wet Bulb Depression (Difference between Dry and Wet Bulb)}
		1	2	3	4	6	8	10	12	14	16	18	20	25	30
0	.038	−7	−20												
5	.049	−1	−9	−24											
10	.063	5	−2	−10	−27										
15	.081	11	6	0	−9										
20	.103	16	12	8	2	−21									
25	.130	22	19	15	10	3	−15								
30	.164	27	25	21	18	8	−7								
35	.203	33	30	28	23	17	7	−11							
40	.247	38	35	33	30	25	18	7	−14						
45	.298	43	41	38	36	31	25	18	7	−14					
50	.360	48	46	44	42	37	32	26	18	8	−13				
55	.432	53	51	50	48	43	38	33	27	20	9	−12			
60	.517	58	57	55	53	49	45	40	35	29	21	11	−8		
65	.616	63	62	60	59	55	51	47	42	37	31	24	14		
70	.732	69	67	65	64	61	57	53	49	44	39	33	26	−11	
75	.866	74	72	71	68	66	63	59	55	51	47	42	36	15	
80	1.022	79	77	76	74	72	68	65	62	58	54	50	44	28	−7
85	1.201	84	82	81	80	77	74	71	68	64	61	57	52	39	19
90	1.408	89	87	86	85	82	79	76	73	70	67	63	59	48	32
95	1.645	94	93	91	90	87	85	82	79	76	73	70	66	56	43
100	1.916	99	98	96	95	93	90	87	85	82	79	76	72	63	52

NAME: _____

SECTION: _____

LABORATORY EXERCISE 5

ISOPLETHING

QUESTIONS TO THINK ABOUT

1. How are weather stations on maps numbered? (Obtain a map from your instructor or from the local weather service.)
2. What is the purpose of any map?
3. What does the prefix *iso* mean?
4. How do we obtain the raw data to draw the isopleth lines on the map?

DISCUSSION

Meteorologists and climatologists make extensive use of maps. It allows them to visualize the total observed data over a very large horizontal area. In order to analyze a map both correctly and efficiently, you must have a feeling for the term you are using. Some of the more familiar terms are:

1. Isobar: Line of equal pressure.
2. Isotherm: Line of equal temperature.
3. Isotach: Line of equal wind speed.
4. Isogon: Line of equal wind direction.
5. Isodrosotherm: Line of equal dew point.
6. Isollobar: Line of equal pressure change.
7. Isohyet: Line of equal precipitation.
8. Isopleth: The general term for a line of equal value.
9. Isohel: Line of equal solar energy.
10. Streamline: Line which is drawn parallel to the wind flow. It has no numerical value associated with it.

RULES FOR MAP ANALYSIS

1. Choose the proper interval you wish to use (so many degrees for temperature, so many millibars for pressure, etc.)
2. Draw the isopleths so that the higher or lower values are always on the same side of each line drawn.
3. Draw lines in a smooth curved flow.
4. Don't go into areas that have no data, for example, off the map surface.
5. Label all lines drawn.
6. Two different value lines can never cross one another.

Materials Needed
1. Soft pencil
2. Good eraser
3. Maps (provided in this exercise)

Procedure
1. Analyze Map I for wind flow (streamline analysis). Start anywhere and draw the lines parallel to the wind at each reporting station. In order to help you draw these lines, imagine that you are standing at each point with your back to the wind.
2. Analyze Map II for sunshine (isohel analysis). Use a 50 langley interval.
3. Analyze Map III for pressure tendency (isallobar analysis). Use a 1 millibar interval and draw for all plus (+), zero, and minus (−) lines.

Questions
1. List areas of diverging and converging winds on Map I.
2. In what general direction does the wind blow in the United States? Why?
3. Why does south Florida experience wind from a different direction than the rest of the United States.
4. What makes the general south to north decrease in the amount of sunshine? How does the Rocky Mountains affect this trend? Why?
5. List all of the maximum and minimum areas of pressure change by geographic region.

Map 1 Streamline analysis. Wind flow

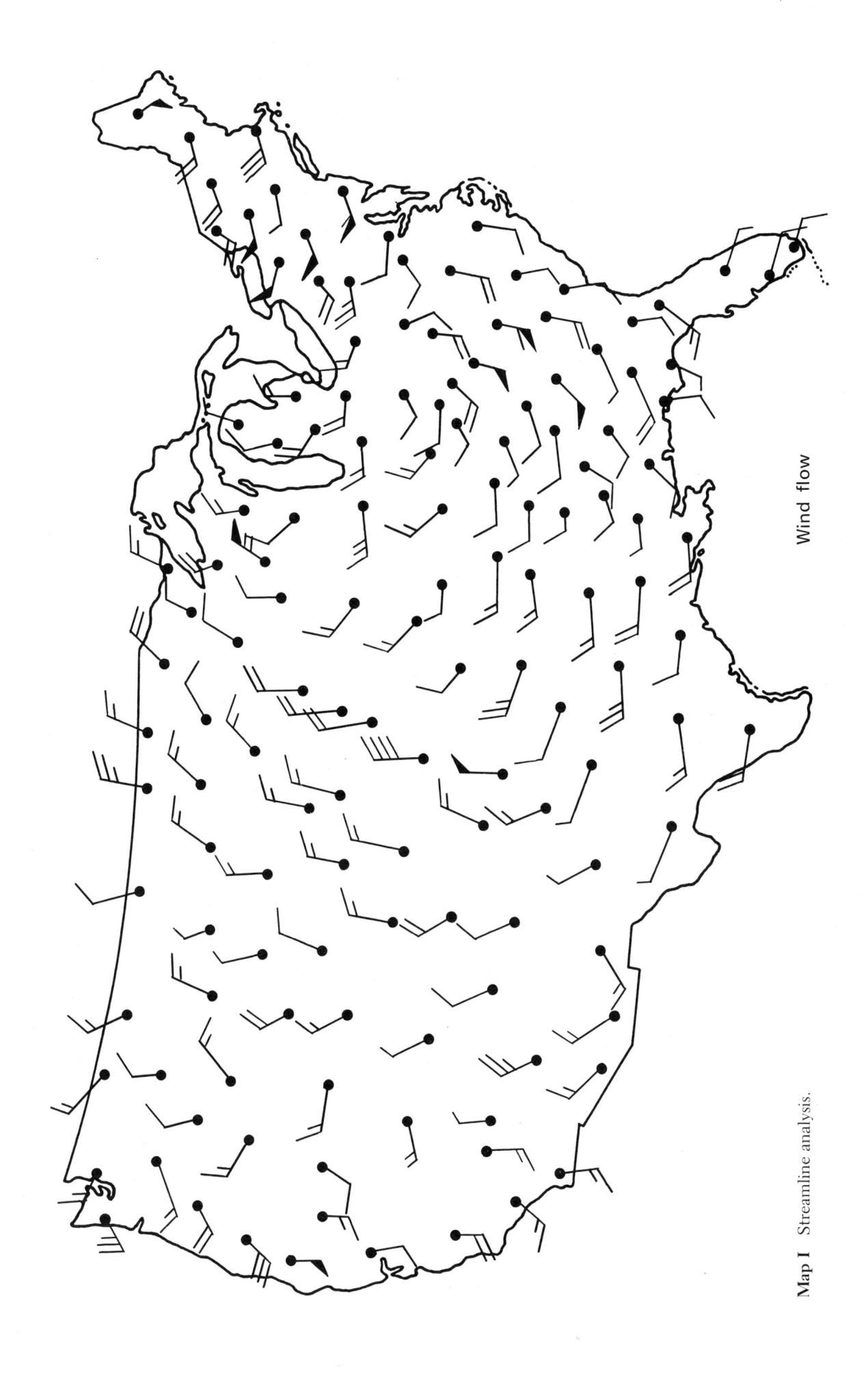

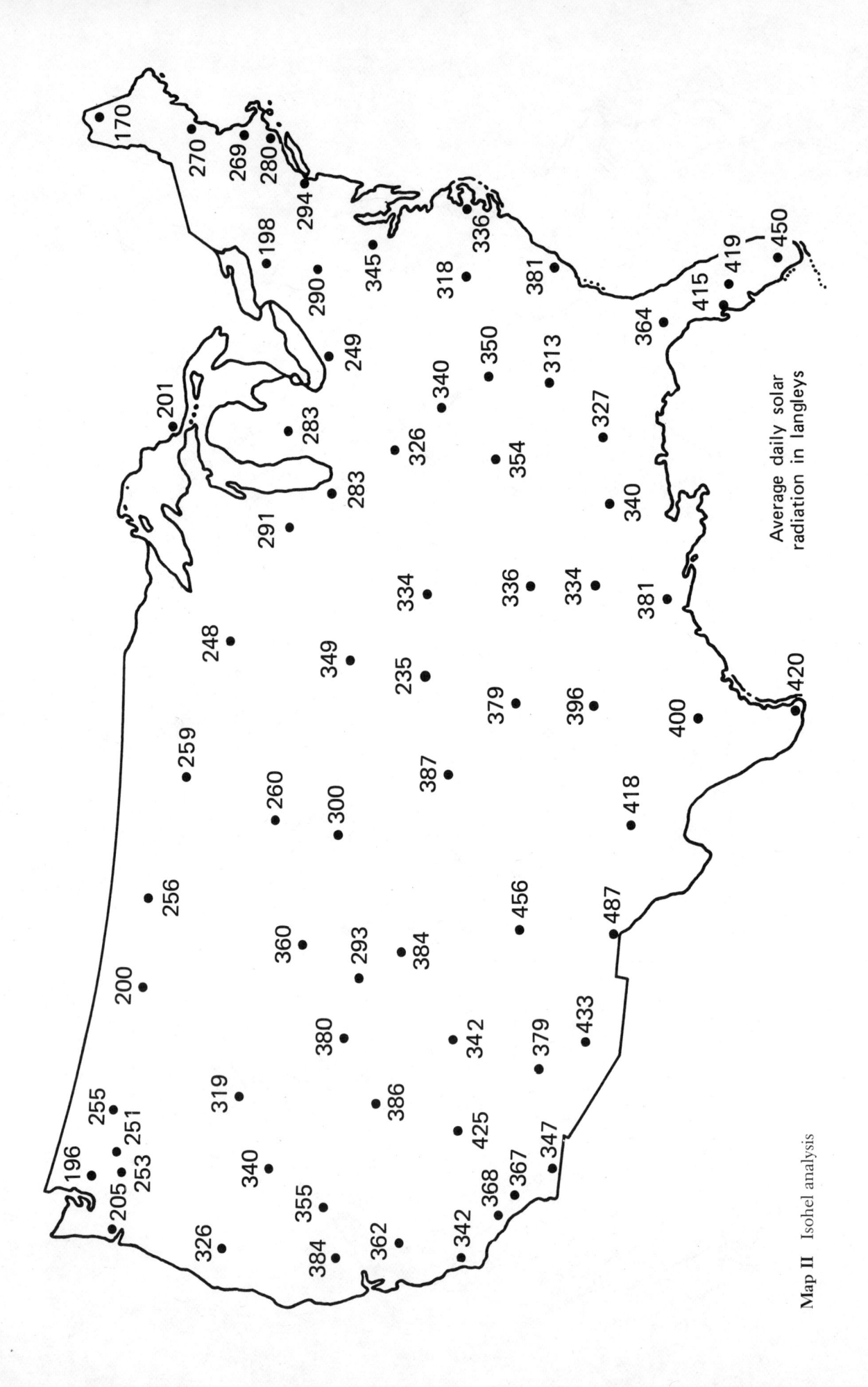

Map II Isohel analysis

Average daily solar radiation in langleys

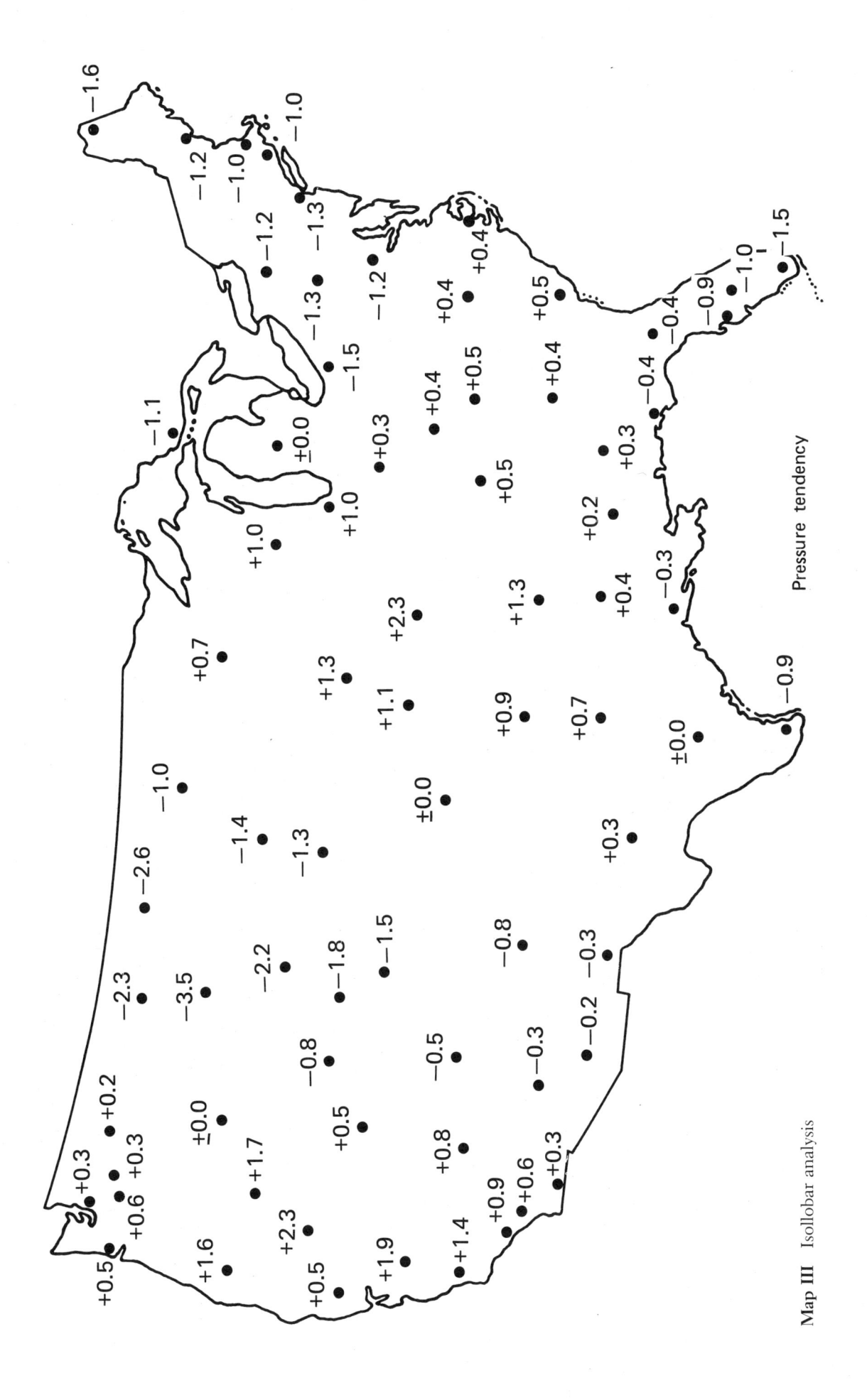

Map III Isollobar analysis — Pressure tendency

NAME: _____

SECTION: _____

LABORATORY EXERCISE 6

OBSERVATION AND MAP PLOTTING
REFERENCE: Chapter 12

QUESTIONS TO THINK ABOUT

1. Why do meteorologists around the world use Greenwich Mean Time to make their observations?
2. How are observations taken in the ocean areas? List at least two ways.
3. How are upper air observations taken?
4. How are satellites and radar used in weather observations?

DISCUSSION

There are four processes a meteorologist must complete to arrive ultimately at a forecast of future weather conditions. These distinct stages are:

1. Observing and transmitting weather phenomena.
2. Plotting a weather map of the observed phenomena.
3. Analyzing the plotted weather map.
4. Interpreting the finished weather map and issuing a forecast.

1. Observing and Transmitting Weather Phenomena

One of the basic duties of many meteorologists throughout the world is to take hourly weather observations. These observations are designed to include all the various known weather elements because these elements are the basic data of weather interpretation. The observations are taken both by instruments and sight. These are then recorded in a definite sequence for ease of handling and encoding. Below is a list of weather elements and the manner in which they are generally observed.

A. Amount of sky cover	Sight
B. Wind direction and speed	Instrument (aerovane)
C. Visibility	Sight and instrument
D. Temperature	Instrument (thermometer)
E. Dew point	Calculation from instrument
F. Relative humidity	Calculation from instrument
G. Type of cloud	Sight
H. Height of cloud	Estimation and instrument
I. Present weather	Sight
J. Pressure	Instrument (barometer)
K. Pressure tendency	Instrument (barograph)
L. Amount of rainfall	Instrument (raingage)

After the weather observation is taken, it is put into a weather code to save both time and space. Weather codes from all over the world are funneled into a central office by short wave radio and teletype. It is at this central office that all the observations can be properly sequenced and used. Any National Weather Service office that has teletype receiving equipment can also receive the encoded weather data and utilize these data for plotting and analyzing their own weather maps. An example of the weather code in both alphabetical representation and actual data is given below. (You need not learn the code.)

ALPHABET REPRESENTATION

iii Nddff VVwwW PPPTT $N_hC_LhC_mC_H$ $T_dT_d$app

ACTUAL DATA

509 83610 32707 01528 864xx 27710

2. Plotting a Weather Map

The second step toward completing a weather forecast is to decode the weather code from all parts of the world and plot the information received on a weather map. This step is quite necessary since it gives the meteorologist a view of the entire section of the globe which he or she is analyzing. Since weather at a particular site is related to global weather conditions, the horizontal two-dimensional view that is obtained by the weather map gives not only a better insight into surface weather conditions but also to the vertical conditions of the atmosphere.

Decoding and Plotting

Each observing station is located on a weather map by a circle designating its geographical location. This station also has a station number assigned to it (designated iii in the weather code) so that the station can be found easily on large weather maps. The encoded weather message is decoded and plotted around the station circle. (*Note:* Not all the weather coding is shown in this model because of the limited time for the lab).

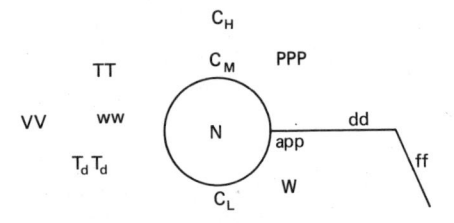

```
                        C_H
              TT       C_M    PPP
    VV      ww                        dd
                       N     app
           T_d T_d                        ff
                        C_L   W
```

Below is a list of the weather symbols and their explanations:

Symbol	Explanation
N	The number of tenths of sky covered by clouds.

○ Clear   ◐ 5/10   ◕ 9/10

◍ 1/10   ◕ 6/10   ● 10/10

◔ 2/10 or 3/10   ◕ 7/10 or 8/10   ⊗ Sky obscured by weather

◑ 4/10

dd Wind direction barb to show the direction the wind is coming from. (*Note:* Calm winds are denoted by a circle drawn around the station circle.)

ff Force of the wind in knots (1 knot is approximately $1\frac{1}{8}$ mph). (*Note:* Always round off observation to nearest 5 kt.)

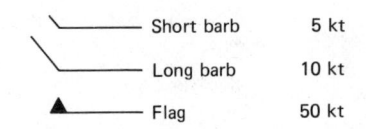

Short barb — 5 kt
Long barb — 10 kt
Flag — 50 kt

VV   Visibility in miles—not plotted if greater than 10 miles.
ww   Present weather—by symbols.

Here are some of the more familiar weather symbols.

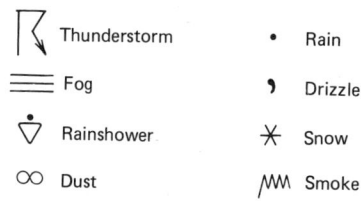

⎡⎨ Thunderstorm   •   Rain
≡   Fog   ,   Drizzle
▽̇   Rainshower   ✱   Snow
∞   Dust   ⋀⋀⋀ Smoke
∼   Freezing precipitation

W   Past weather—occurred in the past six hours.
PPP   Barometric pressure at sea level—plotted in tens, units, and tenths of millibars without the decimal point.
         1010.3——103
         996.3——963
         1014.2——142
TT   Temperature in whole degrees fahrenheit.
C_L Type of lowest cloud.
C_M Type of middle layer cloud.
C_H Type of high layer cloud.
$T_d T_d$ Dewpoint in whole degrees.
app   Pressure tendency—explains how the barometer acted in the past three hours.
         Examples:

\   Falling steadily

_ Falling then rising, lower not than it was three hours ago

—   Steady in the past three hours

A plotted station model of the previously given station 509 (Boston) is shown below.

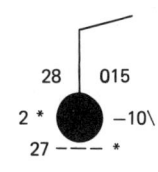

```
        28  | 015
        2 *  ● —10\
        27 --- *
```

Explanation:

Clouds—overcast   Weather—snow
Barometer—1001.5 mbs   Visibility—2 miles
Temperature—28°   Low clouds—stratus layer type
Dew point—27°   Past weather—snow
Wind direction—north at 10 kt
Barometric tendency—barometer falling steadily 0.10 mb in the past three hours.

202 OBSERVATION AND MAP PLOTTING

EXERCISE 1

*Decode the following abbreviated station models by entering the information on the following table.

* Use the tables in Appendix III.

```
        73   054              53   045              21   111
1.  3 , ⊗          5.  10     ●         8.  1 *     ◐
        62                     21                    21

        31   993              58   081              62   172
2.  *   ◐───       6.  3 ⌐⌐   ●         9.  5 ∨   ⊗
        30                     58                    61
                                                     

        78   010              81   143              18   034
3.  5 ⌒⌒  ○        7.       ◐        10.          ◑
        61                    32                    13

        63   090
4. 1/8 ≡  ⊛
        63
```

Number	Cloud Cover	Wind Direct Speed	Pressure	Temperature	Weather	Visibility	Dew Point

… EXERCISE 203

EXERCISE 2

On the accompanying map, plot the abbreviated station model for each station using the following observational data. Use the previously given station model as a guide and the tables in Appendix 3.

Number	Cloud Cover	Wind Direct Speed	Pressure	Temperature	Weather	Visibility	Dew Point
407	Clear	SW—10	1012.8	70	—	12	63
408	1/10	SW—15	1011.0	70	—	15	65
409	1/10	SW—15	1010.9	69	—	15	65
410	10/10	SSE—5	1008.0	67	Fog	1	67
411	6/10	SW—10	1007.2	68	—	10	65
412	5/10	SSW—15	1010.4	71	—	15	65
413	3/10	SSW—5	1011.3	70	—	15	63
414	10/10	NE—15	1005.0	50	—	15	42
415	10/10	S—10	1004.8	66	Rain	5	66
416	10/10	SE—15	1005.3	65	Rain	3	65
417	5/10	SSW—15	1009.5	72	—	12	67
418	2/10	SSW—15	1010.2	70	—	10	65
419	1/10	NE—20	1005.8	50	—	10	39
420	2/10	NE—20	1004.2	48	—	10	39
421	5/10	N—20	999.0	57	—	10	53
422	10/10	S—10	1004.0	68	Rain	2	67
423	1/10	SSW—5	1007.6	71	—	8	66
424	10/10	S—15	999.5	63	Rain	2	63
425	10/10	W—15	1003.2	67	Thunderstorm	7	67
426	Clear	N—10	1004.8	50	—	15	39
427	1/10	NNE—10	1002.7	55	—	15	43
428	2/10	NNW—5	1004.0	57	—	15	50
429	5/10	NW—5	1008.3	58	—	15	55
430	10/10	NW—20	1007.6	65	Thunderstorm	5	65
431	10/10	N—5	1010.0	57	Rain shower	3	56
432	Clear	NNE—10	1009.0	47	—	15	36
433	Clear	N—10	1008.6	52	—	12	35
434	2/10	NNW—5	1010.0	56	—	10	51
435	Clear	N—10	1011.1	46	—	15	32
436	Clear	NNW—10	1012.2	53	—	15	37
437	Clear	NNW—10	1014.6	52	—	10	38
438	Clear	NNE—5	1005.8	54	—	15	36
439	1/10	N—10	1013.8	55	—	15	50
440	3/10	N—10	1011.6	55	—	10	54
441	3/10	N—10	1012.3	57	—	10	53
442	5/10	NNE—20	1014.2	59	—	10	56

Map IV

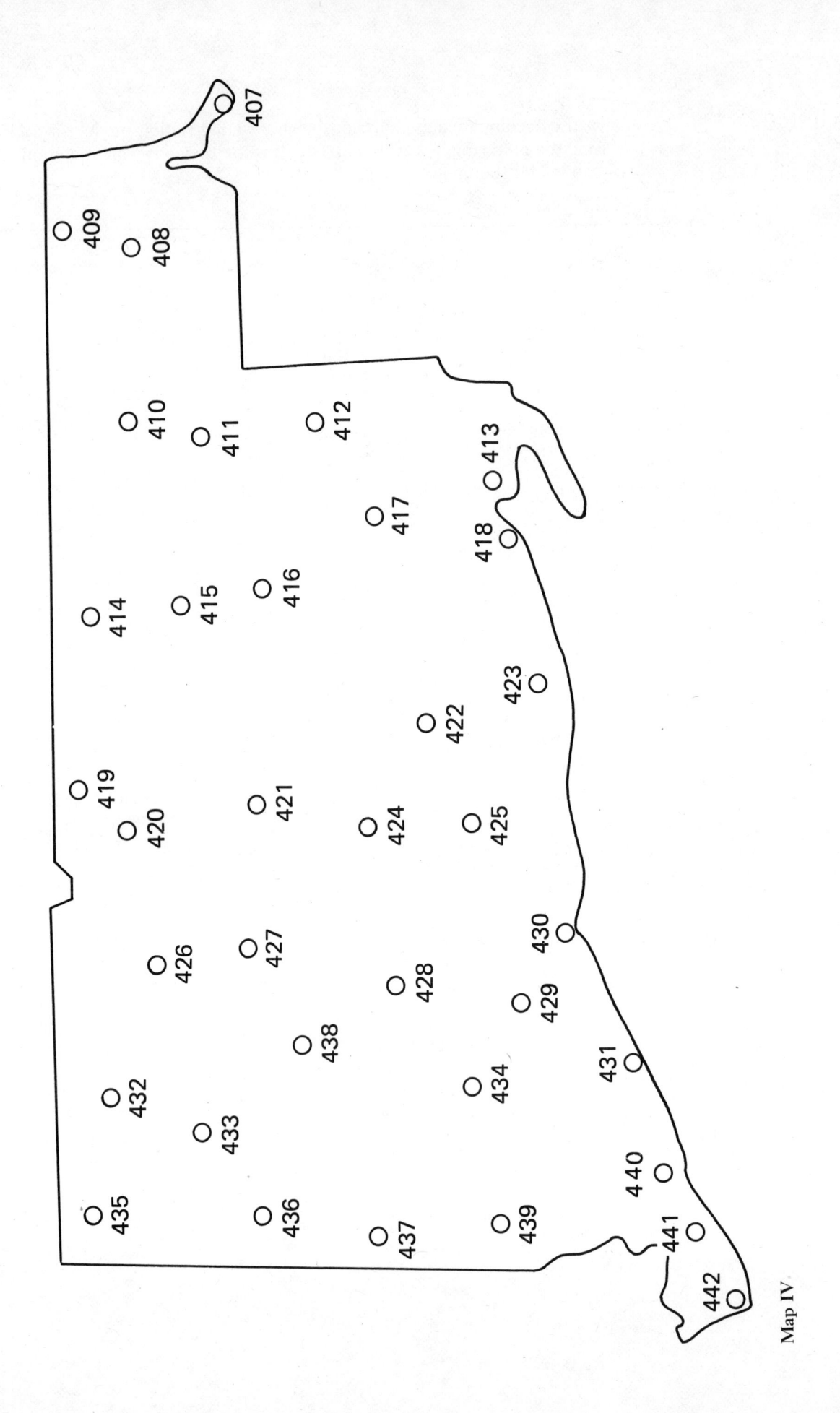

NAME: _____

SECTION: _____

LABORATORY EXERCISE 7

SURFACE MAP ANALYSIS
REFERENCE: Chapter 12

QUESTIONS TO THINK ABOUT

1. Why do pressure systems move?
2. Why are weather fronts only associated with low pressure areas?
3. Why is bad weather associated with low pressure areas and good weather with high pressure systems?
4. Why do pressure systems move in a westerly to easterly direction in the United States?

DISCUSSION

The third step in the completion of a meteorological forecast is analyzing a previously plotted surface weather map. A technique was developed at the turn of the twentieth century in which pressure systems, air masses, fronts, and other existing conditions could be readily identified. (See the end of the exercise for definitions.) By completing a series of pressure analyses, a meteorologist can establish some sort of historical sequence and thus determine how the weather elements are changing and in which way weather systems are moving. This technique is called an *isobaric analysis*, and it entails the drawing of lines of equal pressure called *isobars* (*Iso* means equal, *bar* refers to measure of pressure). These lines are drawn on a map at set intervals parallel to the wind, keeping in mind *Buys–Ballot's law*.

Buys–Ballot's Law
If a person stands with his back to the wind, in the northern hemisphere, the lowest pressure will be on the left. This is a way of readily identifying high and low pressure areas.

Rules of Isobaric Analysis
1. Isobars should be drawn in intervals of 4 mb using 1002 mb as a reference point.

Example

994, 998, 1002, 1006, 1010

2. After the isobar is drawn, it should be realized that the line designates all points on that line as being the same value.
3. Isobars are smooth curved lines and cannot have sharp points. (One exception to this rule is explained in the next lab.)

Example

Right   Wrong

4. Two isobars cannot cross each other but may come very close. If two different-valued isobars did cross, the intersection of the two would have two distinct values, which is impossible.

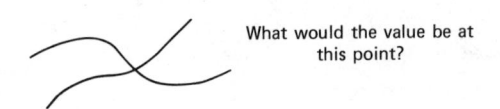

5. Isobars can be drawn between two stations that have different values. This is called interpolation. If the value of the isobar lies between the value of the two stations, then the isobar is drawn between the two.

Example

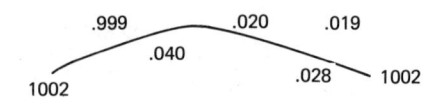

If the value at the station is at the precise interval desired or very close to that interval, then the isobar goes through the station circle.

6. An isobar can *never* stop in the middle of a map, since an isobar is a continuously drawn line of equal pressure that connects itself somewhere on the earth. However, since we are working with a limited map space, the isobars do stop at the border of the map. We stop because we lack the necessary data to proceed further.

Procedures for Isobaric Analysis

1. Pick a point on the map to start. (However, it is easier if you can find the high or low pressure centers first.) The pressure reading can be found in the upper right corner of the station.
2. Decode the true pressure reading and try to fit any chosen interval (1002, 1006, etc.) between any two stations in the area. If the station model has a pressure that is right on the interval, then start at the station.
3. Remember to keep your back to the wind and therefore all lower pressure to the left. Draw the isobar for the selected value parallel to the wind, remembering that these are smooth curved lines.
4. Keep drawing the isobar until it connects into a closed curve or stops at the border of the map.
5. Label the isobar with the proper value.
6. Go up or down four millibars to the next value and start to draw a second isobar.
7. Do not extend the isobar into any area where there are no data.
8. Repeat Step 6 until all values have been used for the data represented.

Materials Needed
1. Soft pencil
2. Eraser

EXERCISE

On the three accompanying maps, perform an isobaric analysis, following the above eight steps.

GLOSSARY

Isobar	A line of equal pressure.
Isallobar	A line of equal pressure tendency.
Isodrosotherm	A line of equal dew point.
Isotherm	A line of equal temperature.
Air mass	A large body of air with temperature and humidity characteristics that are approximately the same over a large horizontal distance.
Front	A zone of transition between two different air masses.
Bar	A measurement of pressure.
Millibar	One one-thousandth of a bar.

Map I

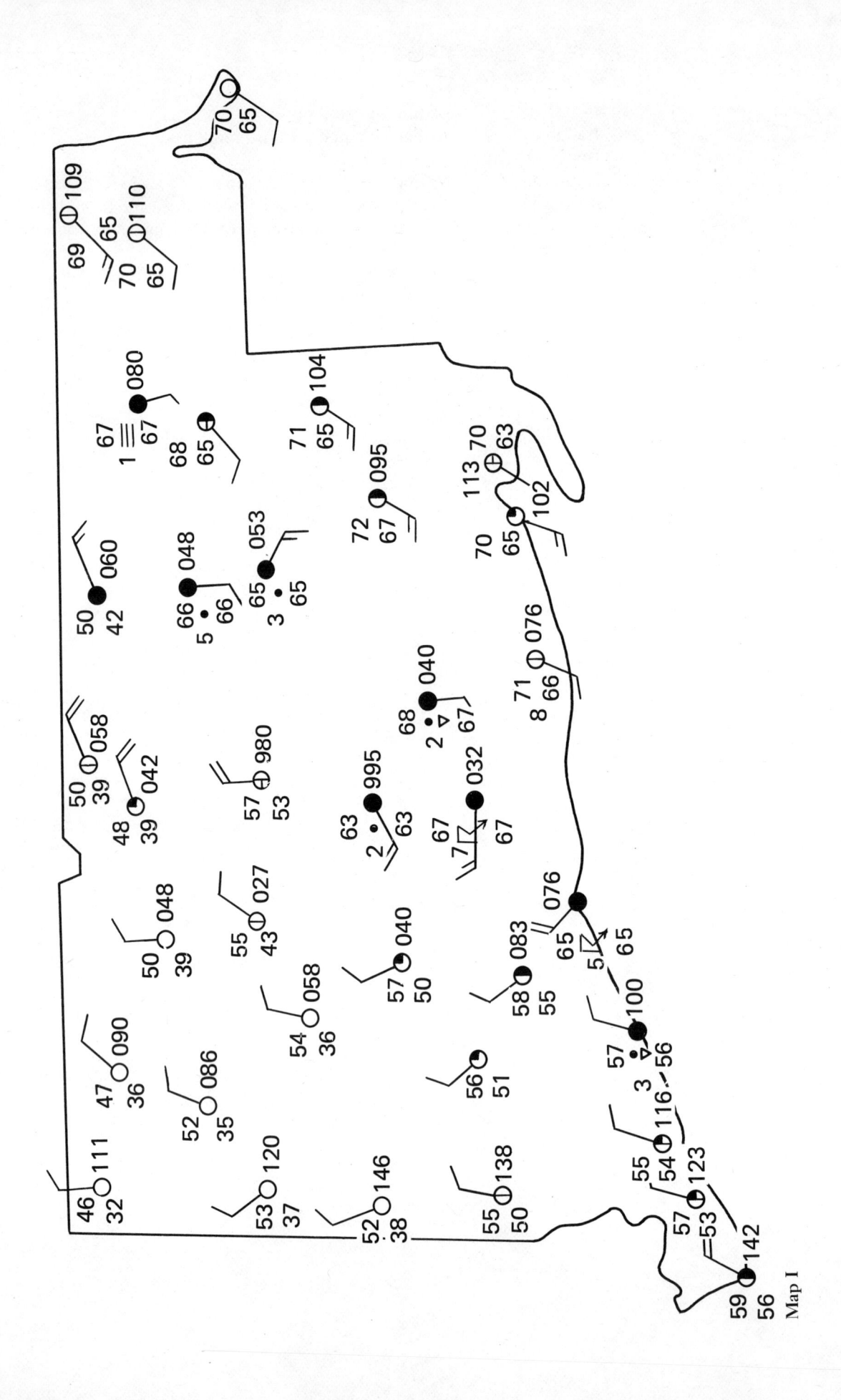

Map II

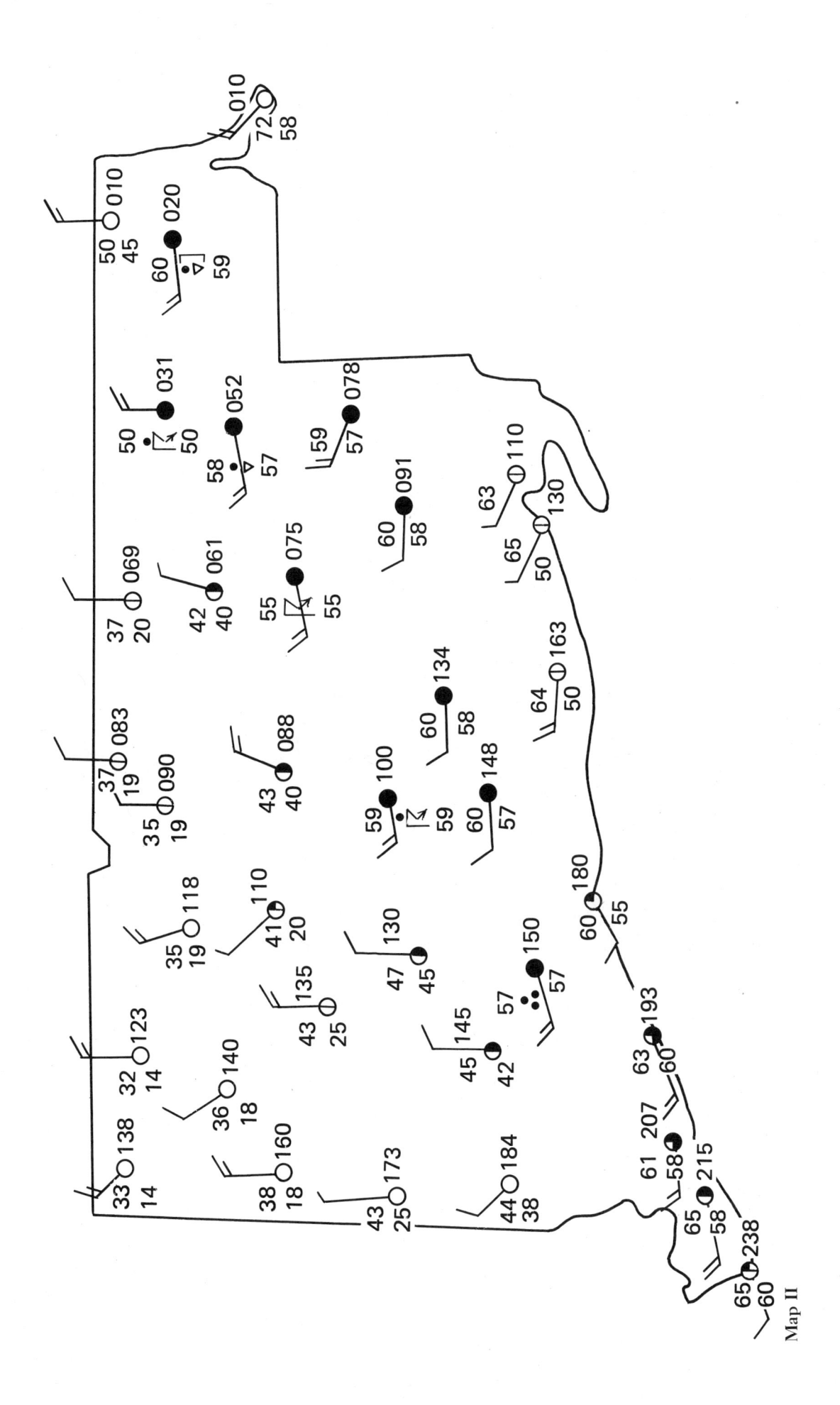

Map III

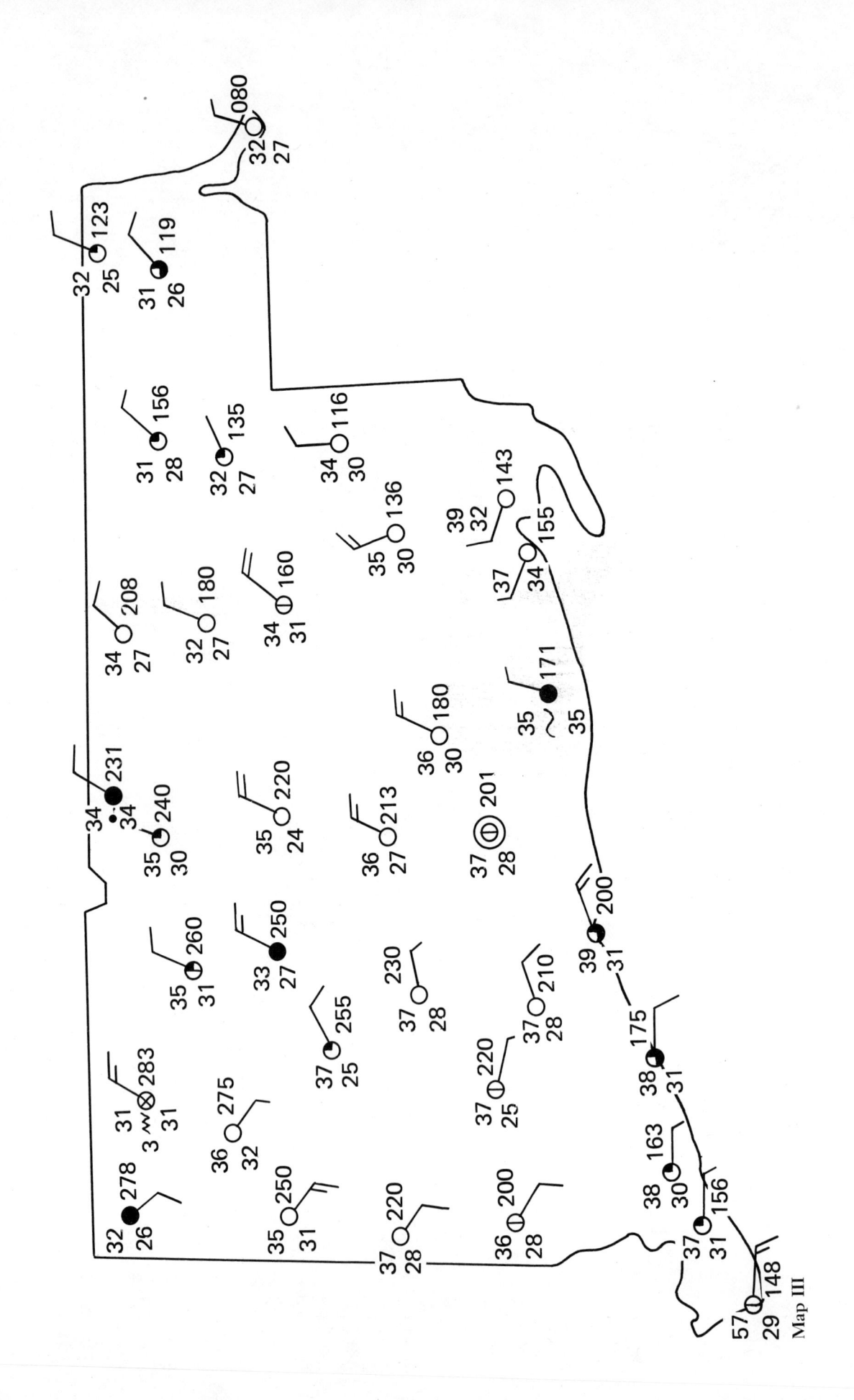

NAME: _____

SECTION: _____

LABORATORY EXERCISE 8

FRONTAL ANALYSIS AND FORECASTING
REFERENCE: Chapters 10 and 12

QUESTIONS
TO THINK ABOUT

1. Why are warm fronts associated with the front part of an approaching low pressure system and a cold front with the back part?
2. Why do cold fronts move faster than warm fronts?
3. What general weather changes would you look for with a frontal passage?
4. How are air masses related to fronts?

DISCUSSION

Frontal Analysis
Front A zone of transition between two distinct air masses. This zone of transition is denoted by a sharp line to show where the leading edge of the invading air mass is located on the surface of the earth. However, it must be noted that fronts extend up into the troposphere for quite a distance.

Every front is associated with a cyclone (low pressure area) and every cyclone is a center of bad weather. Hence, fronts are usually zones of bad weather.

Types of Fronts
Cold Front A line of discontinuity separating the cold air mass that is wedging beneath a warmer air mass.

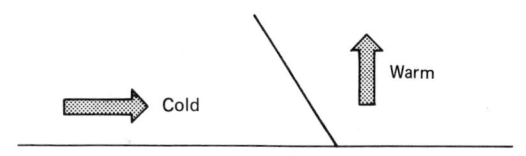

Warm Front A line of discontinuity separating the advancing warm air mass that overides a colder air mass.

Occluded Front A line of discontinuity between two fronts. The cold front moves faster than a warm front and eventually overtakes it.

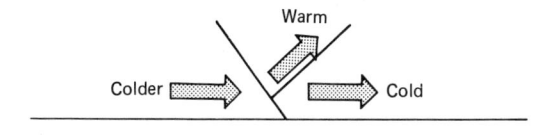

Stationary Front A line of discontinuity between two air masses. However, neither air mass is strong enough to invade the other. These types of fronts are usually identified by past history.

Frontal Identification on Surface Maps

1. Wind. Along the front a sharp contrast between wind directions usually occurs. Look for areas of wind shift:

Example

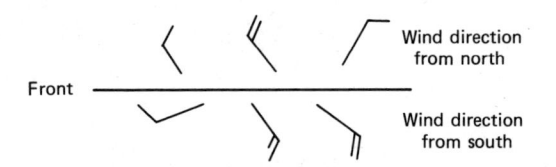

2. Temperature. A large temperature difference will usually result across the front (warm and cold fronts only). Look for areas of large temperature difference.
3. Weather. A front is a bad weather zone. Look for areas of bad weather. Thunderstorm activity or showers usually denote a cold front while steady rain or drizzle a warm front. Along a stationary front a combination of the two fronts exist. Fog can occur ahead of a front.
4. Temperature and dew point range. In back of a cold front the difference between the temperature and the dew point is large while near the front they will be small or they will both be the same.
5. Pressure gradient. Fronts are always located in a trough of low pressure and isobars kink out of a low and into a high. This is the only time in which an isobar is not a smooth curved line.

Example

Although there are other ways to identify fronts, these five steps will suffice as an introduction to the frontal analysis exercise.

Phenomena	Cold	Warm	Stationary	Occluded
1. Wind	Abrupt wind shift from S	Shift from SE to SW	Wind shift very small	Confused
2. Weather	Thunderstorms and showers followed by rapid clearing	Steady rain or drizzle and some fog	Very often no weather	Very large rain area
3. Temperature	Decreases after front passes	Increases after front passes	No significant temperature change	—
4. Dew point	Large spread	Minimum spread	Little difference on either side of the front	Small spread

EXERCISE

Materials Needed
1. Red, blue, green, and yellow pencil
2. Eraser

1. On Maps I and II of lab exercise 7 insert the surface frontal boundaries and high and low pressure systems.
2. Using Map I, complete the attached forecast table using your best guess as to the weather for 12 and 24 hours after map time.

Procedure
a. Locate the front by the five steps previously given.
b. Kink isobars over front so that they kink out of a low or into a high pressure area.

	Present Maptime		
	Map I Station 422	Map I Station 424	Map I Station 426
Sky cover			
Wind speed			
Wind direction			
Visibility			
Weather			
Temperature			
Dew point			
	12 Hours After Map Time		
Sky cover			
Wind speed			
Wind direction			
Visibility			
Weather			
Temperature			
Dew point			
	24 Hours After Map Time		
Sky cover			
Wind speed			
Wind direction			
Visibility			
Weather			
Temperature			
Dew point			

c. Color the fronts with the appropriate color.

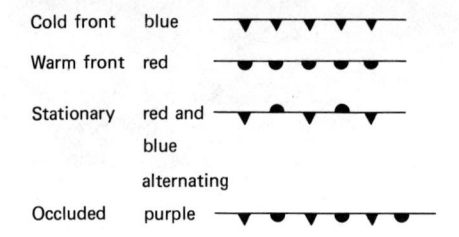

Cold front	blue	▼▼▼▼
Warm front	red	●●●●
Stationary	red and blue alternating	▼●▼●
Occluded	purple	▼●▼●▼●

d. Color in weather areas.
   - Precipitation — Green shading over station with rain
   - Fog — Yellow shading
   - Thunderstorm — Red symbol over station

e. Label high and low pressure areas with colored pencil.
   - High—A blue H
   - Low—A red L

NAME: _____

SECTION: _____

LABORATORY EXERCISE 9

UPPER ATMOSPHERE OBSERVATIONS
REFERENCE: Chapters 4 and 11

QUESTIONS TO THINK ABOUT

1. Who would make use of upper air weather data?
2. How does an air particle react if it is stable, unstable, or neutral?
3. Is the upper atmosphere an important consideration when forecasting what will happen on the surface of the earth?

DISCUSSION

In order to explore the changing upper atmosphere, many weather stations around the world send aloft a helium- or hydrogen-filled balloon with an attached meteorological instrument. This instrument, called a radiosonde, transmits weather data back to the earth's surface via a radio transmitter attached to the instrumented package. These data, consisting of temperature, pressure, and dew point, allows the meteorologist to draw a vertical cross section of weather data above or near the balloon's release point.

After receiving the upper air data, the meteorologist plots the information on a chart called a thermodynamic diagram. Although there are different types of thermodynamic diagrams, we shall use a simplified version for this laboratory exercise.

Description of the Adiabatic Diagram
1. Solid vertical lines—Temperature of the air in degrees Celsius.
2. Solid horizontal lines—Pressure lines from 1050 to 600 mb. Approximate height values based on the standard atmosphere, located to the right of the diagram, can also be applied to these lines.
3. Slanted lines—Dry adiabatic lines representing a decrease (if you go up) or increase (if you go down) of 10°C/km (55°F/1000 ft).

EXERCISE 1
Plotting the Diagram

Materials Needed
1. Pen
2. Ruler
3. Soft pencil
4. Red and blue pencils

Use the data supplied with this exercise to plot the information on the accompanying upper air diagrams. Use a pen to connect the temperature points with a series of solid straight lines. Do the same for the dew point but use dashed lines.

EXERCISE 2
Inversion Locations

1. List all inversions found on the diagram by indicating their bases and heights in millibars of pressure.
2. Indicate the type of inversion you found by using the following information.

216 UPPER ATMOSPHERE OBSERVATIONS

	Data	
Pressure (mb)	Temperature (°C)	Dew Point (°C)
1000	+4.0	+0.0
950	+7.0	+1.0
880	+5.0	+0.0
850	+1.0	−1.0
800	−3.0	−4.0
730	+4.0	+2.0
690	−8.0	−10.0
600	−16.0	−18.0
520	−28.0	−35.0
500	−24.0	−45.0
400	−40.0	−55.0
300	−55.0	Too dry for a reading

(a) Surface inversion—Increase of temperature with height starting at the earth's surface.
(b) Subsidence inversion—Increase of temperature with height aloft in very dry air (temperature and dew point far apart).
(c) Frontal inversion—Increase of temperature with height aloft in very moist air (temperature and dew point close together).

Inversion Base	Inversion Height	Type of Inversion
_____	_____	_____
_____	_____	_____
_____	_____	_____
_____	_____	_____

EXERCISE 3
Freezing Point and Icing Levels

1. Locate the freezing level—level at which the temperature curve crosses the 0°C temperature line.
   Freezing Point _____ °C
2. Locate the icing levels—levels at which the temperature is less than freezing and the temperature–dew point spread is very small (less than 5°C).

Bottom of Icing Levels (mb)	Top of Icing Level (mb)
_____	_____
_____	_____
_____	_____
_____	_____

EXERCISE 4
Stable and Unstable Areas

Locate stable and unstable areas by starting at the surface temperature point and moving up or parallel to the dry adiabatic line. (In this exercise we will assume the air is always dry.)
(a) Whenever this line is to the left of the temperature curve (colder), the air is stable: Color this area blue.

(b) Whenever this line is to the right of the temperature curve (warmer), the air is unstable: Color this area red.

EXERCISE 5
Thunderstorm Forecasting

One method of forecasting thunderstorm activity is to find the surface temperature at which, if the air was forced to rise, it would always be unstable.

Therefore, go along the dry adiabatic lines until you find the first line showing that the air will always be unstable. (The temperature curve is always to the left of the adiabatic line.) Slide down this line to the 1000 mb height and read the temperature.

Record your results here: _____.

This is the surface temperature that must be reached if thunderstorm activity is to be forecast. (Of course, enough moisture must be available in the air.)

EXERCISE 6
Cloud Layers

Cloud layer locations and their thickness can be approximated by locating areas of the diagram where the temperature–dew point spread is very close (less than 1°C). Locate all cloud layers from this observation.

Cloud Bases (mb)	Cloud Tops (mb)
_____	_____
_____	_____
_____	_____
_____	_____

218 UPPER ATMOSPHERE OBSERVATIONS

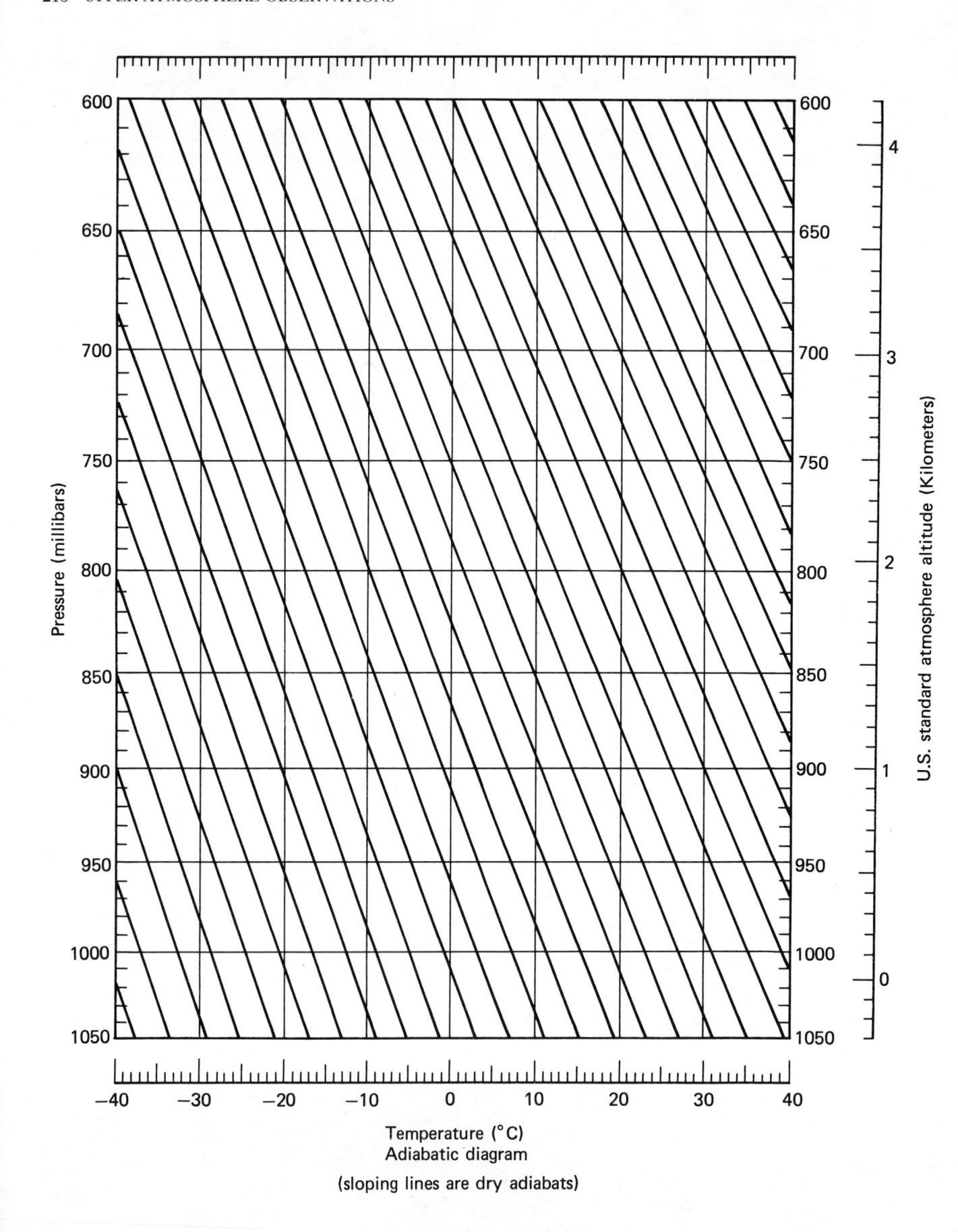

Temperature (°C)
Adiabatic diagram
(sloping lines are dry adiabats)

NAME: _____

SECTION: _____

LABORATORY EXERCISE 10

SUMMERTIME TEMPERATURE ANALYSIS
REFERENCE: Chapters 4 and 13

QUESTIONS TO THINK ABOUT

1. What effect do large water bodies, urban population areas and mountains have on summertime temperatures?
2. What would be the temperature difference if a mountain chain ran north–south as opposed to east–west?
3. What is the warmest part of the United States? Explain your answer.

DISCUSSION

There are many factors affecting the instantaneous distribution of solar radiation, and thus the temperature over the earth. Some of the more obvious factors are the time of the year, the hour of the day, and amount of cloudiness. However, by taking into consideration average temperatures over one specified time period, these specific factors can be ignored or eliminated entirely. Hence, by plotting the mean temperature readings on a map and then completing an analysis of these readings, the student can see the affect of the physical features of the earth on our temperatures. We shall see an example of this application from the temperature distribution map of the United States that will be plotted for this exercise. When finishing the map be sure that the physical features of continentality, oceanity, mountain barriers, and vegetation are realized and compared with the actual temperature distribution.

Materials Needed
1. Weather map of the United States
2. Soft pencil
3. Eraser
4. Red-colored pencil
5. Pen

EXERCISE

Plot the accompanying data in ink on a weather map provided by your instructor. Analyze this data with pencil using a 10°F interval. Retrace the finished product in red pencil. Be sure to label all lines.

Questions
1. Why do the isotherms run parallel around the Pacific coastline and perpendicular to the Atlantic coastline?
2. How do the Rocky Mountains affect the summertime temperature pattern? Do the Appalachian Mountains also affect this pattern? Why?
3. What part of the United States is the warmest? Explain your answer.

220 SUMMERTIME TEMPERATURE ANALYSIS

Data

Station Location	Station Number	July Temperature °F
1. Montgomery, Alabama	226	81.3
2. Birmingham, Alabama	228	80.0
3. Miami, Florida	202	81.7
4. Pensacola, Florida	222	80.6
5. Savannah, Georgia	207	81.2
6. Atlanta, Georgia	219	78.5
7. New Orleans, Louisiana	231	80.1
8. Vicksburg, Mississippi	VKS	81.0
9. Greensboro, North Carolina	317	78.0
10. Mobile, Alabama	223	82.6
11. Shreveport, Louisiana	248	83.7
12. Bristol, Tennessee	TRI	75.4
13. Roanoke, Virginia	411	76.6
14. Parkersburg, West Virginia	413	76.0
15. Richmond, Virginia	401	78.0
16. Memphis, Tennessee	334	80.9
17. Pittsburgh, Pennsylvania	520	74.2
18. Caribou, Maine	712	65.0
19. Greenville, Maine	619	65.0
20. Eastport, Maine	608	60.5
21. Burlington, Vermont	627	69.4
22. Boston, Massachusetts	509	72.4
23. Nantucket, Massachusetts	506	68.0
24. New York, New York	503	74.1
25. Albany, New York	518	70.1
26. Chicago, Illinois	534	75.7
27. Springfield, Illinois	439	78.0
28. Evansville, Indiana	432	79.6
29. Indianapolis, Indiana	438	76.3
30. Kansas City, Missouri	446	80.0
31. St. Louis, Missouri	434	80.2
32. Columbus, Ohio	428	75.4
33. Columbia, South Carolina	310	80.9
34. Sault Ste. Marie, Michigan	734	64.1
35. Houghton, Michigan	744	65.5
36. Traverse City, Michigan	TVC	70.1
37. Detroit, Michigan	537	73.1
38. Alexandria, Minnesota	AXN	69.7
39. International Falls, Minnesota	747	67.5
40. Duluth, Minnesota	745	64.5
41. Minneapolis, Minnesota	658	73.2
42. Miles City, Montana	MLS (near 667)	72.9
43. Cut Bank, Montana	CTB	65.1
44. North Platte, Nebraska	562	75.4
45. Helena, Montana	772	65.7
46. Omaha, Nebraska	553	78.1
47. Bismark, North Dakota	764	70.9
48. Williston, North Dakota	767	69.4
49. Sioux Falls, South Dakota	651	73.6
50. Pierre, South Dakota	PIR	73.3
51. Rapid City, South Dakota	662	71.4
52. Laramie, Wyoming	LAR	63.8
53. Casper, Wyoming	569	72.1
54. Sheridan, Wyoming	666	71.2
55. Idaho Falls, Idaho	IDA	67.1
56. Harve, Montana	777	70.2
57. Green Bay, Wisconsin	645	70.0
58. Milwaukee, Wisconsin	640	70.1
59. Madison, Wisconsin	641	72.1
60. Park Falls, Wisconsin	741	67.2
61. Boise, Idaho	681	74.2

	Data	
Station Location	Station Number	July Temperature °F
62. Dubuque, Iowa	547	74.6
63. Des Moines, Iowa	546	76.3
64. Dodge City, Kansas	451	78.7
65. Wichita, Kansas	450	80.4
66. Goodland, Kansas	465	75.4
67. Little Rock, Arkansas	340	81.2
68. Colorado Springs, Colorado	COS	68.2
69. Denver, Colorado	469	72.5
70. Grand Junction, Colorado	476	77.9
71. Oklahoma City, Oklahoma	353	81.6
72. Amarillo, Texas	363	76.8
73. Dallas, Texas	258	83.7
74. Houston, Texas	243	83.1
75. Brownsville, Texas	250	83.6
76. El Paso, Texas	270	81.4
77. Presidio, Texas	271	84.2
78. Austin, Texas	254	84.6
79. Tonopah, Nevada	485	74.0
80. Winnemucca, Nevada	583	71.9
81. Albuquerque, New Mexico	365	76.7
82. Grand Canyon, Arizona	378	68.8
83. Yuma, Arizona	280	91.0
84. Tucson, Arizona	274	85.1
85. Salt Lake City, Utah	572	77.0
86. Fresno, California	389	81.3
87. Los Angeles, California	295	70.5
88. Sacramento, California	483	73.9
89. San Francisco, California	494	58.9
90. Eureka, California	594	54.6
91. Medford, Oregon	597	72.0
92. Eugene, Oregon	693	66.6
93. Baker, Oregon	BKE	65.6
94. Portland, Oregon	698	66.7
95. Tatoosh Island, Washington	798	55.1
96. Spokane, Washington	785	69.0

NAME: _____

SECTION: _____

LABORATORY EXERCISE 11

WINTERTIME TEMPERATURE ANALYSIS
REFERENCES: Chapters 4 and 13

QUESTIONS TO THINK ABOUT

1. Why are areas close to the water warmer in the winter than those surrounded by a large land mass (at the same latitude)?
2. Why do temperatures get colder as you move away from the equator?
3. What are other factors that alter the temperature of an area?
4. How does the temperature pattern differ from the summertime situation?
5. Where is the greatest drop in temperature within the United States and why?

DISCUSSION

Winter temperatures are largely affected by the distribution of land and water bodies that break up the latitudinal tendencies of parallel isotherms. Since land masses warm up and cool down faster than water bodies, the winter temperatures are much warmer over ocean areas than over the colder land masses. Exceptions to this statement can be found where land is located next to a large water body or mountain barrier.

Since water is mobile and has both vertical and horizontal movements, the heat it absorbs at its surface is distributed over a very wide area. The water is also translucent and therefore energy can penetrate to a much greater depth than it can into the land. For those reasons, the temperature differences that arise can aid in indicating what type of climate prevails over a specific area.

Materials Needed

1. Weather map of the United States
2. Soft pencil
3. Eraser
4. Blue-colored pencil
5. Pen

EXERCISE

Plot the accompanying data in ink on a weather map provided by your instructor. Analyze these data with pencil using a 10°F interval. Retrace the finished product in blue pencil. Be sure to label all lines.

Questions

1. What type of temperature pattern would exist if you analyzed only a large water body?
2. In what ways do you think the temperature pattern of the southern hemisphere differs from that of the northern hemisphere?
3. What part of the country is the coldest in the winter and why?
4. Which way do the isotherms bend when they hit the ocean areas and why?

WINTERTIME TEMPERATURE ANALYSIS

Data

Station Location	Station Number	January Temperature °F
1. Montgomery, Alabama	226	49.5
2. Birmingham, Alabama	228	46.5
3. Miami, Florida	202	68.0
4. Pensacola, Florida	222	53.6
5. Savannah, Georgia	207	52.7
6. Atlanta, Georgia	219	44.0
7. New Orleans, Louisiana	231	53.5
8. Vicksburg, Mississippi	VKS	49.6
9. Greensboro, North Carolina	317	40.6
10. Mobile, Alabama	223	53.0
11. Shreveport, Louisiana	248	47.5
12. Bristol, Tennessee	TRI	38.3
13. Roanoke, Virginia	411	38.1
14. Parkersburg, West Virginia	413	34.7
15. Richmond, Virginia	401	39.1
16. Memphis, Tennessee	334	43.3
17. Pittsburgh, Pennsylvania	520	31.6
18. Caribou, Maine	712	8.5
19. Greenville, Maine	619	12.9
20. Eastport, Maine	608	21.5
21. Burlington, Vermont	627	19.0
22. Boston, Massachusetts	509	29.8
23. Nantucket, Massachusetts	506	32.6
24. New York, New York	503	32.1
25. Albany, New York	518	25.1
26. Chicago, Illinois	534	25.3
27. Springfield, Illinois	439	28.2
28. Evansville, Indiana	432	34.8
29. Indianapolis, Indiana	438	29.5
30. Kansas City, Missouri	446	30.2
31. St. Louis, Missouri	434	32.9
32. Columbus, Ohio	428	30.2
33. Columbia, South Carolina	310	46.0
34. Sault Ste. Marie, Michigan	734	14.3
35. Houghton, Michigan	744	15.2
36. Traverse City, Michigan	TVC	22.3
37. Detroit, Michigan	537	25.5
38. Alexandria, Minnesota	AXN	7.5
39. International Falls, Minnesota	747	3.0
40. Duluth, Minnesota	745	9.1
41. Minneapolis, Minnesota	658	13.1
42. Miles City, Montana	MLS (near 667)	14.5
43. Cut Bank, Montana	CTB	16.8
44. North Platte, Nebraska	562	25.2
45. Helena, Montana	772	20.2
46. Omaha, Nebraska	553	23.7
47. Bismark, North Dakota	764	9.4
48. Williston, North Dakota	767	7.9
49. Sioux Falls, South Dakota	651	15.8
50. Pierre, South Dakota	PIR	17.3
51. Rapid City, South Dakota	662	23.4
52. Laramie, Wyoming	LAR	22.3
53. Casper, Wyoming	569	26.4
54. Sheridan, Wyoming	666	21.3
55. Idaho Falls, Idaho	IDA	15.4
56. Harve, Montana	777	13.9
57. Green Bay, Wisconsin	645	15.7
58. Milwaukee, Wisconsin	640	20.6
59. Madison, Wisconsin	641	16.7
60. Park Falls, Wisconsin	741	8.7

Data

Station Location	Station Number	January Temperature °F
61. Boise, Idaho	681	30.4
62. Dubuque, Iowa	547	19.9
63. Des Moines, Iowa	546	22.1
64. Dodge City, Kansas	451	31.2
65. Wichita, Kansas	450	32.2
66. Goodland, Kansas	465	28.9
67. Little Rock, Arkansas	340	42.6
68. Colorado Springs, Colorado	COS	30.2
69. Denver, Colorado	469	32.0
70. Grand Junction, Colorado	476	25.0
71. Oklahoma City, Oklahoma	353	37.6
72. Amarillo, Texas	363	35.3
73. Dallas, Texas	258	45.8
74. Houston, Texas	243	54.2
75. Brownsville, Texas	250	59.8
76. El Paso, Texas	270	45.4
77. Presidio, Texas	271	48.6
78. Austin, Texas	254	50.4
79. Tonopah, Nevada	485	30.1
80. Winnemucca, Nevada	583	28.0
81. Albuquerque, New Mexico	365	34.1
82. Grand Canyon, Arizona	378	28.9
83. Yuma, Arizona	280	54.6
84. Tucson, Arizona	274	49.6
85. Salt Lake City, Utah	572	30.1
86. Fresno, California	389	45.5
87. Los Angeles, California	295	54.2
88. Sacramento, California	483	45.6
89. San Francisco, California	494	49.8
90. Eureka, California	594	35.4
91. Medford, Oregon	597	47.4
92. Eugene, Oregon	693	39.1
93. Baker, Oregon	BKE	24.9
94. Portland, Oregon	698	39.4
95. Tatoosh Island, Washington	798	41.2
96. Spokane, Washington	785	27.5

NAME: _____

SECTION: _____

LABORATORY EXERCISE 12

ANNUAL TEMPERATURE RANGE ANALYSIS

QUESTIONS TO THINK ABOUT

1. Do you use the coldest and warmest month or coldest and warmest temperatures to calculate temperature range?
2. What are some influences on the annual range of temperatures? Which influences it the greatest? The least?
3. Does the annual temperature range influence how we live in certain areas of the world? What are some of these ways?

DISCUSSION

A Range is defined the difference between a maximum and minimum for any set of data. In this lab average yearly maximum and minimum of temperature are used to calculate range.

Meteorologists use this type of information to help them locate air mass source regions and tracks and to figure the effect mountains and water bodies have on these tracks. Biometeorologists can utilize this same information by applying it to the stresses on a human body over a yearly basis.

Materials Needed
1. Weather map
2. Soft pencil
3. Eraser
4. Pen

EXERCISE

Plot the accompanying data in ink on a weather map provided by your instructor. Analyze this data with pencil using a 5° interval.

Data

Station Location	Station Number	Temperature Range °F
1. Montgomery, Alabama	226	31.8
2. Birmingham, Alabama	228	30.5
3. Grand Canyon, Arizona	378	39.9
4. Yuma, Arizona	280	36.4
5. Tucson, Arizona	274	35.5
6. Little Rock, Arkansas	340	38.5
7. Fresno, California	389	35.8
8. Los Angeles, California	295	16.3
9. Sacramento, California	483	28.3
10. San Diego, California	290	12.4
11. San Francisco, California	494	9.1
12. Colorado Springs, Colorado	COS	38.0

228 ANNUAL TEMPERATURE RANGE ANALYSIS

Data

Station Location	Station Number	Temperature Range °F
13. Denver, Colorado	469	40.5
14. Grand Junction, Colorado	476	52.9
15. Miami, Florida	202	13.7
16. Pensacola, Florida	NAS	27.0
17. Savannah, Georgia	207	28.5
18. Atlanta, Georgia	219	34.5
19. Boise, Idaho	681	43.8
20. Chicago, Illinois	534	49.8
21. Springfield, Illinois	439	49.8
22. Evansville, Illinois	432	44.8
23. Indianapolis, Indiana	438	46.8
24. Dubuque, Iowa	547	54.7
25. Des Moines, Iowa	546	54.2
26. Dodge City, Kansas	451	47.5
27. Wichita, Kansas	450	48.2
28. Goodland, Kansas	465	46.5
29. New Orleans, Louisiana	231	26.6
30. Sault Ste. Marie, Michigan	734	49.9
31. Houghton, Michigan	744	50.3
32. Traverse City, Michigan	TVC	47.8
33. Detroit, Michigan	537	47.6
34. Alexandria, Minnesota	AXN	42.2
35. International Falls, Minnesota	747	64.5
36. Duluth, Minnesota	745	55.4
37. Minneapolis, Minnesota	658	60.1
38. Vicksburg, Mississippi	VKS	31.4
39. Kansas City, Missouri	446	49.8
40. St. Louis, Missouri	434	47.3
41. Miles City, Montana	MLS	58.4
42. Cut Bank, Montana	CTB	48.3
43. Helena, Montana	772	45.6
44. North Platte, Nebraska	562	50.2
45. Omaha, Nebraska	553	54.4
46. Tonopah, Nevada	485	43.9
47. Winnemucca, Nevada	583	43.9
48. Caribou, Maine	712	56.5
49. Greenville, Maine	619	52.1
50. Eastport, Maine	608	39.0
51. Burlington, Vermont	617	50.4
52. Boston, Massachusetts	509	42.6
53. Nantucket, Massachusetts	506	35.4
54. Albuquerque, New Mexico	365	42.6
55. New York, New York	503	42.0
56. Albany, New York	518	44.3
57. Greensboro, North Carolina	317	37.4
58. Bismark, North Dakota	764	61.5
59. Williston, North Dakota	767	61.5
60. Columbus, Ohio	428	44.9
61. Oklahoma City, Oklahoma	353	43.7
62. Baker, Oregon	BKE	40.7
63. Portland, Oregon	698	27.3
64. Pittsburgh, Pennsylvania	520	42.6
65. Columbia, South Carolina	310	34.6
66. Sioux Falls, South Dakota	651	57.8
67. Pierre, South Dakota	PIR	55.7
68. Rapid City, South Dakota	662	48.0
69. Memphis, Tennessee	334	37.6
70. Brownsville, Texas	250	23.8
71. El Paso, Texas	270	36.0
72. Houston, Texas	243	28.9

	Data	
Station Location	Station Number	Temperature Range °F
73. Presidio, Texas	271	35.6
74. Dallas, Texas	258	37.9
75. Amarillo, Texas	363	41.5
76. Salt Lake City, Utah	572	37.9
77. Richmond, Virginia	401	38.9
78. Tatoosh Island, Washington	798	41.5
79. Spokane, Washington	785	41.5
80. Green Bay, Wisconsin	645	34.3
81. Milwaukee, Wisconsin	640	49.5
82. Madison, Wisconsin	641	59.4
83. Park Falls, Wisconsin	741	58.5
84. Laramie, Wyoming	LAR	41.5
85. Casper, Wyoming	569	45.7
86. Shreveport, Louisiana	248	36.2
87. Mobile, Alabama	223	29.6
88. Eureka, California	594	54.6
89. Medford, Oregon	597	35.6
90. Eugene, Oregon	693	31.2
91. Sheridan, Wyoming	666	49.9
92. Idaho Falls, Idaho	IDA	51.7
93. Harve, Montana	777	56.3
94. Austin, Texas	254	34.2
95. Parkersburg, West Virginia	413	41.3
96. Bristol, Tennessee	THI	37.1
97. Roanoke, Virginia	411	38.5

Questions

1. Compare the temperature range of Bismark, North Dakota and Portland, Maine.
2. Compare the temperature range of Portland, Maine and Portland, Oregon.
3. In what part of the United States is the range the largest? Smallest? Explain your answer.

NAME: _____

SECTION: _____

LABORATORY EXERCISE 13

STATISTICAL ANALYSIS OF DATA

QUESTIONS TO THINK ABOUT

1. Why would meteorologists need to use statistical procedures on weather data?
2. What is the difference between meteorology and climatology?
3. What type of weather information would you like to find out if you were planning a trip to Chicago six months from now?
4. How would statistical weather data help the travel agency business?

DISCUSSION

To perform statistical operations on raw weather data you must be familiar with the terms associated with the subject.

Measure of Central Tendency
1. The *arithmetic mean* is the sum of all the values divided by the total number of values.
2. The *median* is the middle value of all the values arranged in increasing order.
3. The *mode* is the value that occurs most frequently.

Measure of Variability
1. The *range* is the difference between the largest and smallest values.

The Frequency Distribution
1. A *frequency distribution* is an arrangement of the values of a series according to magnitude.
2. A *histogram* is the graphic representation of a frequency distribution in the form of a rectangular polygon.
3. If the amount of data is sufficient, a smooth curve may usually be fitted to this polygon.
4. The *normal curve* which is bell shaped results when the fitted curve is symmetrical. (mean, mode, and median are equal to each other).

**EXERCISE 1
Constructing a Histogram (Bar Graph)**

Using the data provided with this lab construct a histogram for all three years of data. (Use a new piece of graph paper for each year.) Steps in constructing a histogram.

1. Use a 10° interval and count the frequency of occurrences in that interval.

2. On graph paper label the intervals along the bottom and the frequency along the side. Plot the data and construct a histogram.

STATISTICAL ANALYSIS OF DATA

	Frequency		
Interval	Year 1	Year 2	Year 3
10°–19°	_____	_____	_____
20°–29°	_____	_____	_____
30°–39°	_____	_____	_____
40°–49°	_____	_____	_____
50°–59°	_____	_____	_____

EXERCISE 2
Central Tendency Calculations

For each year calculate the mean, median, mode, and range. Use the definitions in the discussion section of this lab as a guide.

Mean _____

Median _____

Mode _____

Range _____

		Data	
Day	Year 1	Year 2	Year 3
1	29	36	58
2	40	35	41
3	38	32	44
4	52	38	43
5	43	45	51
6	44	44	43
7	51	43	45
8	43	54	30
9	40	55	29
10	52	30	46
11	32	31	29
12	33	36	26
13	21	38	31
14	21	26	40
15	22	14	38
16	25	15	25
17	34	16	38
18	36	25	35
19	40	22	36
20	44	37	44
21	42	35	40
22	51	47	37
23	47	32	36
24	45	34	35
25	54	37	30
26	56	41	23
27	47	39	32
28	45	37	12
29	36	24	32
30	34	17	32
31	41	27	23

EXERCISE 3
Construction of a Statistical Curve

On a separate piece of graph paper construct a statistical curve for each year. Use the frequency data you calculated in Exercise 1. Label the interval along the bottom and frequency along the side. Plot the frequency number as a point at the intersection of the midpoint of the interval and the frequency number. Connect the points with a smooth curve.

INDEX

Abbot, C. G., 26
Absolute humidity, 53
Absolute scale, 40
Absolute zero, 40
Absorption of radiation, 32
Acceleration, law of, 83
Action-reaction law, 83
Adiabatic lapse rate, 46
   moist, 46
Adiabatic processes, 46
Advection fog, 50, 59, 103
Advection of heat, 29
Aerosols, 20
   effect of, on climate, 158
AFOS system, 143
Agricultural meteorology, 4–5
Agriculture, effect on, of climatic variations, 153–154
Air:
   moisture in, laboratory exercises for, 187, 188
   temperature of, 36–48
   see also Atmosphere; Temperature
Air masses, 98–105
   classification of, 100–101
   general characteristics of, 99
   modification of, 103–105
   source regions for, 99–100
   in United States, 101–103
Air mass thunderstorms, 129
Air mixing, relation to fog dispersal, 61
Albedo, 33
Alberta Low, 115
Altimetry, 77
Altocumulus clouds, 65
Altostratus clouds, 64–65
Amplitude of waves, 29
Anemoscopes, 95
Aneroid barometers, 75
Angle of incidence, 31
Angstrom (unit), 30
Anticyclones, 91, 99, 108
Anvils in thunderstorms, 128
Arctic fronts, 107
Argon, 19
Aristotle, 1, 7
Arithmetic mean, 231
Atmosphere, 16–25
   composition of, 17
   evolution of, 17
   lower, exploration of, 23–25
   stability of, relation to lapse rates, 46–48
   upper, exploration of, 25
   vertical structure of, 21–23
   water in, 49–56
   see also Air
Atmospheric moisture, sources of, 56
Atmospheric pressure, 73–81
   horizontal variation of, 77
   measuring instruments for, 74–75
   semipermanent pressure patterns, 78—81
   terms for, 77–78
   transient pressure patterns, 79
   units for, 75
   vertical variations of, 75–77
Aurora Australis, 22, 28
Aurora Borealis, 22, 28
Aviation meteorology, 4

Baguio, 116
Ball lightning, 127
Baric wind law, 82
Barographs, 75
Barometers:
   ancient, 10
   mercurial, 73, 74–75
Beaufort, Admiral Francis, 94
Beaufort wind scale, 95
Bergeron, Tor, 100
Berti, Gasparo, 10
Biosphere, 19
Bjerknes, Jacob, 10–11, 113
Bjerknes, Vilhelm, 99, 106
Black body radiation, 32, 159
Black out, 155
Blizzards, 115
Bourdon tube, 39
Boyle, Robert, 161
Boyle's law, 161
Buys-Ballot, Christoph, 82
Buys-Ballot's law, 82, 205

California current, 43
Calorie (unit), 50
Capture process of water droplet growth, 69
Carbon dioxide, 19
   effect of on climate, 157–158
Cells in thunderstorms, 128

Celsius, Anders, 40
Celsius scale, 40
Centrifugal force, effect of, on winds, 85–86
Centigrade scale, 40
Centripetal force, 86
Changes of state, 50–51
Charles, Jacques, 161
Charles' law, 161
Chinook winds, 93
Chlorofluoromethanes, 20
Chromosphere, 27
Cirrocumulus clouds, 64
Cirrostratus clouds, 64
Cirrus clouds, 63–64
  and hurricanes, 117–119
Clapeyron, B. P., 49
Clausius, Rudolf, 49
Clausius-Clapeyron equation, 49
Clear air turbulence, 29
Climate, 145–158
  classification of, 145
  controls for, 146–147
  difference of, from weather, 146
  effect:
    of aerosols, 158
    on agriculture, 153–154
    of carbon dioxide, 157–158
    of heat release, 156
    of man, 155–156
    on society, 153–155
  geographic distribution of elements of:
    precipitation, 148–150
    temperature, 147–148
  man's effects on, 155–156
  relation to energy, 154–155
  variability and change of, 150–153
Cloud condensation nuclei, 21
Clouds, 50, 59, 62–67
  classification of, 57
  high, 63–64
  low, 65–66
  middle, 64–65
  types of, 63
  vertical development of, 66–67
Cloud seeding, 5, 72
Cloud streets, 67
Coalescence process, 68–69
Code for weather observations, 139–140
Cold clouds, formation of, 68
Cold fog, 61
Cold fronts, 109–110
Cold-front thunderstorms, 130
Cold-type occlusions, 112
Colorado Low, 115
Computers:
  development of, 12
  role of, in weather forecasting, 142–143
Condensation, 33, 51, 55–56
  nuclei for, 55, 58

Conditionally unstable air, 48, 110
Conduction of heat, 28
Continental air, 100
Continental Arctic air, 101–102
Continental fogs, 60
Continental Polar air, 101–102
Convection of heat, 28–29
  in rising air, 45
Convergence, 91
  in rising air, 45
Conversion systems:
  English/metric, 178
  metric/English, 178
Cooling degree day, 155
Cooling fogs, 59–60
Coreolis force, 84–85
  relation to hurricanes, 116
Corona, 27
Counter radiation, 33
Cumuliform clouds, 63
Cumulonimbus clouds, 67
Cumulus clouds, 66
Cumulus stage of thunderstorms, 125
Cyclogenesis, 115
Cyclolysis, 115
Cyclones, 107, 112–115
  life cycle of, 113–115
  Norwegian model for, 114
  tracks of, in the United States, 115

Daylight, duration of, 31
De Bort, L. P. Teisserene, 16
Degree days, 155
de Saussure, H. B., 10
Dew, 50, 62
Dew point, 52–53, 62
Dimensions, laboratory exercises for, 177–179
Direct capture process of water droplet growth, 69
Disasters, 2–3
Dissipating stage of thunderstorms, 126
Divergence in rising air, 45
Divergent air flow, 108
Doldrums, 89
Drizzle, 50, 69
Dry ice:
  for cloud seeding, 72
  for fog dispersal, 61

Earth, radiation from, 32–35
Earth-sun energy system, 26–35
  diagram of, 34
  net radiation, 35
El Cordonazo de San Francisco, 117
Electrical resistance thermometers, 39
Electromagnetic energy, 30

INDEX

Electromagnetic spectrum, 30
Energy, relation to climate, 154–155
Environmental Data Information Service, 15
Environmental Research Laboratory, 15
Equatorial trough (air pressure), 79
Espy, James, 134
ESSA, 12
Evaporation, 20, 33, 51, 54–55
Evaporation fog, 60
Evaporative techniques for fog dispersal, 62
Eye of tropical storms, 116, 118
Eye wall of hurricanes, 119

Facsimile machines, 141
Fahrenheit, Gabriel, 40
Fahrenheit scale, 40
Fair weather water spouts, 133
FGGE, 144
Findeisen, W., 68
Fitzroy, Admiral R., 98
Flash floods, 3
Föehn wind, 93
Fogs, 50, 59–62
    dispersal of, 61–62
Forecasting, weather, 142–143
    accuracy of, 143
    future outlook for, 143–144
    laboratory exercises for, 213–214
    numerical forecasting, 142–143
    role of computers in, 142–143
Forked lightning, 126
Fractostratus clouds, 66
Franklin, Benjamin, 10, 123
Freezing precipitation, 69–70
Frequencies of waves, 29
Frequency distribution (statistics), 231
Frictional force, effect of, on winds, 86
Frontal analysis, laboratory exercises for, 211–212
Frontal lifting, 45
Frontal systems, 107–112, 142
    types of, 108
    weather map symbols for, 112–113
Frontal thunderstorms, 129–130
Frontogenesis, 107–108
Fronts, see Frontal systems
Frost, 59, 62
Frozen precipitation, 70
Fusion, 50

Galilei, Galileo, 9–10, 36
Gamma radiation, 30
GARP, 144
Gas laws, 161
GATE, 144
Geographic classification of air masses, 100
Geostrophic balance, 87

Geostrophic wind, 86–87
Glacial periods, 150
Global Atmospheric Research Program, 144
Global effects on climate, 156–158
Glossary, 169–174
Glycerine seeding for fog dispersal, 61
GOES satellites, 12, 137, 138
Gradient wind, 87
Gravity winds, 93
Greek meteorologists, 7–8
Greenhouse effect, 33
Ground fog, 60
Gulf Stream, 43
Gusts of wind in thunderstorms, 128

Hadley, George, 88
Hail, 50, 70
    in thunderstorms, 128–129
Hatteras Low, 115
Heat:
    relation to temperature, 29–30
    release of, effect of, on climate, 156
Heat balance, earth and sun, 33–35
Heat lightning, 127
Heat of fusion, 50
Heating degree day, 155
Heavy glaze, 70
Hippocrates, 7, 8
Histograms, 225
Horse latitudes, 89
Howard, Luke, 10, 57, 63
Hurricanes, 115–122
    decaying stage of, 117
    facts of, 120
    formative stage of, 116, 117, 119
    intensification stage of, 117
    mature stage of, 116–117
    naming of, 120–122
    tracks of, 120
    weather associated with, 117–120
Hydrographs, 53
Hydrologic cycle, 50, 56
    Hydrometeors, 69
Hydrosphere, 19, 56
Hydrostatic equation, 75, 161–162
Hygrometers, ancient, 10
Hygroscopic particles, 58

Ice crystal clouds, 50
Ice crystal process, 68
Ice fog, 61
Ice nuclei, 59
Ice pellets, 70
Ice storms, 70
Industrial meteorology, 6–7
Inertia, law of, 83
Infrared radiometers, 29, 39

Inter-glacials, 150
Intertropical convergence zone, 89, 148–149
Intertropical fronts, 107
Inversions of temperature, 44
Ionosphere, 23
Isobars, 76, 78
    on weather maps, 142
Isoplething, laboratory exercises for, 193–197
ITOS satellites, 138

Jet stream, 89, 91

Katabatic winds, 93
Kelvin, Lord, 40
Kelvin scale, 40
Kilometer (unit), 30
Knots (unit), 94
Koeppen, Wladmir, 145

Land breezes, 92
Langley, Samuel P., 31
Langley (unit), 31
Lapse rates of temperature:
    adiabatic, 46
    negative, 44
    normal, 43
    positive, 43–44
    relation to atmospheric stability, 46–48
Latent heat, 33
    of condensation, 116
    relation to hurricanes, 116
    of sublimation, 51
    of vaporization, 51
Law of acceleration, 83
Law of action-reaction, 83
Law of inertia, 83
Length (unit), 177
Lenticular clouds, 65
Level of free convection, 48
Lifting condensation level, 48
Light glaze, 70
Lightning, 123
    process of, 126
    types of, 126–128
Liquid-in-glass thermometers, 37
Liquid precipitation, 69
Liquid propane system for fog dispersal, 61
Longwave radiation, 32, 33
Lower atmosphere, exploration of, 23–25

Macroscale wind systems, 88
Mammatus, 67
Marconi, Guglielmo, 14
Mares tail cirrus clouds, 64

Maritime air, 100
Maritime Polar air:
    Atlantic, 102
    Pacific, 102
Maritime Tropical air:
    Atlantic, 102–103
    Pacific, 103
Mass (unit), 177
Mature stage of thunderstorms, 125–126
Maximum thermometers, 37
Measurements, laboratory exercises for, 175–177
Mechanical lifting in thunderstorm formation, 124
Median (statistics), 231
Melting, 50
Meniscus, 38
Mercurial barometers, 72, 74–75
Mesopause, 23
Mesoscale wind systems, 88
Mesosphere, 21
Meteorographs, 16
Meteorologica, 1, 7
Meteorological satellites, 40, 137–138
Meteorology, history of, 7–13
Metric system:
    conversion factors for, 168, 178
    development of, 177
Micrometer (unit), 30
Micron (unit), 30
Microscale wind systems, 88
Millibar (unit), 75
Millimeter (unit), 30
Minimum thermometers, 37–38
Mist, 69
Mode (statistics), 225
Moisture:
    in air, 51–53
    effect of, on air masses, 105
        laboratory exercises for, 187–188
Moisture classification of air masses, 100–101
Monsoon effect, 150
Monsoon winds, 92–93
Mother of pearl clouds, 22
Mountain breezes, 93

Nacreous clouds, 21
Names for tropical storms, 120–122
National Environmental Satellite Service, 15
National Hurricane Center, 143
National Meteorological Center, 143
National Oceanic and Atmospheric Administration, 15
National Oceanographic Data Center, 15
National Service Storm Forecast Center, 143
National Weather Records Center, 4, 15

National Weather Service:
  forecasting accuracy of, 143
  history of, 13–15
Neutral atmospheres, 46–48
Newton, Sir Isaac, 53
Newton's laws of motion, 83
Nimbostratus clouds, 66
Nimbus, 12–13
Nitrogen, 18
Nitrogen oxides, 19
Noctilucent clouds, 23
Nor'easters, 102, 115
Normal atmospheric pressure, 75
Normal curve (statistics), 231
Northeasters, 102, 115
Northeast semicircle of hurricanes, 117
Northeast trade winds, 89
Northern lights, 28
Northers, 101
Notation, scientific, 175–176
Numerical weather prediction, 142–143

Observations of weather, 135–138
Occluded fronts, 111–112
Oceans, effects of, on temperature, 42, 43
Orographic lifting, 44–45, 60, 103
Oxygen, 18–19
Ozone, 20
Ozonosphere, 23

Paine, Congressman Halbert, 13
Pauses, 23
Periods of waves, 29
Photolytic dissociation, 17
Photosphere, 27
Photosynthesis, 17
Pilot balloons, 94
Pilot leader (lightning), 126
Pilot reports of weather, 137
Polar easterlies, 89
Polar front jet stream, 89
Polar fronts, 89, 106, 107
Polar outbreaks, 101
Precipitation, 56
  forms of, 69–70
  geographic variations of, 148–150
  measuring devices for, 70–72
  process of, 67–69
Prefrontal fog, 60
Pressure gradient, 78, 84
Pressure systems, airflow around, 91–92
Prevailing westerly winds, 43, 89
Priestly, Joseph, 18
Ptolemy, 8

Radar observations of weather, 137
Radiation fog, 59
Radiation of heat, 28, 29

Radiometers, infrared, 39
Radiosondes, 23
  development of, 11
  for moisture measurement, 54
Rain, 50, 69
  in thunderstorms, 128
Rain forests, 149
Rain gauges, 70–72
Rainsqualls in thunderstorms, 128
Range (statistics), 231
Rawinsondes, 136–137
Rayleigh Fountain experiment, 68
Reflection of radiation, 32
Refraction of radiation, 32
Relative humidity, 53
Remote sensing of precipitation, 72
Ridges (air pressure), 78
Rising air, 44–45
  temperature variations in, 44–46
River Forecast Centers, 143

Satellites:
  development of, 12
  meteorological, 40, 137–138
Saturation of air, 52
Saturation vapor pressure, 52
Scattering of radiation, 32
Scientific notation, 175–176
Sea breezes, 92
Self-recording thermographs, 38–39
Seneca, 8
Sensible heat transfer, 34
Sheet lightning, 127
Silver iodide, 72
Sleet, 50, 70
  in thunderstorms, 125
Sling psychrometers, 53
Small hail, 70
Smog, 44
Snow, 50, 59, 70
Snow pellets, 50
Solar constant, 26, 30–32
Solar flares, 28
Solar radiation, 30
Solidification, 50
Solid particles in atmosphere, 20
Solute effect, 59
Southern lights, 28
Specific heat, 29–30
Specific humidity, 53
Spectrophotometers, 39
Squall lines, 109, 128
Squalls in thunderstorms, 128
Stable air, 46–48, 110
Stationary fronts, 111
Station model (for weather maps), 142
Statistical analysis of data, laboratory exercises for, 231–233

Steam fog, 60–61
Stefan-Boltzman law, 159
Storm surges, 116, 120
Stratiform clouds, 63
Stratocumulus clouds, 66
Stratopause, 23
Stratosphere, 18, 21
Stratus clouds, 66
Streak lightning, 127
Sublimation, 19, 51, 125
Subsidence, 89
Subsidence inversion, 90
Subtropical lows, 115
Sulfur dioxide, 19
Sulfuric acid, 19
Sun, 27–35
Sun-earth energy system, 26–35
   diagram of, 34
   net radiation, 35
Sunspot cycle, 27
Supercooled water, 51
Supercooled water droplets, 59
Superior air, 103
Super saturation, 56
Surface observations of weather, 135–136
Synoptic meteorology, 3–4
Synoptic observations, 136
Synoptic weather code, 138–139

Temperature, 36–48
   annual range of, laboratory exercises for, 227–229
   effect of, on air masses, 103
   geographic variations of, 147–148
   inversions, 44
   relation to heat, 29–30
   summertime, laboratory exercises for, 219–221
   upper air, 39–40
   variations in, 41–48
      earth's irregularities affecting, 42–43
      environmental air, 43–44
      horizontal, 42
      latitude effect, 41–42
      ocean currents, 43
      oceanic effect, 42
      in rising air, 44–46
      seasonal, 41–42
      vertical, 43
   wintertime, laboratory exercises for, 223–225
Temperature classification of air masses, 100
Temperature conversions, 182
Temperature measurements, laboratory exercises for, 181–183
Temperature scales, 40
   conversions, 40–41
Theodolites, 94

Thermal convection cells, laboratory exercises for, 185–186
Thermocouples, 40
Thermographs, self-recording, 38–39
Thermometers, 36–40
   ancient, 9
Thermosphere, 22–23
Thunder, 128
Thunderstorm clouds, 67
Thunderstorms, 124–130
   formation of, 124–126
   phases of, 124, 125–126
   types of, 129–130
   weather associated with, 128
Time (unit), 177
Tipping bucket rain gauges, 71
TIROS satellites, 12, 137, 138
Topography, effect of, on air masses, 104
Tonadoes, 130–131
   effects of, 131–132
   facts about, 132–133
Torricelli, Evangelista, 10, 73
Towering cumulus clouds, 66–67
Trade winds, northeast, 89
Transducers, 40
Transpiration, 20, 56
Tricellular wind model, 89
Tropical storms, 116–122
Tropopause, 23
Troposphere, 18, 21
Troughs (air pressure), 78
   relation to hurricanes, 117
Typhoons, 117

Ultraviolet radiation, 30
Unstable air, 46–48, 109
   in thunderstorm formation, 124
Updrafts in thunderstorms, 125
Upper atmosphere:
   exploration of, 25
   observations of, 136–137
      laboratory exercises for, 215–218
      satellite observations, 137–138
   temperatures of, 39–40
Upslope fog, 59, 60

Valley breezes, 93
Vapor pressure, 50, 52
Vertical development clouds, 66–67
Virga, 66
Visible range (radiation), 30

Wake capture process of water droplet growth, 69
Warm clouds, formation of, 68–69

Warm fog, 61
Warm fronts, 110–112
Warm-front thunderstorms, 129–130
Warm-type occlusions, 112
Water droplets:
  shape of, 59
  size of, 58
Water in atmosphere, 49–56
Waterspouts, 133
Water vapor, 20
  measuring instruments for, 53–54
Wavelengths, 29
Weather, difference from climate, 146
Weather Bureau, history of, 13–15
Weather data:
  coding of, 138–139
  transmission and collection of, 138–140
Weather forecasting, 142–143
  accuracy of, 143
  future outlook for, 143–144
  laboratory exercises for, 213–214
  numerical forecasting, 142–143
  role of computers in, 142–143
Weather maps:
  analysis of surface maps, 142
    laboratory exercises for, 191–193, 205–210
  frontal symbols for, 112–113
  observations for, 199
  plotting of, laboratory exercises for, 199–204
  preparation of, 140–142
    surface maps, 141–142
    upper air maps, 141
  symbols for, 163–167
Weather modifications, 5–6
Weather radar, development of, 11
Weather systems, 106–122
Weighing rain gauges, 71
Wien displacement law, 159
Willy-willy, 117
Wind-chill factor, 97
Wind gusts, 94
Wind patterns, general, 88–90
  local, 92–93
  worldwide, 88
Winds, 82–97
  balance of forces on, 86–88
  directions of, 93–94
  flow of, around pressure systems, 91–92
  measurements of, 95
  primary forces affecting, 83–84
  secondary forces affecting, 84–86
  speed of, 94–95
  in thunderstorms, 128
Wind shear, 4
  in clouds, 64
Wind vanes, 95
World Weather Watch, 143